NATO ASI Series

Advanced Science Institutes Series

A series presenting the results of activities sponsored by the NATO Science Committee, which aims at the dissemination of advanced scientific and technological knowledge, with a view to strengthening links between scientific communities.

The Series is published by an international board of publishers in conjunction with the NATO Scientific Affairs Division

A Life Sciences B Physics	Plenum Publishing Corporation London and New York
C Mathematical and Physical Sciences D Behavioural and Social Sciences E Applied Sciences	Kluwer Academic Publishers Dordrecht, Boston and London
F Computer and Systems Sciences G Ecological Sciences H Cell Biology I Global Environmental Change	Springer-Verlag Berlin Heidelberg New York London Paris Tokyo Hong Kong Barcelona Budapest

PARTNERSHIP SUB-SERIES

1. Disarmament Technologies	Kluwer Academic Publishers
2. Environment	Springer-Verlag/Kluwer Academic Publishers
3. High Technology	Kluwer Academic Publishers
4. Science and Technology Policy	Kluwer Academic Publishers
5. Computer Networking	Kluwer Academic Publishers

The Partnership Sub-Series incorporates activities undertaken in collaboration with NATO's Cooperation Partners, the countries of the CIS and Central and Eastern Europe, in Priority Areas of concern to those countries.

NATO-PCO DATABASE

The electronic index to the NATO ASI Series provides full bibliographical references (with keywords and/or abstracts) to about 50000 contributions from international scientists published in all sections of the NATO ASI Series. Access to the NATO-PCO DATABASE is possible via a CD-ROM "NATO Science & Technology Disk" with user-friendly retrieval software in English, French and German (© WTV GmbH and DATAWARE Technologies Inc. 1992).

The CD-ROM can be ordered through any member of the Board of Publishers or through NATO-PCO, Overijse, Belgium.

2. Environment – Vol. 45

Springer
Berlin
Heidelberg
New York
Barcelona
Budapest
Hong Kong
London
Milan
Paris
Singapore
Tokyo

Environmentally Devastated Areas in River Basins in Eastern Europe

Edited by

Alfons Georges Buekens

Free University of Brussels
Pleinlaan 2
B-1050 Brussels, Belgium

Vasily Victorovich Dragalov

MC Moscow-Consult
Marchlewskogo
Moscow 13-309, Russia

With 84 Figures and 60 Tables

Springer

Published in cooperation with NATO Scientific Affairs Division

Proceedings of the NATO Advanced Research Workshop on Integrated Rehabilitation of Environmentally Devastated Areas in River Basins in Eastern Europe, held at Moscow, Russia, September 1994

Library of Congress Cataloging-in-Publication Data

Environmentally devastated areas in river basins in Eastern Europe / edited by Alfons Georges Buekens, Vasily Victorovich Dragalov.
p. cm. -- (NATO ASI series. Partnership sub-series 2, Environment ; vol. 45)
"Proceedings of the NATO Advanced Research Workshop on Integrated Rehabilitation of Environmentally Devastated Areas in River Basins in Eastern Europe, held at Moscow, Russia, September 1994."
ISBN-13:978-3-642-72268-4
1. Water quality management--Europe, Eastern--Congresses. 2. Water quality management--Former Soviet republics--Congresses. 3. Water quality management--Case studies--Congresses. 4. Water--Pollution--Europe, Eastern--Congresses. 5. Water--Pollution--Former Soviet republics--Congresses. 6. Watersheds--Europe, Eastern--Congresses. 7. Watersheds--Former Soviet republics--Congresses. I. Buekens, Alfons Georges, 1942- . II. Dragalov, Vasily Victorovich, 1950- . III. NATO Advanced Research Workshop on Integrated Rehabilitation of Environmentally Devastated Areas in River Basins in Eastern Europe (1994 : Moscow, Russia) IV. Series.
TD255.E58 1998
363.739'4'0947--dc21 98-36220
CIP

ISSN 1431-7141
ISBN-13:978-3-642-72268-4 e-ISBN-13:978-3-642-72266-0
DOI: 10.1007/978-3-642-72266-0

Softcover reprint of the hardcover 1st edition 1998

Typesetting: Camera-ready by authors/editors
SPIN: 10683606 31/3137 – 5 4 3 2 1 0 – Printed on acid-free paper

Preface

The fall of the iron curtain revealed widespread environmental and safety shortcomings in relation to industrial operations, mining and other human activities.

It was the purpose of this Advanced Research Workshop, sponsored by NATO, to bring together a group of eminent experts from East and West in order to present and analyse environmental data, assess environmental damage, establish priorities, strategies and methods of remediation and compare these with case studies from the Western World.

In this volume a number of important contributions have been gathered, that reflect the presentations and discussions held.

Alfons Georges Buekens
Vasily Victorovich Dragalov

Preface

Table of Contents

I

GENERAL CONSIDERATION

Scientific and Technological Policy of the Ministry of Science of the Russian Federation in application to Water Treatment.

V.N.Novoseltsev
The Ministry of Science
of the Russian Federation
Tverskaya Str. Moscow
Russia

The Ministry of Science of the Russian Federation is the central organ with executive authority responsible for the preparation and eventual implementation of the State Scientific and Technical Policy of Russia.

The definition of this policy is organized as follows:

- establishment of priorities;
- competitive selection and expert review of programs and projects;
- material and financial support from Federal sources.

Another aspect is to devise the required incentives to encourage the practical realization of the projects considered; such incentives may be economic (monopoly), fiscal (levels of taxation) or other.

These projects must also have their commercial value, their characteristics must be equal to or exceed the current state-of-the-art at worldwide level. The projects may be introduced into a State Directed scientific and technical program, a State science center Program, or be independent - such as externally sponsored large international projects. The "Chemistry and Technology of Pure Water" was considered as priority among other scientific and technological topics. The solving of this problem includes various economical, methodological and technological tasks. A number of important methodological problems dealing with the protection of surface waters in large river basins, are in the sphere of interests of scientists and businesspersons from different countries, and are to be solved in the framework of "Oka - clean river" project. As a part of this project, an international project "Oka-Elba" is being carried out.

NATO ASI Series, Partnership Sub-Series, 2. Environment – Vol. 45
Environmentally Devastated Areas in River Basins in Eastern Europe
Edited by Alfons Georges Buekens and Vasily Victorovich Dragalov

Some of results of these projects are illustrated in the ARW. The study of technological aspects of "Chemistry and Technology of Pure Water" also received a high priority. It includes six groups of projects such as:

- creation of new types of equipment, optimization of traditional and development of new water treatment technologies;
- drinking water water production and purification methods;
- sorption methods in natural and drain water treatment;
- biological methods of water treatment;
- purifying and recuperation of galvanic drainage, with creation of a closed water circuit and removing of heavy metals;
- analysis and remote sensing of complex water systems.

Then there is a wide variety of project groups, as listed above, which an all devoted to the main aim: the realization of a technological scheme or a process, which shall provide at minimal expense an acceptable water treatment to be applied for pre-defined conditions and demands to water quality. In practice, the systems of computer optimization for water treatment are already close to their realization.

Already, there are numerous examples of separation technological schemes and equipment, for example - membrane separation and combined schemes, which provide a reduction of energy expenses by 20-50% related to usual industrial practice; inexpensive schemes of chicken litter treatment without liquid effluence; ecologically pure flocculants and coagulants with reduced discharge levels and finally numerous; methods and pieces of equipment, which makes increase the efficiency of water treatment plant, etc.. Numerous progressive technologies of water treatment are being further developed and marketed via commercial structures, some of which supported by different non-government funds.

It can be concluded that the Ministry of Science and Technical Policy of the Russian Federation provides an organizational, juridical and financial support for these developments, new structures, and infrastructure dedicated to Russian R&D-efforts in applications in Water Treatment.

TARGET FEDERAL PROGRAMME - REVIVAL OF THE VOLGA

Valentin V. Naidenko
State Architecture & Construction Academy
Nizhniy Novgorod
ul. Gorkogo 152189, Nizhniy Novgorod
RUSSIA

RUSSIAN FEDERATION

Current critical ecological situation in the Volga basin, one of the largest regions of the Russian Federation, calls for urgent measures to provide a radical improvement of socio-ecological situation.

The Volga basin, incorporating 6 economic zones, 26 provinces and 6 republics, is inhabited by people of more than 100 nationalities. The total population of the region is 60.9 mln. For many decades the Volga basin concentrated the main industrial, power engineering and agricultural potential of this country. The real ability of the ecological load was ignoored all the time. Specialists now observe a catastrophic distruction of ecological systems with complete anihilation of some precious species of its flora and fauna and considerable growth of infantile disease and mortality rate.

According to the instructions by the President and Government of Russia a decision was adopted in 1994 to work out a target federal program - *Improvment of ecological situation in the Volga river, its tributaries; rehabilitation and prevention of degradation of natural complexes of the Volga basin (Revival of the Volga).*

NATO ASI Series, Partnership Sub-Series, 2. Environment – Vol. 45
Environmentally Devastated Areas in River Basins in Eastern Europe
Edited by Alfons Georges Buekens and Vasily Victorovich Dragalov

This work is aimed at a radical improvement of ecological situation in the region, conservation and development of ecological systems, preventing degradation and restoring natural complexes, providing favourable conditions for people's living, reducing disease incidence and death rate, extending life expectancy.

This conception is based on the principle proclaimed by the UN Conference on Environment and Development (July 3-4, 1992, Brasil), viz. providing balance between solving socio-econominie problems and conservation of environment, satisfying the vital needs of the present generation and providing the same opportunities for the future generations.

The programme includes 15 basic directions - economic, education; ecology of bodies of water; rational use of fish resources; water management; urban ecology; ecology and health service; ecologically safe industrial production; providing radiation and chemical safety; ecologically safe agricultural production; air pollution control; ecologically safe environmental engineering, forestry and wild life; special protected nature reserves; ecological monitoring; processing domestic and industrial wastes; ecological problems of geology and hydrogeology.

The programme is of an applied character and is oriented at introduction of well-established methods and means capable of improving in a radical way the ecological situation in the Volga region.

II

RIVER BASINS AND THEIR CHARACTERIZATION

River Basin Management In The Upper And Middle Odra River

Richard M Ashley
Wastewater Technology Centre
Department of Civil Engineering, Surveying & Building
University of Abertay Dundee
Bell Street
DUNDEE DD1 1HG
Scotland

S Czaban
Academia Rolnicza, Institut Inzyuierii
Stodowiska, Wroclaw,
Poland

INTRODUCTION

The problems of applying integrated approaches to manage water resources in river basins in Western Europe and the United States are well known. The difficulties of reconciling the anthropogenic demands of water consumers and users (for supply and disposal) have led to a number of development programmes in various countries, which address the tools required to make technical studies and also propose methodologies for holistic assessment taking due cognisance of environmental needs. Notwithstanding these initiatives, a truly integrated approach to water resource management, which includes all significant factors, has never been undertaken anywhere in the world, although various 'action programmes' have allowed transnational initiatives to develop in a coherent way (e.g. Dieperink, 1993). Conspicuously absent from integrated approaches are elements of:

- real public participation in prioritisation and valuing;
- true cost-effectiveness assessments;
- life-cycle accounting of all mass and energy flows.

In the latter case, the manifestation of the accounting omission is apparent in terms of resource utilisation and a false accounting for both natural resources (energy and matter) and waste disposal. In general the analysis of the *'system'* in terms of profitably (or cost-benefit) has a false boundary as illustrated in Figure 1. The most often quoted expression of this false accounting is ***'the polluter must pay'***, which is assumed to mean the manufacturer or extractor of a commodity or a resource. In fact, as the commodity or resource is required by the society - then it is the society who are the *'polluters'* and

NATO ASI Series, Partnership Sub-Series, 2. Environment – Vol. 45
Environmentally Devastated Areas in River Basins in Eastern Europe
Edited by Alfons Georges Buekens and Vasily Victorovich Dragalov

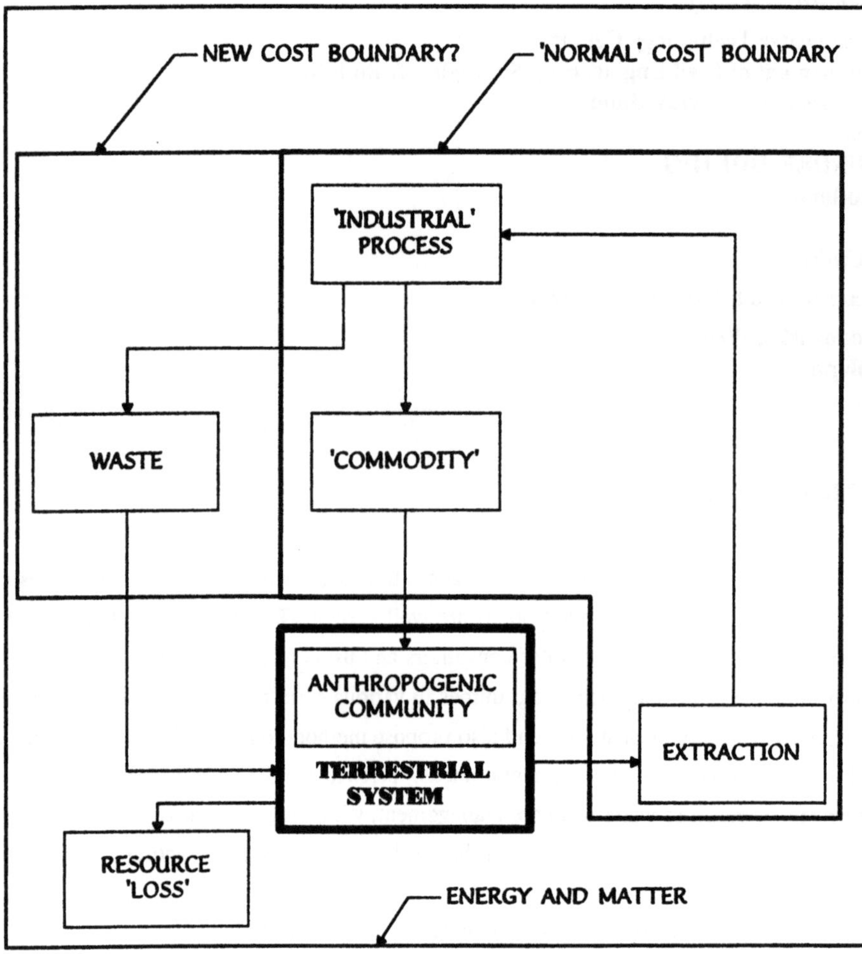

Evaluation system - environment vs costs depends upon system boundaries
Figure 1

consequently the *'payers'*. This more realistic approach would of course, render some commodities (such as gold) so expensive that they would no longer be *'economic'* to mine. Clearly, even the proper formulation of the *'best'* approach to manage water resources in a particular river basin will depend upon the perspective, or point of view, adopted. Classically the two extremes are *'top-down'* and *'bottom-up'*. In the former, a global overview is taken and decisions will be based on a holistic perspective, whereas in the latter, local strategies are adopted without necessarily being cognisant of the wider perspective. Public participation can occur with either approach, although in the *'top-down'* case, this means via (often remote) politicians. To be truly *'bottom-up'* an approach must directly involve local people, not politicians as surrogates. Interestingly, the problems of perspective when balancing anthropogenic and environmental needs have been addressed in terms of morals by economists in the past such as Adam Smith [Smith, 1976] and selectively ignored in terms of limiting accounting boundaries to *'local'* factors [Figure 1]. Waste and *'resource loss'* costs have traditionally been overlooked. The concepts leading to the logical conclusion that life-cycle assessment and accounting in all mass and energy flows are required are thus not new, but because the earth's resources have now been recognised as not inexhaustible, such concepts have only recently assumed a greater significance.

A number of recent thinkers have questioned received approaches to questions of sustainability of resources and of pollution *'sinks'* within terrestrial systems, eg Semple, 1991 and, specifically related to Water Resources, MacGarvin and Johnston, 1992 and Beck et al, 1994. In each case, calls have been made for new ways of thinking and approaching the overall *'sustainability'* problem. MacGarvin and Johnston describe how the EU countries' attitude to wastewater management has undergone a *'paradigm-shift'* to focus on source management and control rather than *'end-of-pipe'* technologies. Of course, the USA and Japan have been in the vanguard of these new approaches. Semple describes the difficulties of accounting for environmental benefits and the false accounting which fails to include life-cycle assessments which take account of *'lost resources'* used not for the *'best'* purposes. His theme exploring the law of diminishing returns on environmental protection is also taken up in a later paper by Tyson et al, 1993 at the INTERURBA [Interactions between sewers, treatment plants and receiving waters in urban areas] Conference in Wageningen in 1992. In each case, it is clearly demonstrated that false accounting can also apply to environmental protection measures, where unreasonable excessively stringent legislation can lead to unrealistic protection measures causing improvements in one sphere, eg water pollution control, to be negated by a consequent increase in pollution elsewhere in the terrestrial system. This results from the protection measures causing an increase in demand for resources and fossil fuel consumption - thus affecting the system as a whole.

Lessons continue to remain unlearnt, and despite a belief in *'clean production'* techniques, *'false accounting'* continues in the developed countries. Countries which are undeveloped or in transition, such as in Eastern Europe, with vast problems, cannot afford to re-learn by making the same mistakes. There is thus a need to adopt holistic and integrated approaches to the rehabilitation of environmentally devastated river basins, which where possible, are based on clean technologies, life cycle accounting and as a consequence, result in realistic environmental standards.

THE ODRA RIVER

The Odra River is one of the two primary waterways in Poland, the other primary river basin draining to the Vistula (Figure 2). The Odra originates in the Czech Republic, south-west of Poland and flows north to the Baltic at Szczecin. Along its lower course, the Odra River is a natural border between Poland and Germany. The river passes through Wroclaw and several industrial areas, eg Karwina-Orawa and the Rybnik District of Silesia. As a result of industrial activity and also in some instances, because of natural contamination by groundwater seepage, many of the rivers in Poland are highly polluted. The Odra is heavily contaminated by municipal wastewaters, chlorides and other suspended and dissolved compounds in the Czech Republic and is further polluted in the mining areas by coal in Rybnicki and Slaski (Silesia) and from the copper mining district in Legnic Glogow, where for example, up to 10% of the total pollutant load carried by the river originates.

Figure 2 shows the classification of all Polish rivers for a water quality survey carried out in 1991/92 based on physical, chemical and bacteriological quality. Figure 3 shows these rivers and the corresponding levels of salinity for the same survey. The problem of improving river water quality in the Upper Silesian region by coal mine management utilising a variety of techniques has been discussed (Suschka, 1991) in the context of various national remediation programmes designed to bring the water quality to Class II in terms of salinity. This standard requires chlorides not to exceed 300mg/l, sulphates to be limited to 200mg/l and total dissolved solids (TDS) to be less than 1000mg/l. It is argued (ibid) that it will be necessary to close coal mines even to achieve river quality standards which are not as good quality as Class II and which are in any case considered unecessarily stringent.

Municipal wastewater from Wroclaw, Zielona Gora and industrial wastes from industrial plants (inefficiently treated) enter the Odra River. In some areas the flow in the tributaries, eg Klodnica, Bytomka, comprise almost entirely raw sewage, with BOD levels of 175mg/l, TDS up to 2900mg/l, AmmN of 19.8mg/l and Chlorides up to 1024mg/l (Suschka, 1991, Suschka et al, 1994).

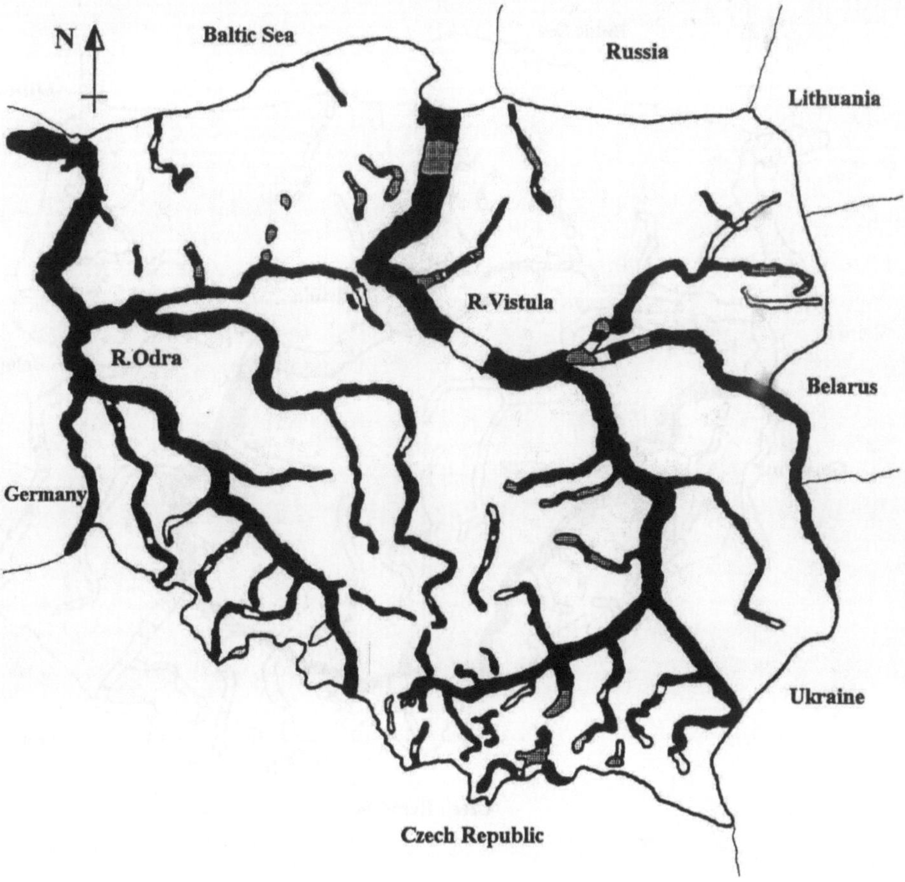

Class	Suitable for :-
I	drinking water abstraction
II	animals, bathing, recreation use
III	industrial and agricultural use
	water quality below standards

Classification of water quality of all Polish rivers 1991/92 survey - Physical, chemical and bacteriological.

Figure 2.

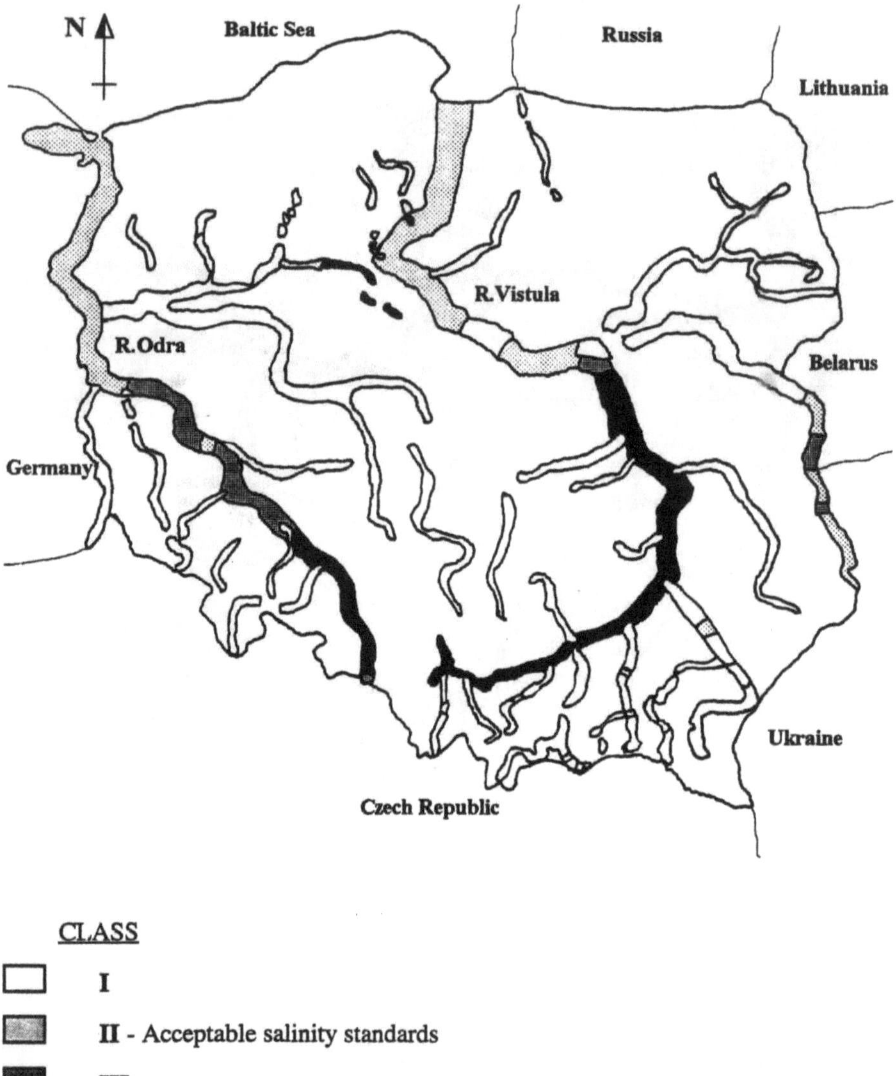

CLASS

- I
- II - Acceptable salinity standards
- III
- water quality below standards

Classes of water purity of all Polish rivers - Based on salinity survey 1991/1992

Figure 3.

Ironically, despite the richness of the mineral resources, water resources are scarce in Poland and hence the Odra River is a primary artery for navigation and potentially a potable water source. Unfortunately rainfall has been very low for a number of years and the controlling upland reservoirs used for river regulation have been too low to facilitate river levels sufficient for navigation to be possible. In addition, low river flows have been inadequate to ensure appropriate dilution of the various pollutant discharges.

Thus there is an intrinsic inter-relationship between the hydrology of the river basin and the water quality in the Odra, and subsequently as it affects the quality of the discharge into the Baltic Sea. There is clearly a need for an integrated approach to be taken to resolve the needs of river regulation, water supply and wastewater disposal. Countries in the EU are currently facing increasingly stringent water quality standards relating to all aspects of potable waters and potentially contaminating discharges. For example for drinking water, the EU guideline level for chlorides is 25mg/l, contrasting with WHO recommendations of 600mg/l. Despite a high level of controlling national legislation having been in place for many years in countries such as the UK, the new EU standards are imposing high cost penalties to achieve these new clean water requirements. Poland, and in particular the Odra river catchment, borders the EU and there is a temptation to envisage that EU standards should be applied to the Odra river. Such an approach would require an unrealistically massive investment necessitating:

- river regulation
- discharge treatment and control
- remediation of contaminated aquatic sediments
- concurrent air quality control

Whilst steps must be taken to improve the water quality in the Odra, these should be at a level appropriate to the state of development of the Polish economy and must ensure that industrial pollution is controlled by a staged implementation of controls at a pace that can be afforded.

The approach recommended is a holistic one which deals with the whole of the Odra basin and will thus include a joint action involving the Czech Republic as well as Poland. The response developed in the UK to deal with river basin pollution management and which is tackling problems of a much smaller scale than those of the Odra river basin, is sufficiently flexible to be used as a methodology for the implementation of an integrated approach to the problem. This is encapsulated in the BATNEEC [Best Available Technology Not Entailing Excessive Cost] and BPEO [Best Practicable Environmental Option] principles, and relies on effective computational tools to determine the 'best'. The selection of appropriate environmental standards is then open depending upon individual legislation or other constraints.

This paper considers the specific problems caused by the high levels of salinity in the Odra basin, and in particular that arising from the copper mining industry in Silesia. These problems are considered in the context of the overall river basin quality problems.

WATER MANAGEMENT IN THE UPPER AND MIDDLE ODRA RIVER BASIN

At present, the water authorities in Poland are involved in the implementation of new storage water reservoirs. The management of the Upper and Middle Odra River has its headquarters in Wroclaw (Regional Board of Water Management). The total area of the region is 42,850 square kilometres, including:

57%	arable land
31 %	forests
2 %	waters
4 %	urban areas
4 %	roads and railroads
1.7%	wasteland
0.3%	others.

The region is rich in natural resources, such as hard coal, brown coal (lignite), copper ore with traces of precious metals, as well as zinc, lead and nickel ores, granite, sandstone, marble, limestone and malm. The natural resources of the region have encouraged the development of the coal industry, power stations and steelmills. Chemicals, electrical machinery and paper and cellulose industries also play a significant role in the region.
For three decades (1961-1990), the level of annual rainfall averaged 739 mm and ranged from 545 mm in Legnica to 1380 mm in Jakuszyce. In recent years, however, the region has suffered a long-term drought. In the last decade, the drought caused a 200 mm drop in annual precipitation. As a result, the water supplied to industry and, in some places also to households, had to be limited. The annual flow on the Odra averages 16,123 billion cubic metres, which accounts for 27.3% of the outflow from Poland. There are 19 storage water reservoirs of total capacity of 740.6 million cubic metres which function as water supply, flood control facilities, alimentation of waterways and also as a source of energy.

The density of storage reservoirs is a quotient of their service capacity and annual mean flow of many years. The low percentage (2.7%) shows that the number of artificial storage reservoirs does not satisfy the needs of the region.

The Czech Republic has seven large storage water reservoirs whose total capacity is 168.3 million cubic metres, including 118.4 million cubic metres of service capacity. When the Czechs open the reservoir Slezka Hora, these will rise to 385.6 and 314.4 million cubic metres, respectively. Five storage water reservoirs are located in the Ostrawica, one in the Opawa and one in the Olza basin. Their task is to provide households and industry with water and power, to control floods, and also for recreation. There are eight storage water reservoirs on the Polish side of the Upper and Middle Odra basin. Their total capacity amounts to 629.72 million cubic metres, including 372.35 million cubic metres of service volume. Navigation on the Odra is mainly affected by the following reservoirs: Otmuchow-Nysa, Turawa, Dzierzno Duze and Mietkow, which are able to increase low flows. The total service capacity of the reservoirs is 323.94 million cubic metres (Table 1).

Table 1: Main characteristics of the storage reservoirs in the catchment

No.	Reservoir	River	Capacity million m^3					Year of opening
			Total	Dead	Reserve	Service	Flood	
1	Otmuchow	Nysa Kl	124.46	9.64	9.74	66.44	38.64	1933
2	Nysa	Nysa Kl	113.6	6.0	15.28	64.39	27.92	1992
3	Turawa	Mala Panew	106.2	4.0	5.20	83.30	13.70	1948
4	Dzierzno	Klodnica	94.0	40.5	-	46.50	7.00	1963
5	Mietkow	Bystrzyca	69.91	2.06	-	63.31	4.54	1986

POLLUTION

There are 1116 individual sources of pollution in the Upper and Middle Odra basin. The total effluent discharged to surface waters in 1991 reached 562.6 million cubic metres, of which 202 million cubic metres accounted for the wastewater treated physically, 28.2 million cubic metres were treated with chemicals and 228.8 million cubic metres were treated biologically, leaving 103.6 million cubic metres of non-treated wastewater.

The effects of industrial pollutants on water contamination are characterised by the following values for the discharges into the Odra basin: 328,000 tons of chlorides/year from 17 coal pits and 175,000 tons of chlorides/year from copper ore mines. The salinity of the Odra waters is of major concern for the water management of the region. Excessive salinity is mainly caused by the coal pits (72%), chemical industry (10%) and copper industry (12%) (Chabar et al, 1993, Suschka et al, 1994).

COPPER MINING WASTES

Since 1968, mining for copper ore in Legnica-Glogow Copper District has been accompanied by high inflows of saline waters to the pits. The industrial safety regulations require that such waters are removed. Essential mining for copper ore is undertaken in four mines: Lubin, Polkowice, Rudna and Sieroszewice. Subsequent technological processing is done in the ore processing plants of Lubin, Polkowice and Rudna. Table 2 shows the annual volume of pumped water, excavated ore and copper concentrate, the ratio of pumped out water to the amount of processed ore and copper concentrate obtained in the ore processing plants.

Table 2: Annual indices of water pumped out from copper pits and plants

Year	Mining water m^3 (million)	Ore tons (million)	Concentrate tons (million)	2/3 m^3/t	2/4 m^3/t
1	2	3	4	5	6
1988	30.318	29.38	1.705325	1.03	17.78
1989	30.158	26.11	1.543110	1.16	19.54
1990	28.995	24.29	1.409986	1.19	20.56
1991	29.625	23.78	1.434072	1.25	21.07
1992	30.189	24.10	1.432616	1.25	21.07
1993	30.528	25.48	1.560178	1.20	19.57

The hydrotransport of copper ore needs 4.5 m^3 of water per 1 tonne of the processed material. The sources of service water for this are as follows:

- mining water (ca. 29% of demand)
- water from stored wastes (ca. 62% of demand)
- water from ore concentrate (ca. 8% of demand)
- rain and industrial water from yards, roofs, irrigated roads, treated surface waters, refrigeration systems and water from sewage treatment plants (ca. 1% of demand).

The by-products of this process make 90% of the processed ore. These are conveyed to the storage reservoir Zelazny Most by pump and pipe systems (Figures 4 & 5).

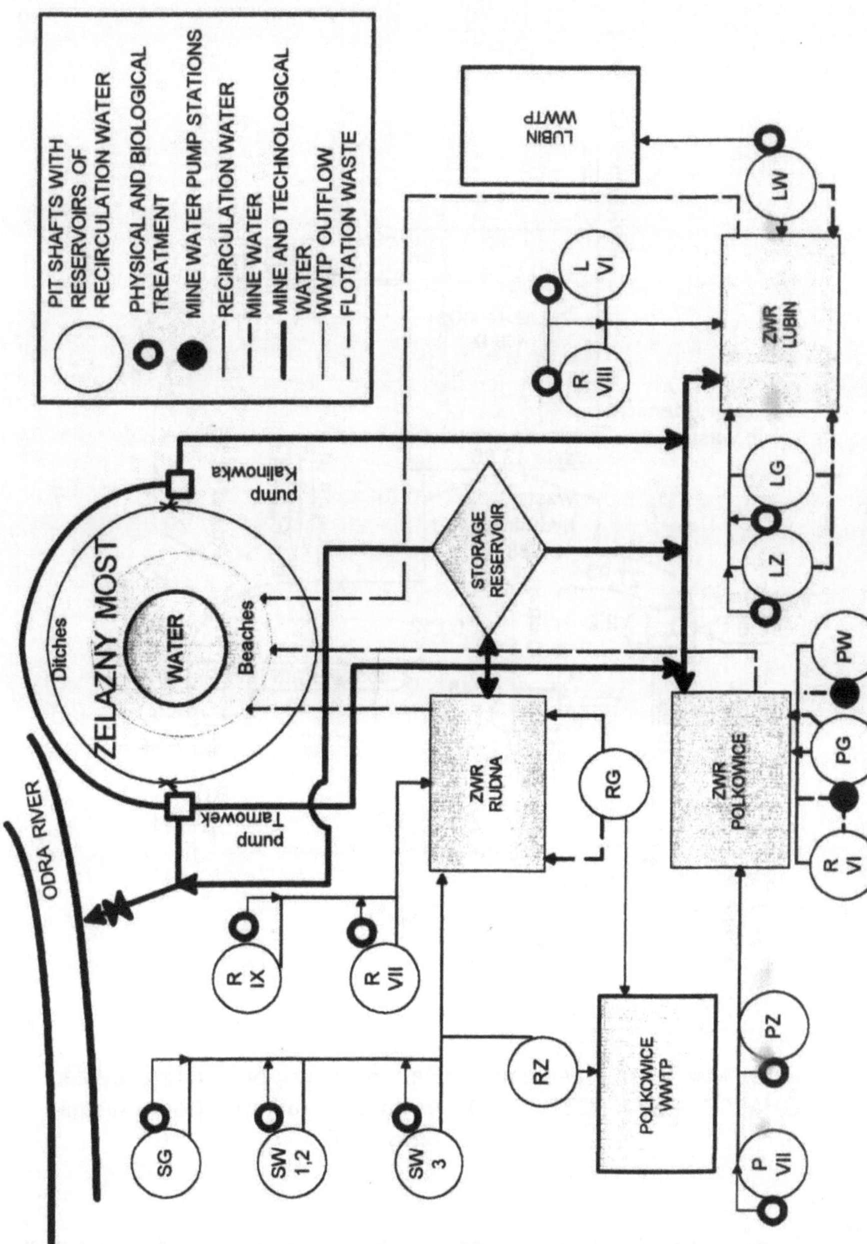

Diagrammatic representation of copper mining system (KGHM "POLSKA MIEDZ" S.A)

Figure 4

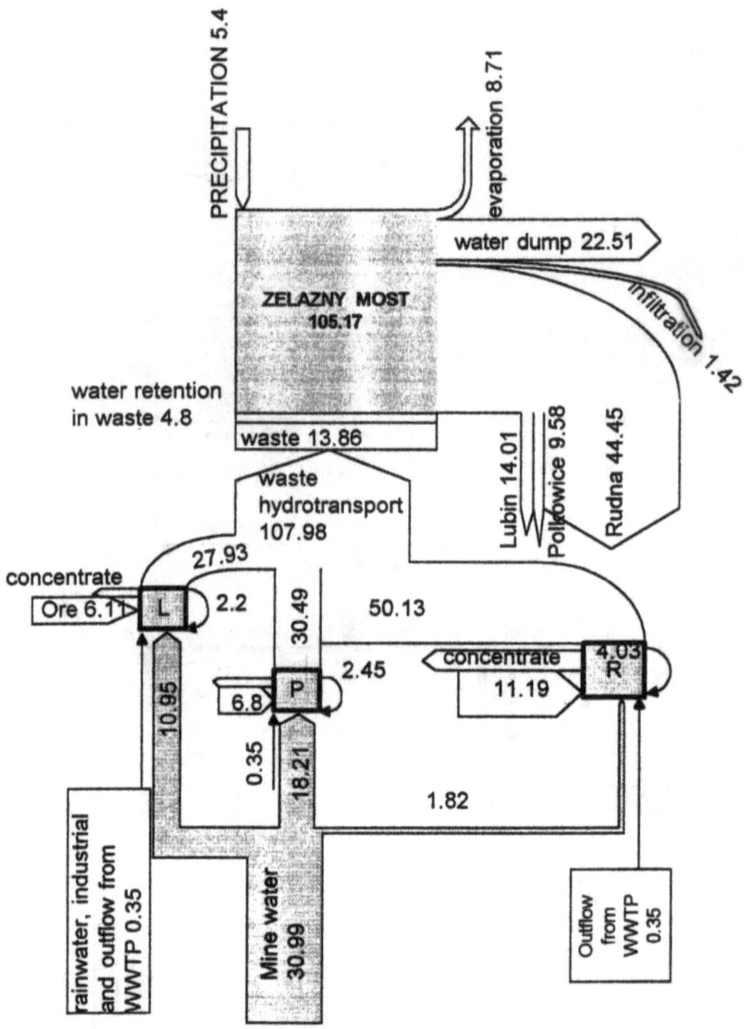

Annual water and wastewater balance for copper mining system (KGHM "POLSKA MIEDZ" S.A.) in units of million cubic metres
Figure 5

When these pass through alluviums (beaches), the processes of sedimentation and granular segregation occur. Fine particles pass to the reservoir, where they deposit to form sludge.

The supernatant water is discharged by two pump stations of Kalinowka and Tarnowek, which transfer it to ore processing plants or discharge it to the Odra at Glogow. It is extremely difficult to avoid excessively deep suspensions in the reservoir. Stratification of the flow resulting from the differences in density of the water reaching the tank significantly affects the inflow of suspended solids to the water tower system. Stratification, depending on the factors affecting water density, is observed in the reservoir Zelazny Most due to:

- thermal (differences in density resulting from variations in temperature)
- physical (differences in density due to suspension concentration)
- chemical (differences in density depending on dissolved chemicals).

The stratified flow is also influenced by winds which are responsible for sludge movements and abrasion of the banks. The forces which enable free movement of the water vary, but the impact of winds is substantial. The wind generates random vibrations on the surface of the water (whirl). The flotation and transport of the suspended particles is also affected by the location and capacity of the various water towers and pipes containing small openings.

POLLUTION CONTROL STANDARDS

The Ministry of Environmental Protection issues discharge permits via District Council Officers/Inspectorate. Currently the combined chloride and sulphate concentration must not exceed 500 mg/l in the Odra river basin. Once this level is reached all discharges are then prohibited. Unfortunately for the copper mines, the salinity is already very high in the Odra due to the discharges upstream in the Czech Republic (ca. 240,000 tonnes pa) and from the Polish coal mine drainage. As a consequence, discharges are frequently prohibited, particularly during low rainfall periods. One such occurrence happened when there was no rainfall for seven months (1991). This caused dire problems for the copper mining industry as no discharges were allowed from the spoil reservoir. In 1992 there were 162 days when the water level in the Odra was too low to discharge any wastewater and in 1993, the number of prohibited days rose to 209. It is believed, following recent French studies (in 1987 and 1992) that overall the copper mine drainage adds only some 10% to the saline pollution of the Odra river. Currently there are uniform emission standards of 35 g/l chloride and sulphate for all discharges provided the resultant river concentration does not exceed 500 mg/l.

There are frequently unrealistic expectations by the communities in the Odra basin following the major upheavals in the Polish mining industries accompanying the political moves towards a more enterprise-driven economy. Derogations by central government relaxing mining pollution controls have often been opposed vigorously by local people despite the potential stress this may apply to mining associated jobs. The expectation of *unjustifiably* high quality standards (Johnstone & Horan, 1994), is common to many parts of the world in the wake of emerging environmental awareness. Whilst it is apparent that there are sound scientific reasons for implementing improved pollution control standards for the Odra basin, the framework in which these are imposed must be taken into account. Usually, the imposition of a rigid restraint, in this case on effluent discharges to the Odra during low flows or high upstream pollution, results in a consequential pollution problem elsewhere. For the copper mines, this is a build-up in stored drainage water and deposits in the Zelazny Most reservoir, which currently occupies an area of some 16 square kilometres.

APPROACHES TO MANAGE WATER QUALITY IN THE ODRA BASIN

Individual pollutant sources can be managed by the application of BATNEEC principles, which can in many cases be environmentally sensitive in terms of land use and environmental impact, eg the use of wetland systems (Bavor & Mitchell, 1994), associated with infiltration provided groundwater is protected if it is used for other purposes. The application of BPEO however, is much more sensitive as 'best environmental option' is often difficult to define, but studies in the UK (Dempsey, 1994) have developed methodologies which can be used as the backbone for integrated catchment pollution control management strategy development. Not only can the approach be used for remediation, but it is also the basis for implementing a 'sustainable development' system (Gardiner, 1994) within a river catchment. One approach that may be used is to balance the environmental benefit against cost using the 'law of diminishing returns' as illustrated in Figure 6 (Semple, 1991).

In a recent review paper (Beck et al, 1994) questioning 'standardised' approaches to urban drainage problems, it is clear that many western countries such as the UK, Holland, USA, but also Japan, are interested in new 'sustainable' approaches to waste management, which may also include life-cycle assessments incorporating all vectors of energy and materials. Importantly, major factors such as climate change are also identified as of critical significance. Clearly this broader perspective is directly relevant now to the problems of the Odra basin. Unfortunately a fully detailed and all-pervasive methodology for

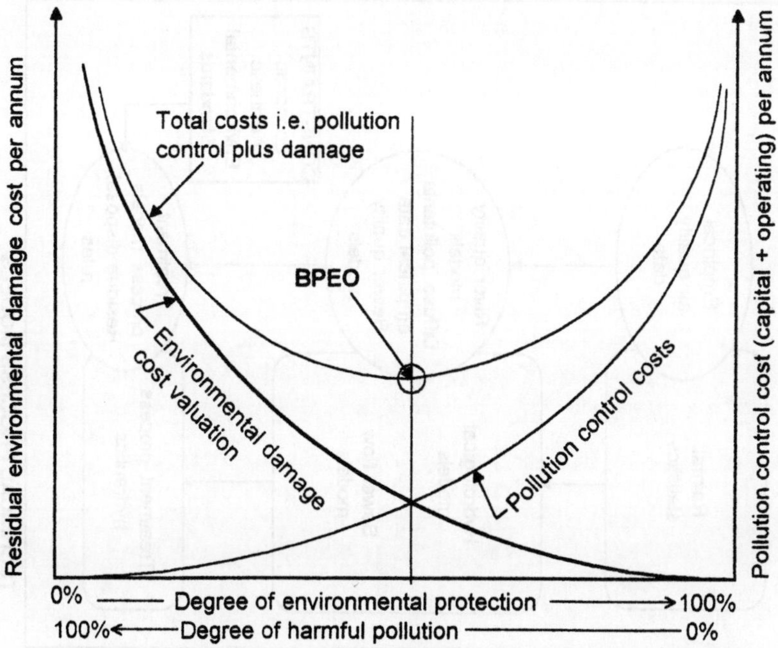

Best Practicable Environmental Option (BPEO) in holistic economic terms (from Semple, 1991) Figure 6

application is not yet devised and so far only some aspects of the various interacting system can be dealt with.

Essentially the current UK approach is as summarised in Figure 7, and incorporates technical (computational) models to simulate the natural and unnatural (pollution) processes linked to economic and socio-political constraints. These aspects of the 'system analysis' could be readily applied to the Odra basin problems.

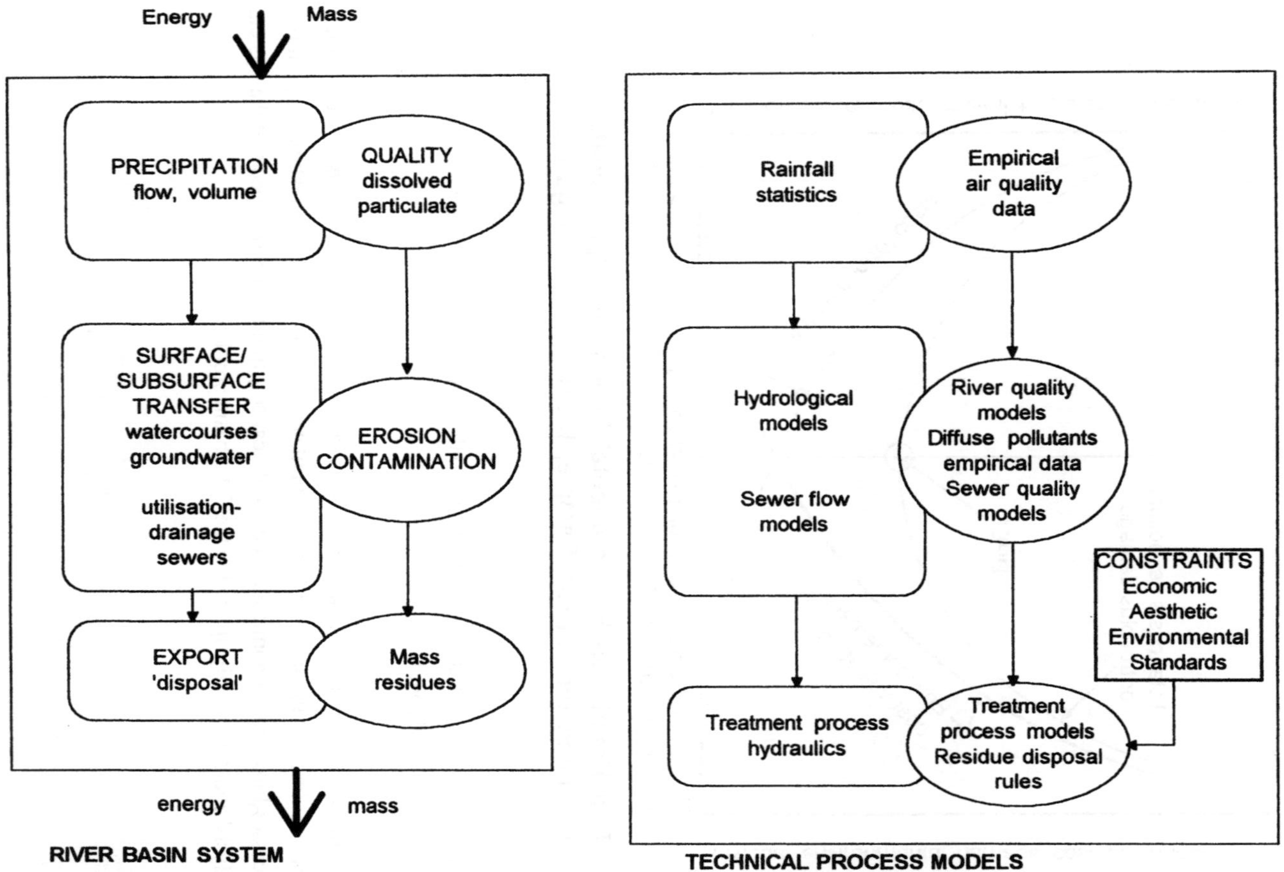

River catchment systems and process models avaialable for simulation Figure 7

SPECIFIC POSSIBILITIES TO CONTROL THE SALINITY IN THE ODRA BASIN

The available options comprise selection of some or all of the following:

- find new storage areas for effluents
- install treatment systems
- develop river regulation systems based on existing and new reservoirs
- introduction of integrated river basin management and control

Hydrogeological methods (depending on the way of discharging the water) include:

➔ Water recirculation and pumping. Using this method, the saline water is pumped into an irrigated water-bearing complex; part of it reaches the pit, whereas the remaining part migrates in the opposite direction, thus limiting the flow of water from the lower parts of the stratum.

➔ Funnelling the waters into remote geological structures, eg gas-bearing areas. These methods (depending on local and hydrological conditions) require less energy compared with desalination. Pumping of saline waters may be disrupted during transfer and this may cause a threat to the environment. Each breakdown and subsequent pollution of surface and ground waters as well as the soil with chlorides and sulphates are potential hazards to the environment.

In Polish mines, the most common are hydrotechnical solutions (Suschka, 1991). Mining water can be discharged to the nearby rivers provided that its physico-chemical composition meets the standards prescribed by Polish law. If the requirements cannot be met, the following solutions are used instead:

➔ Pipe systems carry saline water along the course of the river to the place below the last plant abstracting water. The wastewater may be discharged directly to the sea.

➔ The use of saline water dispensers. Relatively small water tanks could be built which would be able to reduce the peak of salinity, or cause long-term quality balance, in order to avoid any adverse effects of saline waters on the environment.

New and innovative localised wastewater treatment processes are being developed (Beck et al, 1994); for the copper mining wastes a radical method of treatment is desalination. The simplest methods of separation, among all well-known processes of desalting (distillation, cold treatment, electrodialysis, ion exchange and reverse osmosis) are membrane processes. These methods are mainly used for desalting sea water, especially in countries where water deficits are high, and access to the raw materials for power stations is not a problem. Generally, the condensate (brine, post-crystalline lye) is returned to the sea. Demineralisation of mining waters, whose chemical composition varies depending on the pollutant, is a more complex process and would result in high power consumption, residues and an increase in the already polluted atmosphere in the area where desalination plants would have to operate.

CONCLUSIONS

Whichever of the technical options for controlling the pollution in the Odra basin is utilised, the solution to the problem also requires the implementation of the 'broader' aspects encompassing river regulation and the development of an integrated catchment management plan. Unfortunately this is a transnational river basin which also affects at outfall to the Baltic, a polluted sea, which is subject to international pollution reduction agreements (*Red list* substances). Without a holistic approach involving all national and international interests it will be impossible to arrive at a solution to the problem which is equitable in both economic and environmental terms to both Poland and the Czech Republic.

The search for truly integrated and 'sustainable' systems for managing waterborne wastes is gathering pace, incorporating life-cycle assessments of material and energy flows (e.g. Kop, 1993). As these new methodologies emerge, they should be applied to the major areas of pollution degradation existing in eastern europe, such as the Odra basin, to ensure that clean-up measures are economically sound and result in long-term sustainable solutions, not simply stop-gap expedient measures to bring such areas up to 'western' standards.

ACKNOWLEDGEMENTS

The authors are grateful to Professor P. Blaszczyk of the Institute of Environmental Protection, Warsaw for his advice on the paper.

REFERENCES

Beck M B, Chen J, Saul A J, Butter D (1994) Urban Drainage in the 21st Century: Assessment of new technology on the basis of global material flows. Wat.Sci.Tech. Vol 30, No.1

Bavor H J, Mitchell D S [Eds] (1994) Wetland Systems in Water Pollution Control. Wat Sci Tech, Vol 29, No 4, 1994

Chaber M, Krogulski K (1993) Saline water in the catchment of the Odra River. Konferencja Naukowo-Techniczna "Odra i jej Dorzecze", 26-28 maja 1993r Wroclaw. Zeszyty Naukowe Akademii Rdniczej we Wroclawiu, Nr 232

Dempsey P (1994) Urban Pollution Management Manual - A planning guide for the management of urban wastewater discharges during wet weather. UK Foundation for Water Research.

Dieperink C (1993) Salmon, sustainability and transboundary river basin management, some reflections. International syposium on *Transboundary river basin management and sustainable development*, Delft, 1992. International Hydrology Programme: UNESCO Paris 1993.

Gardiner J L (1994) Sustainable Development for river catchments. Journal UK Institution of Water and Environmental Management, 1994, 8, June

Johnstone DWM, Horan NJ (1994) Standards, costs and benefits: an international perspective. Journal IWEM (UK) Vol 8, No.5

Kop J H (1993) Principles underlying integrated development and control of water resources in river basins. International syposium on *Transboundary river basin management and sustainable development*, Delft, 1992. International Hydrology Programme: UNESCO Paris 1993.

MacGarvin M, Johnstone P A (1993) On precaution, clean production & paradigm shifts. Wat.Sci.Tech. Vol 27, No. 5-6

Semple N G (1991) Principles of Environmental Protection. Proc UK Inst Civ Engrs, Part 1, 1991, 90, Aug

Smith A (1976) Theory of moral sentiments. Oxford, Clarendon Press, Oxford (first published circa 1760).

Suschka J (1991) Political, economic and ecological impact of coal mine saline water discharges in Poland. Environmental pollution 1 - ICEP.1 Inderscience Enterprises.

Suschka J, Ryborz S, Leszczynska I (1994) Surface water and sediment contamination in an old industrial region of Poland - two critical examples. Wat Sci Tech, Vol 29, No 3, 1994

Tyson J M, Guarino C F Best H J Tanaka K (1992) Management and Institutional aspects. Wat. Sci. Tech. Vol 27, No.12

Water Quality and Environmental Degradation in the Tom River Basin (Western Siberia): the need for an integrated management approach

Oleg F.Vasiliev,
Institute for Water and Environmental Problems
Siberian Branch of the Russian Academy of Sci.
ul. Papanintsev, 105
656099 Barnaul
Russia

Introduction

From 1990 through 1992 the Institute for the Water and Environmental Problems (IWEP), and other institutes of the Siberian Division of the Russian Academy of Sciences, has conducted a research program devoted to assessing water management and ecological state of the Tom River and its basin. Several environmentally devasted urban and industruial areas are located in the river basin. River stretches lain therein are also under heavy pollutant load and sometimes experience shortages of water.

The Tom River basin contains the Kuzbass Industrial Complex and is also the locus of agricultural activities. Kuzbass is the largest center of coal mining in Russia, one the centers of metallurgical and chemical industries. Cities such as Kemerovo, Tomsk, Novokuznetsk, Mezhdurechensk, and others, - are located on the river (Fig.1). The Kemerovo region is one of the most densely populated and heavily industrialized areas in Siberia. The region plays a key role in the economic development of Russia.

The Tom River is the main source of water supply for the Kuzbass region and the ultimate collector for the industrial, municipal, and agricultural waste waters. Over a thousand industrial plants and human settlements discharge waste waters with low efficiency of treatment into the Tom River and its tributaries. The wash-out of the pollutants from the municipal territories and industrialized areas, waste from the cattle-breeding complexes and the run-off with surface waters from the agricultural lands are also very substantial pollutants. In addition our recent data of field observations suggest that

NATO ASI Series, Partnership Sub-Series, 2. Environment – Vol. 45
Environmentally Devastated Areas in River Basins in Eastern Europe
Edited by Alfons Georges Buekens and Vasily Victorovich Dragalov

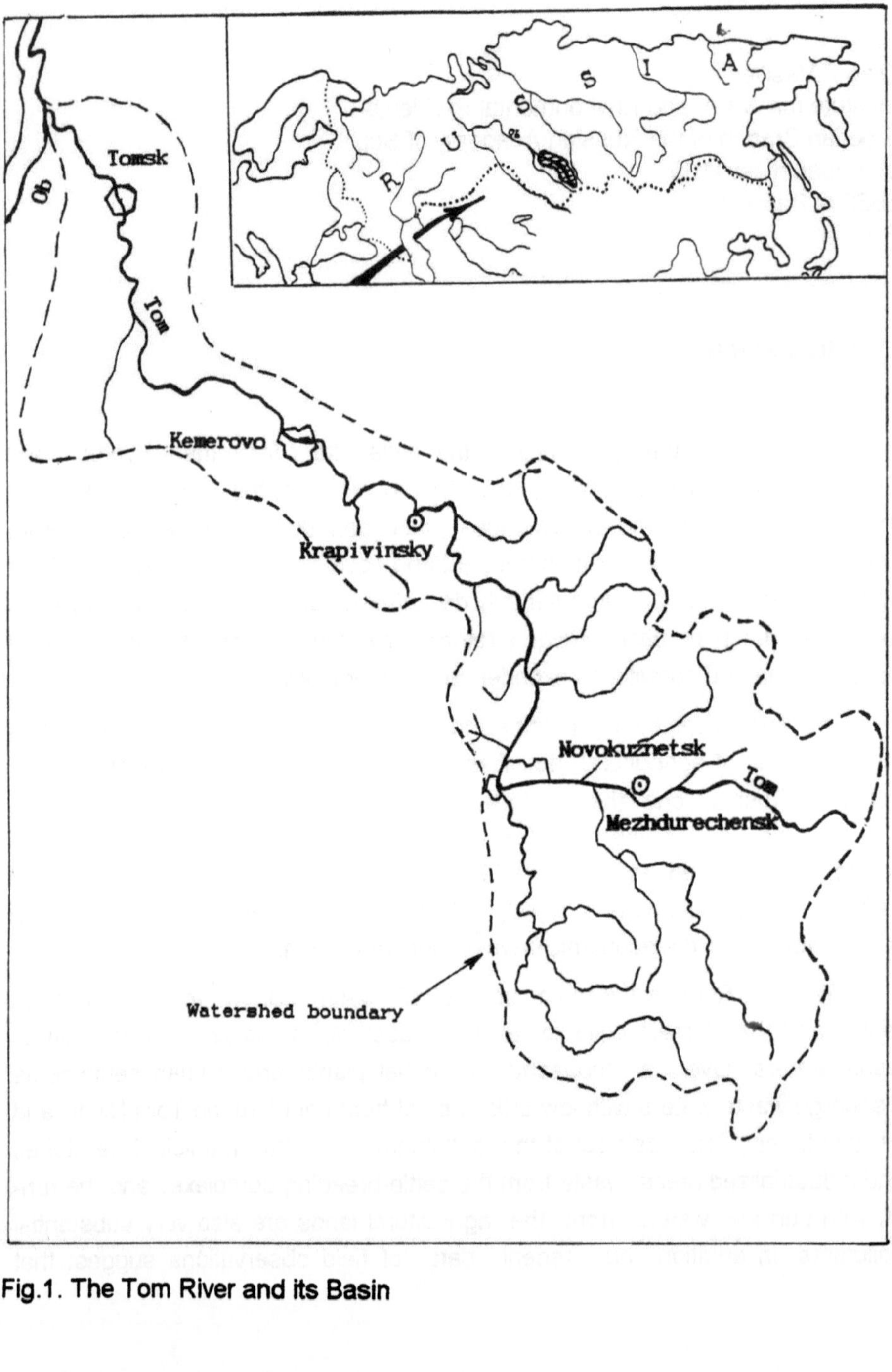

Fig.1. The Tom River and its Basin

airborne pollutants play an essential role in contaminating the watershed area and its surface waters. As a result the river waters contain practically all types of pollutants (a variety of organic compounds, heavy metals and others). In order to understand the complexity of environmental interrelations in the river basin it is necessary to consider avariety of biogeochemical processes involved and their interactions.

At the same time a rapid response is required since the unfavorable environmental conditions aggravate the tense socio-economic and political situation in this region making the environmental problems in the Kuzbass (primarily - the qualitative state of urban air and river waters) ones of the most urgent in Siberia.

Krapivino project

The Tom River run-off has a very nonuniform annual distribution. During periods of low flow (particularly in winter) hydrological conditions regularly occur most unfavorable both for water supply (lack of water and low levels) and for the quality of river water. The latter is due to the fact that the concentrations of certain pollutants released mainly from point sources reach high levels because of low dilution. Expansion of industrial activity without adequate technological changes (low waste processes; pollution abatement) will make situation only worse.

About 20 years ago, in the seventies, a project has been proposed being aimed at mitigating the water quality state and the hydrological regime along the lower part of the river: near Kemerovo and downstream. The concept of the project was to create a rather large river reservoir by constructing a dam near Krapivino (Fig.2), a settlement located upstream from Kemerovo, and to regulate river flow (run-off) providing sufficient discharges to dilute the pollutant load at the low-water periods.

Construction works started in 1975 but in 1989 it was stopped according to a governmental decision: the project had to be considered once more with the view of making a new environmental impact assessment.

IWEP has been proposed to do and lead such a study. Therefore, a number of research organizations, working under the supervision and together with the institute, have fulfilled an investigation on the effects of the Krapivino project on the hydrology of the Tom River, its water quality, river ecology and

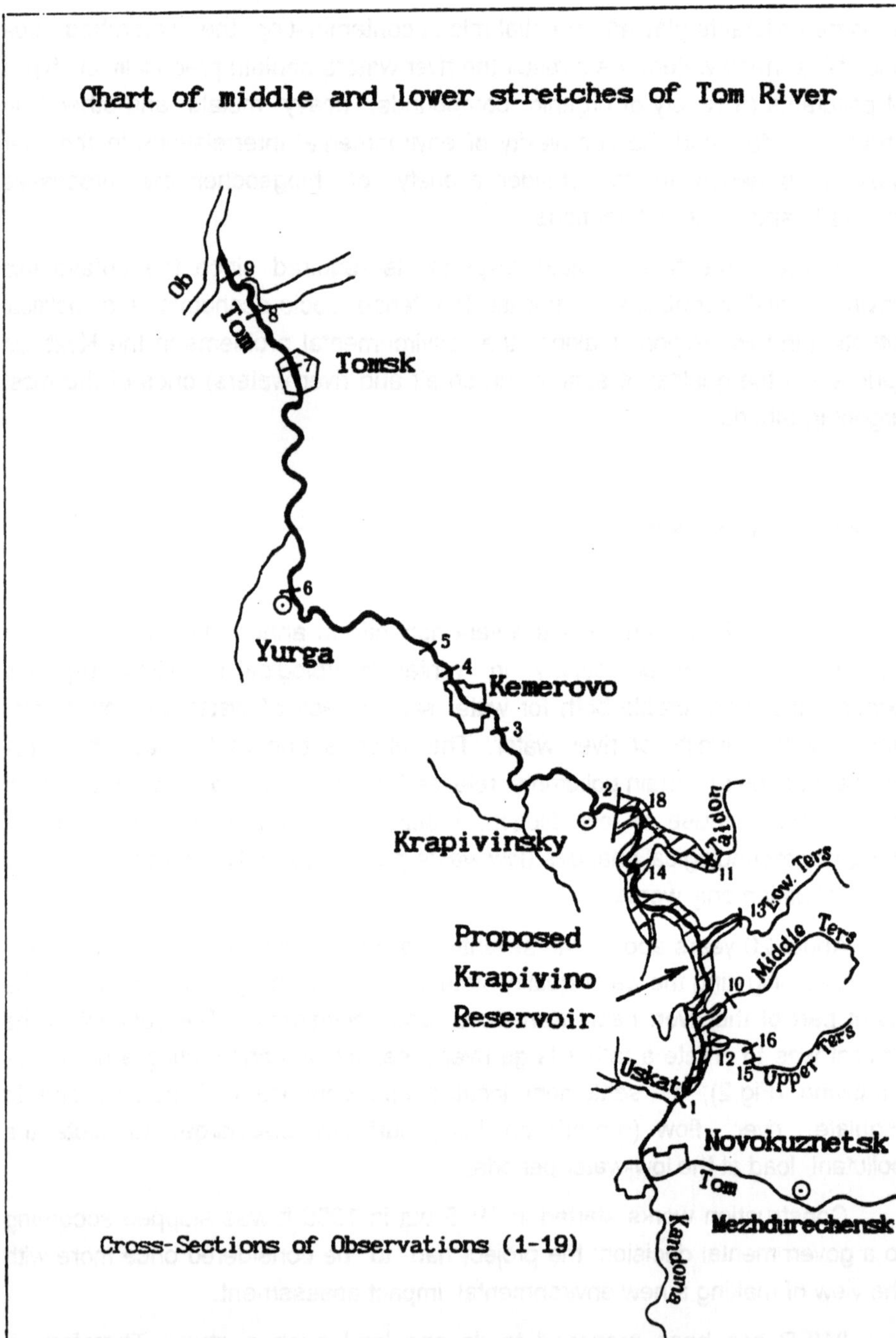

Fig.2. Chart of middle and lower stretches of the Tom River (with the location of the Krapivino Reservoir and the points of observation)

generally on assessing of possible environmental changes on the surrounding area.

Study area description

The Tom River is a right tributary of the Ob River, one of the largest rivers of Siberia as well as the world. The river is 827 km long and drains a watershed area of 62000 km^2 .The total mean annual run-off of the river is 34.1 km^3/year (1080 m^3/sec). Its seasonal distribution reflects the mixed type of formation where input of snow melting is prevailing (50-85% , in average about 70%). Near Novokuznetsk city (situated upstream of Krapivino reservoir) the mean annual run-off is 20.4km^3. The maximum and minimum values observed were 29.9 km^3 and 13.8 km^3 , respectively. In the vicinity of the dam location (near Krapivino settlement) the mean annual flow is 29.6 km^3. This amount of water exceeds the total volume (storage) of reservoir by a factor of 2.5 and its operating storage by a factor of 3 (the storages are 11.7 and 9.7 km^3 , respectively).

There are some groundwater resources prospected in the area under consideration. They have been partially developed. However, these resources are not sufficient to meet the total water demand in the region, especially in large industrial centers (even for drinking water supply). And still it is recommended to consider the use of groundwaters for drinking water supply as a priority.

The pollutant load differs along the river. Annual discharges of pollutants from point and non-point sources into the Tom River according the preliminary and rather rough evaluation by Prof.I.V.Gordin (VODGEO R&D Institute) for the period of 1988-1990 years are showed in Table 1 for Northern and Southern Kuzbass (upstream and downstream of Krapivino Reservoir).

Research works and results

A wide-range multidisciplinary research program of the study has included the collection and analysis of available hydrological, hydrochemical, ecological and other data on the river system and its watershed area, field investigations for enlarging this information and mathematical modelling as a tool for assessing environmental mpacts of the Krapivino project. Hydrological and hydrochemical data were collected at 19 points of observations (Fig.2).The field investigations have been pursued mainly on extensive survey on necessary hydrochemical, biogeochemical and hydrobiological data specifying the present state of the Tom River basin: quality of surface and ground waters; pattern of sediment, soil and snow pollution; species diversity under condition of anthropogenic stress and bioaccumulation of chemicals by the river inhabitants. An environmental assessment of the Krapivino project and its ability to ameliorate water quality was based on worldwide experience as well as mathematical modeling of hydrophysical, hydrochemical and some hydrological processes in the proposed reservoir and in a river stretch downstream of it. For this purpose the set of mathematical models was developed (or adopted): models of thermal and ice regimes of river and reservoir; models of sediment, pollutant transport and water quality in river and reservoir (including the models of nutrient, namely-phosphorus and nitrogen cycles in the reservoir). Research works were carried out by several scientific groups from different institutions during a period over two years.

The major results of this investigations on water resources and environmental state of the Tom River and its basin in brief are as follows:

1. Currently the total supply of water resources in the Tom River basin is sufficient to satisfy the regional water demand. However lower river discharges during low water periods cause water shortages as well as deterioration in water quality during these periods. Moreover, groundwater supply in the basin is limited and does not allow to satisfy the demands of the large industrial centers (even for drinking water supply).The seasonal regulation of river run-off by developing the Krapivino reservoir is the most pragmatic solution of the problem. This will enable to increase the low discharges in the river during low water periods.

2. The hydrochemical study of water quality reveals that the Tom River and some of its tributaries are heavily exposed to anthropogenic contamination especially immediately downstream from the large industrial centers. Main pollutants are the organic compounds (petroleum compounds, phenols, polycyclic aromatic hydrocarbons, formaldehyde, aniline, organic chlorine compounds, some amines, naphthalene and its derivatives, dibutylphtalate and its derivatives), nitrate and ammonia nitrogen, sulfates, chlorides, as well as some heavy metals (cadmium, zinc, chromium, copper et.al.).

Concentrations of the above substances often exceed drastically the national standards for water quality in natural water bodies.

The pattern of water contamination of the river changes during annual hydrological cycle. At the time of winter low water the pollution by petroleum compounds is dominant everywhere along the river. Some stretches of the river are highly polluted by volatile phenols and ammonia nitrogen. Although the spring flood water dilutes the petroleum compounds, water pollution by pesticides increases everywhere along the river. Also the level of contamination by formaldehyde, phenols and nitrate increases locally during the high water period in spring.

3. Heavy metals content in the water and bottom sediments does not exceed as a rule the normal background. However there are contaminated stretches in the locality of industrial centers (e.g.,Novokuznetsk and Kemerovo).

4. During the last decades an essential alteration in the populations of many different aquatic species occurred in the Tom River. Downstream from the large industrial centers the populations of river inhabitants is dramatically depressed. This is ubstantiated with elevated levels of bioaccumulation of some heavy metals and other chemicals in aquatic species. These data reflect a high degree of contamination of abiotic components of river ecosystem. The mean content of cadmium and lead in tissues of all fish pecies exceeds national safety limits for fish consumption. In some fish species mercury was also found to exceed safe limits.

5. Presently the quality of surface waters in the Tom River basin is determined to a great extent by pollution from different sources; including industrial plants, urban sewerage systems, agricultural lands and urban and industrial areas. Therefore its essential improvement is impossible without significantly reducing of emissions to air, of water and soil pollution from industrial and urban sources as well as without a more careful use of fertilizers, pesticides and other agrochemicals.

In more general terms we may conclude that the hydrochemical and hydrobiological investigations have revealed a general pattern of pollution in the Tom River and its tributaries and allowed us to identify priority pollutants for different stretches of the river. More work will be required, however, to construct mass balances of key pollutants for river sections. It is necessary to achieve a clear understanding of the relative importance of the various point and non-point sources of river pollution. Without such an understanding it is impossible to develop any substantiated strategy of water quality management.

The evaluation of non-point sources of river basin diffuse pollution also needs to be considered as a total spatial and temporal environmental context at the river basin scale. This must include the transport of airborne pollutants, their deposition on the watershed surface and soil pollution.

It is worth to tell about another an interesting phenomenon concerning the pattern of the chemistry of the river waters. Observations have revealed rather high content of petroleum compounds and phenols along the total stretch of the Tom River and its tributaries. Even on the upper section of the river, upstream of Mezhdurechensk, such a pattern is observed. The question arises whether the origin of these chemical is anthropogenic or natural? Certainly the answer is of importance for developing the strategy of water quality management.

As far as it is concerned the assessment of environmental effects caused by Krapivino project they are the following:

1. The flooded land area of the proposed Krapivino reservoir extends over an area of 650km^2 , consisting mainly of alluvial soil with local accumulations of peat throughout the area. Heavy metal content in soils and peat deposits is close to normal background and no radioactive anomalies on the flooding area have been revealed. After reservoir filling there is a possibility of peat rising to the water surface (the volume of such a peat can be assessed at about $3.4 \cdot 10^6 m^3$ that is 6% of total peat deposits). According to experience of Novosibirsk reservoir, however, this should not seriously ffect the hydraulic structures operation and would not significantly affect water quality.

2. Mathematical modeling of hydrothermal and ice regimes of the Krapivino reservoir (with the maximum depth of 54,2m) indicates two distinct periods of thermal stratification, one occurring during the summer period and another during the period of ice cover. The behavior of the ice cover is expected to be slightly different from the one under the present river condition. The stretch of open water in ice downstream from the dam (the polynia) will be approximately 60-80km long. The use of selective withdrawal of water from reservoir could shorten its size up to 20- 40 km.

The impact of reservoir on the climate of the area would be limited to a 1-2 km distance from its shoreline. During the winter the frequencies of fog occurence will increase (up to 3-4 per month) along the polynia stretch.

3. The oxygen regime of reservoir is predicted to be rather satisfactory. During the typical year the average concentration of oxygen will fluctuate within the 6-12 $mg \cdot l^{-1}$ range (the lower limit for protection of aquatic organisms is 4 mg

$\cdot l^{-1}$). Depletion of oxygen could possibly occur only during the winter locally at the bottom of the reservoir.

The gross content of various organic biodegradable constituents (described by biochemical oxygen demand, BOD) in the main part of reservoir and immediately downstream from the dam is lower than in the river water entering the reservoir except in the period of intensive phytoplankton growth during the summer. During all periods of the year it will be the same for dissolved inorganic nitrogen content as well.

4. The Krapivino reservoir would be an oligomezotrophic water bodysimilar to the Novosibirsk reservoir in relation to phytoplankton development. In the summer the expected average volume content of phytoplankton biomass is 0.3-1.8 $mg \cdot l^{-1}$.

5. The impact of the proposed reservoir on the water quality downstream from the dam relating to labile dissolved organic content and nutrients may be considered as a whole to be positive. It is stipulated by magnification of natural treatment (or the selfpurification, including settling of contaminated suspended sediments) under reservoir condition as well as by the dilution of pollutants with increasing discharges of the water during low water periods. Moreover, the reservoir could play a positive role in case of accidents at the treatment plants and the tailings situated up stream by dampening the high concentration of chemicals in the water and providing favorable conditions for settling of suspended matter.

6 The change of hydrogeological regime will impact on down-stream water users but it would only occur for a limited distance (tens of kilometers) below the dam along the river channel.

Conclusion

Research conducted was aimed at establishing the scientific basis to develop a strategy for improving the ecological state of the Tom River basin. The results obtained thus far have demonstrated conclusively that the water quality problem can be considered separately from other environmental issues of the region. It can be done only within the total environmental context of the river basin.

There is a need to develop a comprehensive overview of the sources, ways of migration (or transport) and accumulation of contaminants on the watershed and in river system. The result of such an analysis could serve as a base for developing a strategy for integrated management of water and environmental quality in the river basin scale. In elaborating the strategy, it is necessary to establish priorities in the management of environmental quality.

Applying for the system approach to solve the problem it is supposed to carry out the research works on the assessment of present and expected water quality in the Tom River Basin. The study must include an analysis and appraisal of possible ways for transition to new environmentally friendly technologies which reduce waste loads, in particular - to closed technological systems to decrease waste discharge into the environment.

Table1. Annual discharges of pollutants into the Tom River

Type of pollutants	Point sources ton/year	Non-point sources ton/year
	Southern Kuzbass	
Suspended Sediments (matters)	15 069	153 696
Petroleum Compounds	385	2 072
COD	25 578	72 208
BOD	4 737	29 307
Nitrogen	13 213	28 386
Phosphorus	5 81	41 176
	Northern Kuzbass	
Suspended Sediments (matters)	11 537	29 805
Petroleum Compounds	244	369
COD	10 765	16 504
BOD	2 153	8 459
Nitrogen	6 852	8 214
Phosphorus	284	1 644

NATO ADVANCED RESEARCH WORKSHOP
ARW

INTEGRATED REHABILITATION OF ENVIRONMENTALLY DEVASTATED AREAS IN RIVER BASINS IN EASTERN EUROPE

september 12-16, 1994, Moscow, Russia

the Hron River Basin, Slovakia
Danube Environmental Programme

by Peter J.M. Kerssens, Project Manager
DELFT HYDRAULICS, the Netherlands

1 INTRODUCTION

The regional environmental programme for the Danube river basin has been jointly established by the riparian states of the Danube at a meeting in Sofia in september 1991. The Sofia meeting, also attended by supporting financing organizations such as the Global Environmental Facility of the World Bank, the Commission for the European Community and the European Bank for Reconstruction and Development (EBRD), as well as bilateral sponsors, adopted a broad framework for a 3 year-action programme to develop a pre-investment programme and for human resource development activities that establish a foundation for further cooperation.

The aim of the pre-investment programme is to establish a phased, strategic investment programme for improved environmental management and pollution control in the basin. As such it is considered a key activity in the development of a strategic action plan for sustained environmental management in the Danube river basin, effectively meaning the preparation of "bankable" projects to solve environmental problems at somentioned "hot spots". A second, but not less important objective of the pre-investment studies is an identification of the reforms required to establish an institutional and policy framework that will enable sustained investment and environmental management. Technical assistance should be supplied to the local authorities in order to train them in the field of comprehensive environmental planning and programming, setting up the institutional framework for such and develop an environmental management system.

The subject study forms the Dutch input/contribution to the pre-investment studies for the mentioned Danube Environmental Program (DEP) and preparation of the strategic Action plan. In september 1992, in discussion with the Program Coordination Unit in Bruxelles, a project proposal was prepared for the Hron river basin in Slovakia and presented to the Netherlands government. The project was approved in october 1992 and started accordingly. Project executing agency from the Netherlands side is a joint venture of DELFT HYDRAULICS as project leading organization and Rijkswaterstaat/RIZA. The major local partner in the project execution was the Water Research Institute, Bratislava. Preparatory meetings between the Slovak Ministry of the Environment, the Slovak Focal Point Coordinator, a representative of the Netherlands Government, and the executing agency took place in November 1992 in Bratislava. The Diagnostic Mission phase of the project was executed between November 1992 and March 1993. In April '93 the first activities of the actual Pre-Investment Studies started with a mission to Slovakia for data collection, identification of the Decision Support System setup and initial work for computer modelling. This abstract describes the project set-up, the various subprojects for pre-investment studies, the overall environmental study "surrounding" the subprojects, the function of the mentioned Decision Support System and the supporting study on economic/financial and institutional aspects.

NATO ASI Series, Partnership Sub-Series, 2. Environment – Vol. 45
Environmentally Devastated Areas in River Basins in Eastern Europe
Edited by Alfons Georges Buekens and Vasily Victorovich Dragalov

Figure 1 Hron River Basin, Slovakia, and its location within the Danube basin

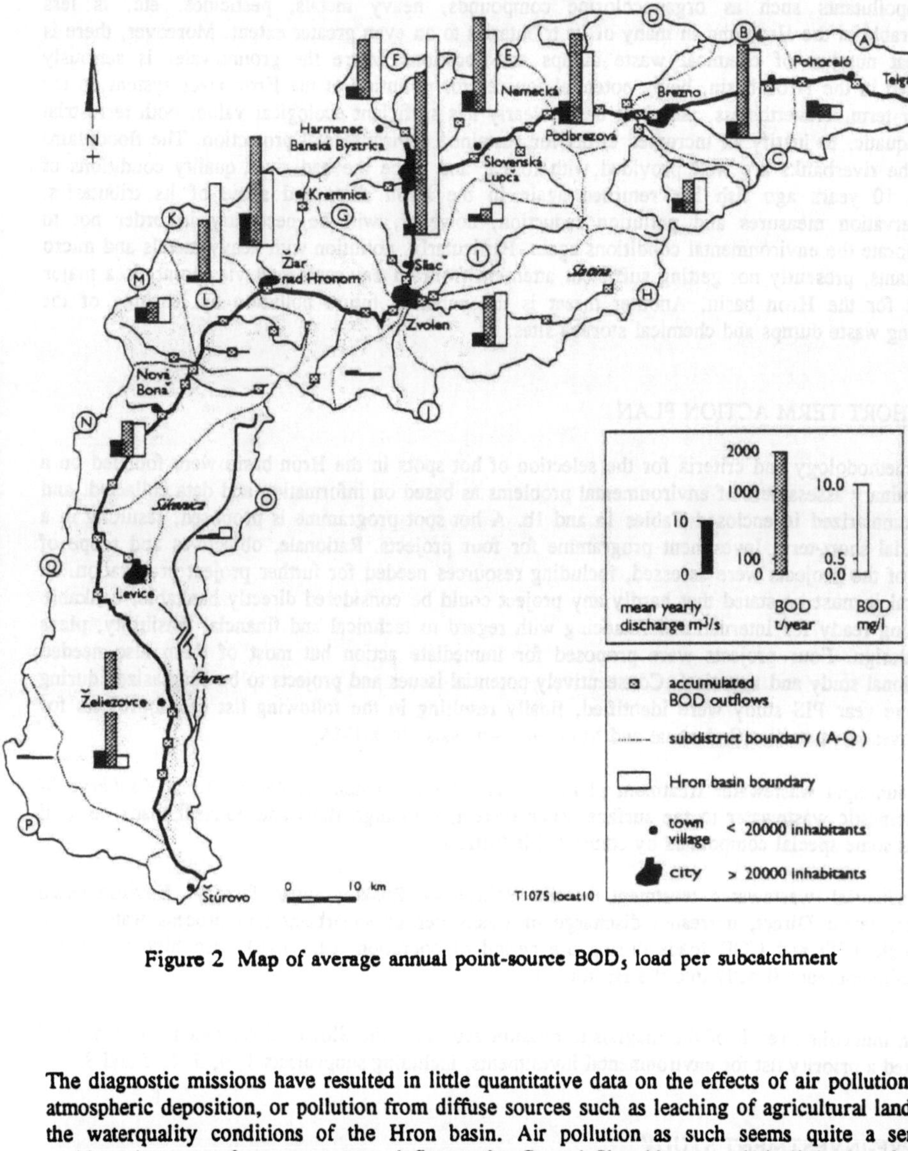

Figure 2 Map of average annual point-source BOD_5 load per subcatchment

The diagnostic missions have resulted in little quantitative data on the effects of air pollution, via atmospheric deposition, or pollution from diffuse sources such as leaching of agricultural lands on the waterquality conditions of the Hron basin. Air pollution as such seems quite a serious problem, but apart from some general figures for Central Slovakia as a whole little data for the Hron basin itself have been found. The same applies for the pollution by diffuse sources. It is clear that leaching of agricultural land, historic dumping sites for chemical waste as well as factory/plant sites polluted by accidental spilling and leakage, contributes significantly to the water pollution of the ground- and surface watersystem in the Hron basin. However, apart from an overview of the most serious chemical dumps, a list of mine and ore processing locations and estimates on pesticide use in the basin, no data on this subject can be produced.

Overall it can be stated that the Hron basin, despite heavy bacterial pollution of the river itself and high amounts of contaminants at selected locations, presently is in a reasonable state regarding environmental conditions. However, this seems to be a too optimistic view. Although there are not sufficient data for any firm conclusions, there is evidence that the situation concerning micropollutants such as organochlorine compounds, heavy metals, pesticides, etc. is less favourable in the Hron and in many of its tributaries to an even greater extent. Moreover, there is a great number of chemical waste dumps and locations where the groundwater is seriously polluted in the Hron basin, being potential sources for pollution of the Hron river system on the longer term. Nevertheless, the river basin clearly has sufficient ecological value, both terrestrial and aquatic, to justify an increased effort for sustained environmental protection. The floodplains and the riverbanks are well provided with forest, and since the bad river quality conditions of about 10 years ago fish has returned again in the Hron river and some of its tributaries. Conservation measures and pollution reduction, however, will be necessary in order not to deteriorate the environmental conditions again. Particularly, pollution with heavy metals and micro pollutants, presently not getting sufficient attention from an environmental viewpoint, is a major threat for the Hron basin. Another threat is the potential future pollution by leaching of the existing waste dumps and chemical storage sites.

4 SHORT TERM ACTION PLAN

The methodology and criteria for the selection of hot spots in the Hron basin were founded on a preliminary assessment of environmental problems as based on information and data collected, and are summarized in enclosed Tables 1a and 1b. A hot spot programme is proposed, resulting in a potential short-term Investment programme for four projects. Rationale, objectives and scope of each of the projects were assessed, including resources needed for further project preparation. In general it must be stated that hardly any project could be considered directly bankable, bankable meaning ready for international financing with regard to technical and financial feasibility, plans and design. Four projects were proposed for immediate action but most of them also needed additional study and appraisal. Consecutively potential issues and projects to be emphasized during the one year PIS study were identified, finally resulting in the following list of subprojects for pre-investment studies (Industrial and Municipal Analysis block/IMA):

1) Municipal wastewater treatment plant at Ziar. Environmental problems: Direct discharge of domestic wastewater to the surface water system, with high BOD and nutrient loads, as well as some special compounds by connected industries.

2) Industrial wastewater treatment at the Pohronsky Ruskov sugar factory. Environmental problems: Direct, untreated discharge of washwater of sugarbeets and process water, with high BOD and COD loads during the period of operation (campaign), also high suspended solid contents directly into the Hron river.

As an immediate result of the diagnostic mission activities, the Slovak Government in May 1994 adopted a priority list for environmental investments, including subprojects 1, 4, 5, 6, 7 and 8.

5 PRE-INVESTMENT STUDY

As indicated before, the Pre-Investment Study (PIS) phase consists of a number of feasibility/ appraisal studies on specific subprojects in order to assess bankability for international financing. In addition a subproject has been executed on the preparation of a Decision Support System for the Hron river basin, consisting of a Database, a Water Quality Simulation model for the Hron and relevant surface water system, a Case Analysis and Case Management Tool, graphical presentation techniques, an economic evaluation model and the necessary interfaces to allow communication between the various computer modules. The Hron DSS allows an evaluation of the effects and/or impacts of pollution reduction measures on the Hron water system, and to weigh the efficiency against costs for implementation. It simulates the fate and transport of pollutants throughout the Hron system and its discharge to the Danube river. By means of a waste load

Table 1a Characteristic parameters of main pollution sources used as criteria for hot spot selection

Criteria / Potential Hot Spots	Industrial Waste Water/plants/processing							Soil Containination Sites				
	Bucina wood Zvolen	Pohronský Ruskov sugar	Biotika Slovenska Ľupča	Harmanec paper	ZSNP alum.	Levitex textile	Petro. Dubova	Sliač airfield	ZSNP factory	Bučina Zvolen	Mining Tailings	ČSD complex Brezno
1 Environmental conditions												
a emission load (quantity)	●●●	●●	●	●	●●	●●	●	●●●	●●●	●●●	●●	●
b type of pollutants												
- BOD/COD	•	•	•	•	•	•	•	•				•
- heavy metals	•				•	•	?		•	•	•	
- N and/or P												
- micro pollutants	•		?	?	•	•	•	•	•	•	•	•
c effect on human health and ecosystem												
- short term		-	-	-	-	-	-	-	-	-	-	-
- long term (sustainiability)					-	-	--	--	--	--	--	--
- after emission reduction	n.r.	-	-	-	-	-	?	?	?	?	?	?
d possibilities for emission reduction												
- short term (non point & point sources)	+	+	+	+	+	+	+	+	+	+	-	+
- long term	+	+	+	+	+	+	n.r.	n.r.	n.r.	+	+	+
2 Economic/financial aspects												
a economic feasibility (industry)	??	?	+	+	+	?	+	n.r.	+	?	?	-
b availability of cost recovery system	-	-	•	-	•	?	•	-	-	-	-	-
c availability of financial (investment) plan	-	•	•	•	•	?	•	(*)	•	-	-	-
3 Technical aspects												
a availability of technical solutions	•	•	•	•	•	?	•	•	•	-	-	(*)
b availability of technical plans	?	•	•	•	•	?	•	(*)	•	-	-	-
4 Institutional aspects												
a availability of local organization to execute the project (technical, financial responsibility, political aspects, structure for cost recovery)	?	•	•	•	•	?	•	?	•	?	?	-
5 Various remarks	simple treatment in operation	new plant in tender phase	3-stage treatment plant in operation	biological treatment in planning	connection to Žiar WWTP in planning	more analysis needed	new biological treatment plant in operation					

●	moderate	•	present	n.r.	not relevant
●●	high	+	positive	?	questionable
●●●	very high	-	negative		
		o	no significant effect		

Table 1b Characteristic parameters of main pollution sources used as criteria for hot spot selection

Criteria / Potential Hot Spots	Municipal Waste Water discharge/plants					Miscellaneous			
	Banská Bystrica	Žiar nad Hronom	Brezno	Kremnice	Zvolen	Sewerline BB	Fish-ladder VK	VVO Budca pig-farm	chemical dumps
1 Environmental conditions									
a emission load (quantity)	●●●	●●●	●	●	●●	●●●	n.r.	●●●	●
b type of pollutants									
- BOD/COD	*	*	*	*	*	*	n.r.	*	
- heavy metals	?	?	?	?	?	?			*
- N and/or P	*	*	*	*	*	*		*	
- micro pollutants	?	?	?	?	?	?			*
c effect on human health and ecosystem									
- short term	-	-	-	-	-	-	-	-	o
- long term (sustainiability)	-	-	-	-	-	-	-	-	-
- after emission reduction	o	o	o	o	o	?	n.r.	?	?
d possibilities for emission reduction									
- short term (non point & point sources)	+	+	+	+	+	+	n.r.	+	?
- long term	+	+	+	+	+	+		+	
2 Economic/financial aspects									
a economic feasibility (industry)						n.r.	-	-	?
b availability of cost recovery system	*	*	+	*	*	*	-	-	-
c availability of financial (investment) plan	*	*	?	?	?	*	*	*	-
3 Technical aspects									
a availability of technical solutions	*	*	*	*	*	*	*	*	-
b availability of technical plans	*	*	?	*	?	*	(*)	*	-
4 Institutional aspects									
a availability of local organization to execute the project (technical, financial responsibility, political aspects, structure for cost recovery)	*	*	*	*	*	*	*	*	-
5 Various remarks	Various common issues to be studied in PIS							treatment plant in construction	serious loads to be expected in future

● moderate
●● high
●●● very high

* present
\+ positive
\- negative
o no significant effect

n.r. not relevant
? questionable

model simulating the inputs within the basin, including point source inputs such as effluent from existing or planned industrial or domestic wastewater plants as well as diffuse non-point sources from viz. agriculture or mining activities, various input scenarios can be analyzed. The DSS system has been transferred to the Slovak counterparts to be utilized for future policy development and water quality management studies, and training courses on use of the DSS were given.

An overall environmental/ecological substudy has been carried out to look at overall environmental conditions and potential effects of sanitation measures and/or expected effects of existing solid/liquid waste dumps (with/without sanitation), on the potential for nature protection, restoration and development, on recreation potential, and related environmental aspects. Finally, economic/financial aspects as well as institutional issues have been addressed explicitly in order to put the Action Plan in the right development perspective, to identify the local and international financing capabilities as well as the implementation potential.

The PIS part of the study started in April '93 and was finished in October '94 with the final report as well as a final workshop for transfer of the results to the Slovak governmental authorities, related industries and water management authorities and the Danube Project Coordination Unit. Major objectives of the Workshop were to report and discuss the results of the subproject studies in the field of Industrial and Municipal Analyses, to transfer the Decision Support System to the various interested parties including training and transfer of knowledge, and to discuss follow-up and financing potential for the most urgent sanitation projects. For international financing existing contacts with the WorldBank as well as the European Bank for Reconstruction and Development have been explored. In 1994 already an EBRD-loan of 110 million US $ has been granted to the Slovak Government for environmental investments for the ZSNP Aluminium Factory, and the EBRD commissioned further appraisal studies to determine the necessary technical measures.

6 THE DECISION SUPPORT SYSTEM

An integrated approach of water management on a regional scale, in this case a complete river basin, increases the complexity of policy making dramatically. A computerized decision support system (DSS) can be used to *support* the process of selection of hot spots, evaluation of environmental impacts of alternative measures and the effectiveness of investments from an economical point of view.

The PC decision support system developed for the Hron by DELFT HYDRAULICS and VÚVH constitutes a specific *application* of *existing software* that was developed in the course of 15 years of national policy support in The Netherlands. The system has been adapted and implemented in many other countries. The DSS for the Hron consists of three sub systems:

a. the *information* system containing system specific databases (geographic data, environmental standards, monitoring data etc.), and a database management system;
b. a set of *simulation models* for:
   - hydrology and hydrodynamics (flow);
   - quantification of waste loads;
   - water quality prediction of nutrient dynamics, primary production, BOD, bacteria, oxygen, heavy metals and organic micro-pollutants;
c. an *analysis system* to define and compare management strategies, to perform risk assessments and evaluate costs.

The user communicates with these three modules via a User Interface and much of the output can be presented using graphical techniques (maps, bar-charts, graphs etc.). In the user interface one can switch between two languages, it this case between Slovak and English.

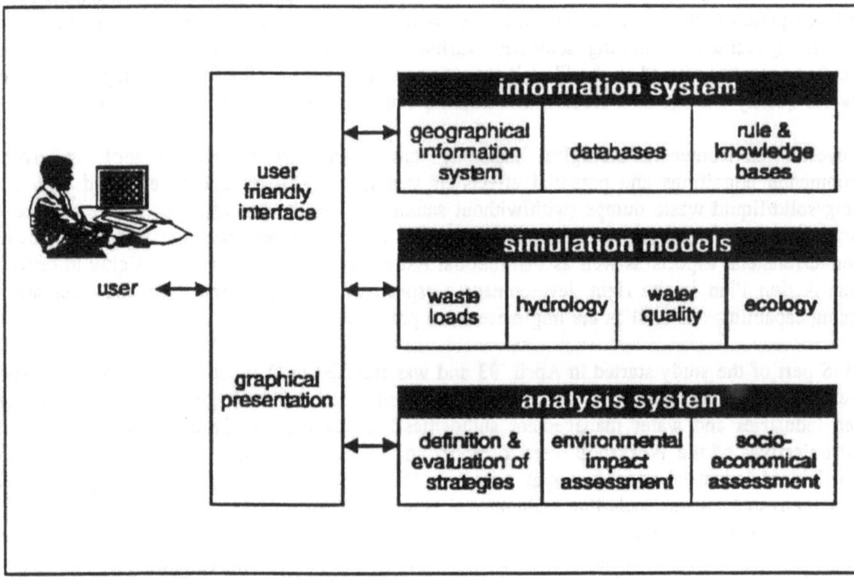

Figure 3 General outline of a Decision Support System.

The Hron DSS software has been implemented at VÚVH, Bratislava, and is also available to other governmental institutes that are involved in environmental policy planning for the Hron.

The DSS has been calibrated with data from the calender year 1990, using the measured discharges in the Hron as a starting point. The schematization for both the water quantity and the water quality simulation comprises 26 segments covering the entire basin of the Hron river and the major tributaries such as the Slatina.

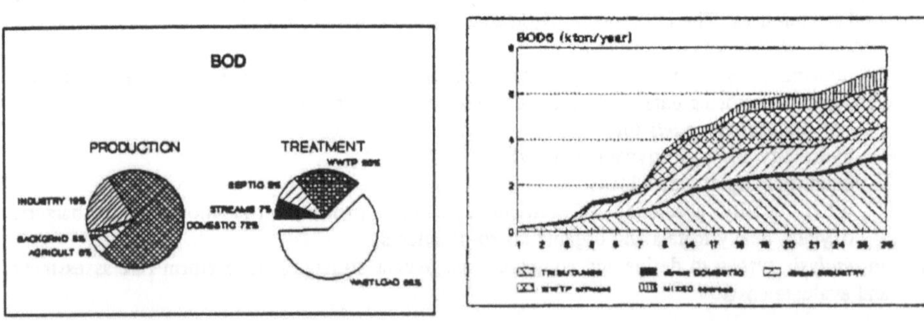

Figure 4 Relative contribution of various sources of BOD waste loads (production) and purification (treatment) in the Hron river basin (left) and cumulative waste load profile of discharges to the Hron river (right).

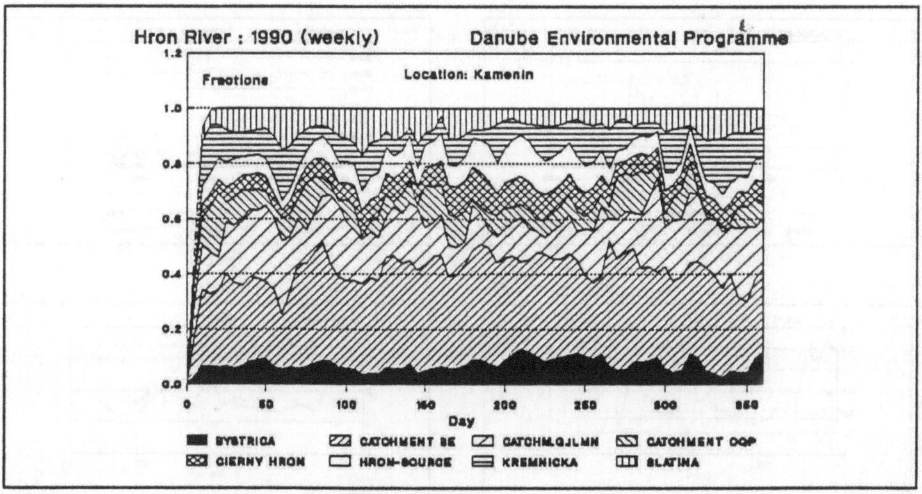

Figure 5 Origin of water in the Hron river at Kamenín (1990)

Based on the available data, both the water distribution in the system and the residence time for the Hron have been calculated, resulting in a residence time for the main part of the Hron ranging from 2 days during high flow situations and 6 days in low flow periods (1990). Furthermore, the origin of the surface water at Kamenín has been analyzed, as is indicated in the figure. In the same way, the various water quality substances have been simulated (calibrated) with the DSS.

A number of scenarios have been identified for analysis with the DSS. The scenarios mainly include the construction of various waste water treatment plans at specific locations, i.e. major cities along the Hron river, connection of the Levitex factory to the WWTP in Levice, extension of the sewer system and prevention of direct discharge in Banská Bystrica, connection of the Bučiná wood factory to the WWTP in Zvolen, and finalization of the bentonite wall at the Aluminium factory in Žiar to prevent pollution via the groundwater to the Hron surface water system. An extensive description of the individual measures and the selected scenarios is presented in the DSS report.

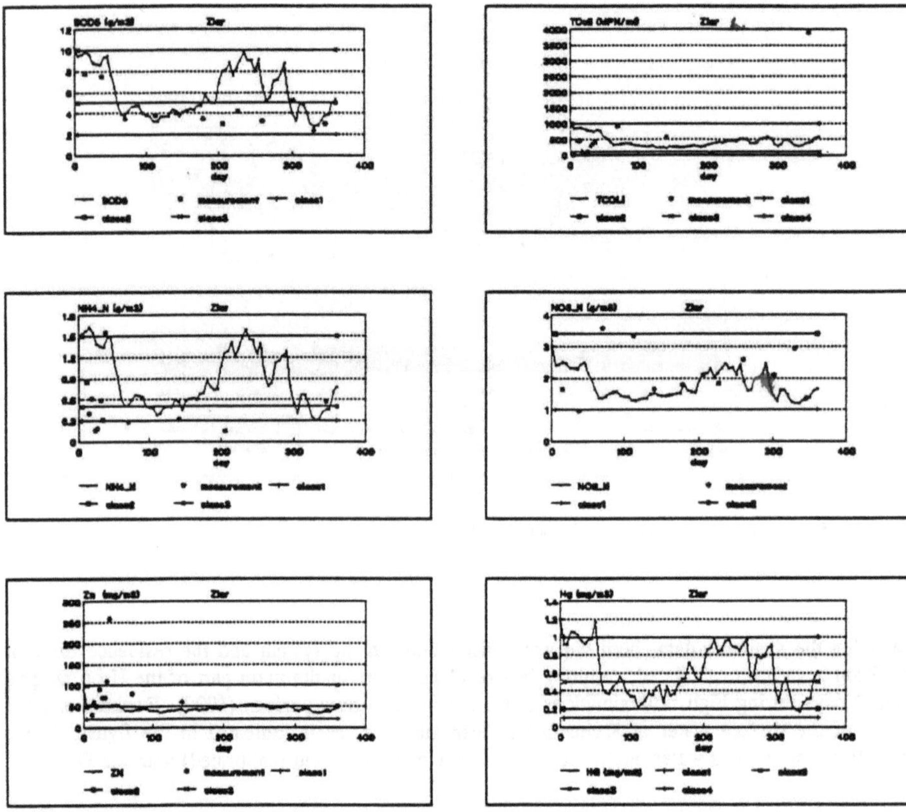

Figure 6 Monitoring and calibration results for BOD_5, Coliform bacteria, ammonium, nitrate, zinc and mercury compared to Slovak standards at Žiar (1990)

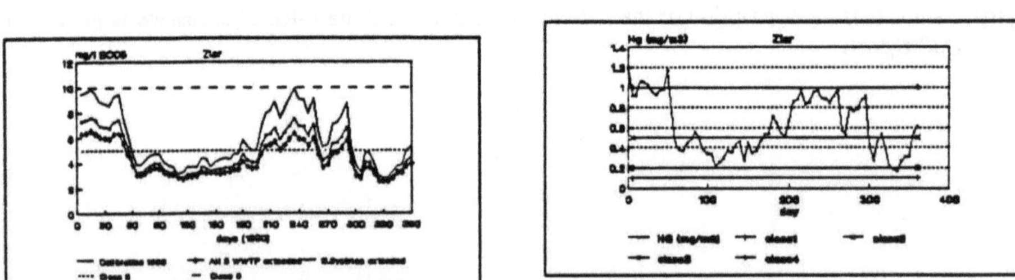

Figure 7 Effect of proposed measures on the BOD_5 concentration at Žiar and Kamenín (Danube)

Major conclusions from the application of the decision support system are:

- **The DSS in general and the Waste Load model in particular gives a good overview of inputs for various substances at the different locations within the Hron river basin;**
- **The simulated results generally compare reasonably well to the monitoring data available;**
- **The present water quality situation in the Hron system varies from moderately till severely polluted on most substances: BOD_5, bacteria, nutrients, micro-pollutants and heavy metals;**
- **Due to the relative low residence time of the water in the Hron river the self-purification capacity of the Hron is small;**
- **A significant decrease of BOD_5 (20%) and coliform bacteria (20%) can be achieved with the construction of WWTP's in selected cities. Locally the reduction can be substantially higher, particularly in low flow situations. Reduction of nutrients and heavy metals is hardly effectuated by such measures;**
- **Reduction of micro-pollutants and heavy metals should be implemented at the (point) source of origin, i.e. end-of-pipe or in process technology at factory level;**
- **The pollution of diffuse sources, such as agriculture, atmospheric deposition, traffic, mining, etc. should also be tackled at source level as hardly any reduction of concentrations takes place by water quality processes in the river system.**

7 WASTEWATER ANALYSIS SUBPROJECTS/HOT SPOT ANALYSIS

7.1 Municipal waste water treatment plant Banská Bystrica

A proposal was made for a two-stage extension of the treatment plant to optimize the use of those extension parts which have already been constructed. In phase 1 the plant capacity will be made equal to the capacity of the current main sewer, to avoid overflow of untreated water. These activities and the renovation of the existing plant should be completed at the earliest possible. In phase 2 a second main sewer from Banská Bystrica will be constructed. However, in comparison with earlier plans this sewer will be much smaller, resulting in lower costs. After phase 2 there will be no more overflow to the surface water of untreated waste water from the sewer system of Banská Bystrica and surrounding area and it will be possible to meet the Slovak effluent standards for nitrification and the 2005 standards for the removal of phosphorus.

Total investment costs for the total extension works of the treatment plant and sewerage system are estimated at 392 million Sk (US $ 12.3 M). This results in a rise of operational and maintenance costs per m^3 of treated water from Sk 2.6 in 1993, to Sk 2.9 in 1995 and Sk 3.2 in 2005.

7.2 Municipal waste water treatment plant Zvolen

Contrary to Banská Bystrica, only a small part of the planned extension has been realized in Zvolen so far. The phase 1 and 2 extension works started both in 1991. In phase 1 the present treatment plant is upgraded and an additional sewer is connected. In phase 2 a new treatment is to be build resulting in a drastic increase in hydraulic and biologic treatment capacity. To prevent unnecessary investment related to over-capacity it is recommended to discontinue the extension works immediately and to first adjust the design to the new projections of required capacity. The necessary investment costs for the treatment plant and a main sewer are estimated at 120 million Sk (US $ 3.8 million).

7.3 Other municipal waste water treatment plants

Apart from the WWTP's at Banská Bystrica and Zvolen there are two more plants under expansion- or reconstruction works in the Hron basin, namely at Žiar and at Brezno.

At Žiar the domestic waste water, with a high BOD and nutrient load, directly discharges into the surface water system. In addition, effluents from industries connected to the sewer system are polluting the surface water. During a short visit to the treatment plant at Žiar the civil part of the construction works appeared of a good quality and was nearly finished. The mechanical equipment about one third of the cost. The Ministry of Environment also looks for funding possibilities. Although in the present situation it is not easy to raise funds in Slovakia, it may be feasible in this case due to the relatively small size of this project.

8 INSTITUTIONAL ASPECTS

The study on institutional aspects has resulted in the following conclusions and recommendations:

- The responsibility for policy formulation, management and control with respect to the water sector is divided over two Ministries, i.e. the Ministry of Soil management and the Ministry of Environment. The financial resources of the Ministry of Soil Management to invest in the surface and groundwater infrastructure are very limited. The Ministry of Environment coordinates and supports expenditures for public water supply and sanitation from the Environmental Fund, but the required investment costs greatly exceed the available budget;

- Decentralization of the water supply and sanitation company (StVAK) to limited stock companies at municipal level will complicate foreign investment in waste water treatment plants as a) the projects will be relatively small, and b) the Central Government may not guarantee the international loans. Clustering of projects may solve the first problem but will work against the trend of decentralization;
- Water and waste water charges do presently not cover the costs involved. A study and policy for cost recovery in the Water Sector will be urgently needed.

Through lines of communication, cooperation, regulation and financing a lot of institutions are involved in the national and regional decision making process with respect to the environmental investments that should take place. Currently many of these lines are being evaluated in the framework of proposed legislation (the new Water Act), decentralization and privatization. The institutional situation therefore is far from clear. Principles of government guaranteed free access to vital commodities such as drinking water may be partly in conflict with the principles of the free market economy, full cost recovery and "the polluter pays".

In the final report of the pre-investment study a number of comments is being made to the situation of september 1994. Also an economic assessment of different investments in municipal waste water treatment plans and the consequences of full cost recovery in consumer prices is presented.

9 GENERAL ENVIRONMENTAL ISSUES: NATURE REHABILITATION POTENTIAL

Essential to nature rehabilitation is to restore natural habitat patterns and processes as a prerequisite for the sustainable life of particular species and communities. The restoration of habitats and natural processes instead of nature conservation may not seem very efficient when at endangered locations the last endemic species are under heavy pressure. As shown in the picture, there is quite a large area with landscapes currently under pressure. However, there is an increasing awareness among scientists and policy makers that species protection has no use when the corresponding habitats are degrading.

Habitat rehabilitation never can be understood as reconstruction of any former state. Although the historical state can serve as a reference, the current boundary conditions put by both nature and culture are different. New objectives have to be found. This is a challenge for landscape planners, engineers and environmentalists. Because natural succession is, apart from human management, mainly dependent on habitat conditions, i.e. abiotic conditions of the site, geographical aspects play a substantial role. Although a certain level of uncertainty is inherent to the process of habitat rehabilitation, a proper analysis method makes clear where decisions have to be taken at specific time steps during the project process. By presenting clear comparisons of alternatives on criteria defined by the interest groups, an integrated approach should circumvent long discussions between interest groups. Generally, and specifically in Central Europe, such discussions are still in a process of evolution towards fruitfulness. Involvement of the local inhabitants in the process is therefore crucial to create enough platform for rehabilitation plans, which otherwise may be encountered as culture- and nature-alien.

Nature rehabilitation may lead to new forms of economical land use as well, such as nature-friendly meat production, cropping and ecotourism, thus supporting the implementation of sustainable rehabilitation plans. Because in many instances little expenses are required, or even substantial benefits can be gained from the rehabilitation projects, habitat rehabilitation can evolve into a turning point and model issue in the discussion on improving the environment and quality of life in Slovakia. From interviews it is concluded that the Slovak Ministry of Environment is ready to take an initiating role regarding this topic.

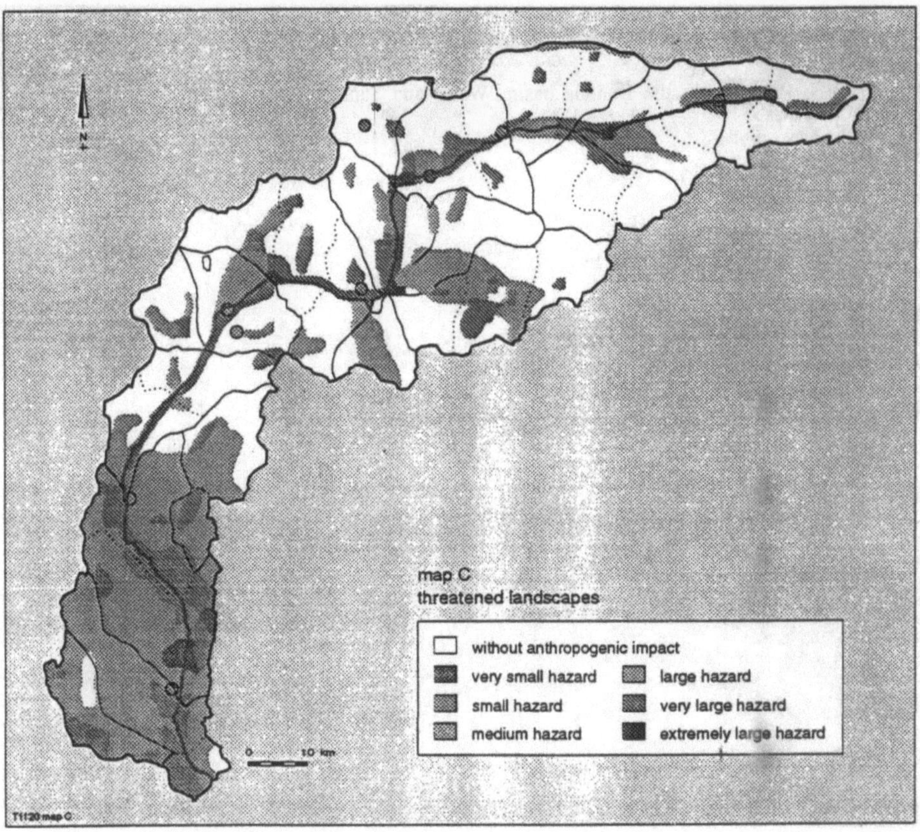

Figure 8 Map of threatened landscapes in the Hron river basin

One example to be mentioned of a rehabilitation issue combined with economical profitability, is the plan to realize a number of small hydropower plants on the lower Hron, currently under preparation under the supervision of the Povodie Hrona, Banská Bystrica. Main aims would be groundwater recovery. side stream rehabilitiation and power generation. If well designed, such plan could indeed contribute significantly to the rehabilitation of diversity in riverine ecosystems. However, it should be stressed that a sound integrated policy analysis of the complete cascade system should be carried out, followed by Environmental Impact Assessment, to guarantee that economic argumentations will not unduely dominate the planning process.

REFERENCES:

Danube Environmental Programme, Hron river basin pre-investment study. Final report of the Diagnostic mission. DELFT HYDRAULICS/RIZA/WRI, April 1993.

DANUBE river study for the Slovak water catchment area; Water and Sanitation for Health Project, Source SHMU. 1992. Ministry of the Environment.

Water quality/watermanagement balance study for the Hron Subcatchment and Levice Industrial-Residential Complex. Ing. Anna Horváthová and Ing. Dása Madarásová, WRI Bratislava, 1990.

Statistical Yearbook of Slovakia 1991, Slovak Statistical Office

Hydrology of the river Danube. A. Stancik and S. Jovanovic et al, 1988

Point Source Pollution in the Danube basin. Water and Sanitation for Health (WASH) Project, USAID. July 1992.

GLOBAL REGULARITIES IN THE OCCURRENCE OF THE ENVIRONMENTALLY DEVASTATED URBAN AREAS AS A FUNDAMENTAL BASIS FOR URBO-MONITORING (METHODOLOGY OF SUSTAINABLE DEVELOPMENT OF TOWNS IN RIVER BASINS)

Irina V. Lazareva
Central Research and Design Institute for Town Planning
pr. Vernadskogo 29
Moscow
RUSSIA

The current decade of the 20th century at the turn of the century and transition to the 21st century and the 3rd millenium is characterized by numerous natural and man-induced disasters of various scale which give grounds for UN to evaluate this period as International Decade for Natural Disaster Reduction (IDNDR). The frequency and intensity of catastrophes are governed by the global temporal and spatial laws. Most frequently they are observed in the vicinity of big rivers in the basins of which a great number of towns are located. As a rule, such towns have a long history and many valuable historic and architectural monuments, some of them included in the UNESCO list of the worldwide heritage. Therefore, in order to ensure sustainable development of the above towns it is important to conserve both their cultural heritage and urban objects that support their normal functioning. This fact makes it necessary to adopt a strategy of preventive measures. In this case special emphasis is to be placed on the urbo-monitoring as an instrument to regulate the activity of various experts' groups under risk situations (RS).

Urban monitoring is to be based on due consideration of the temporal and spatial regularities in the occurrence of the natural disasters that become

NATO ASI Series, Partnership Sub-Series, 2. Environment – Vol. 45
Environmentally Devastated Areas in River Basins in Eastern Europe
Edited by Alfons Georges Buekens and Vasily Victorovich Dragalov

evident actually against the background of evolutionary natural processes. In most cases this factor determines the location of the man-induced risk situations.

For that reason it is important to develop an integrated list of ecologically significant qualities of territories in addition to the engineering-geological conditions also including geo-chemical and geo-physical conditions (provoking unmotivated behaviour of operators, cancer diseases, etc.). The author proposes the so called *formula of territories* including 0-7 factors which makes it possible to evaluate the type of risk situation, the actual risk for a human being or environmental object, the size of the risk area. The formula dictates the need to take into account both local and regional geological conditions within different elements of the river basin such as watershed - plateau - slope - valley.

Analysis of urban territories considered as urbo-geo-systems shows that in the recent years risk situations are occurring with higher frequency.

Time periods are mainly associated with the centennial cycles of activation of the geodynamic processes that are also related to the most important events in human history including ups and downs in the worldwide development. The frequency of the catastrophic manifestations of geo-cosmic interactions is 42 and 666 year for the mountainous and valley areas known from the Bible, and about 2000 years on the plains. Similar regularity is observed in the spatial distribution of the catastrophes. Thus, catastrophic eartquakes and floods occur approximately every 80-100 years in the same place. Coordinates of geoactive spots of the plant are known for each particular country.

Man-made disasters as well as the sources of military conflict that affect the normal life of people of the whole planet and of regions, such as burning of oil in the Persian Gulf, are mainly related to the tectonic floating spots - the junctions of the structural blocks of the Earth crust. It should be emphasized that many old historic settlements, including many world capitals, are located within the above ecologically-hazardous areas. (The evaluation of 35 capitals of the Central and Western Europe was made with the participation of the author).

In this respect of principal importance is : (a) to identify the regularities in the occurrence of the risk situations (RS) for a concrete town, and (b) to take into account the regional specific conditions in the activity of the local authorities. Thus, there is good reason to bring to the notice of the National Committee on Catastrophes and to the appropriate UN Commissions (such as Habitat, UNDP, etc.) their own doctrine to evaluate the RSs) the violation of the population rights to an adequate environment. In this case priority is given to the value of human life. So, it seems more reasonable to replace the traditional concept of the logical safety by the risk concept which is the only reasonable under the limited time and resources and to come to the development of an integrated programme of preventive and operational measure.

The author has developed a special form based on N-index which was actually tested to evaluate the quality of the urban environment taking into account the period of preventing a risk situation for the population and vitally important elements of the urban environment. The form could contribute to better coordination of the activity of all authorities responsible for the control and to the identification of the degree of potential risk.

Thus, an evaluation of environmentally devastated urban areas taking into account the global regularities in the occurrence of risk situations will be used as a basis to formulate programes for sustainable development of towns. Sustainable development of towns is ensured by the integration of data about the history of towns beginning from their pre-history to post-present realities.

Principles of Integrated Water Management for the Revival of Old Industrial Areas

by Professor Dr.-Ing. W.F. Geiger
University of Essen, Institute of Environmental Engineering,
D-45117 Essen, Germany

Table of Contents

Acknowledgements

The material presented draws on work done as part of

- UNESCO-IHP Project M-3-3a, Objective 1 "Integrated Water Resources Management in Urban and Surrounding Areas", and
- projects contributing to the revival of the Emscher area.

The author wishes to acknowledge the contributions to the UNESCO-IHP Project M-3-3a by the participants of two workshops held from June 15 to 18, 1992 in Essen and from May 28 to June 2, 1994 in Gelsenkirchen, Federal Republic of Germany. Especially honoured are the rapporteurs of the second workshop Professor Herricks, Urbana-Champaign, Illinois, USA, Mr. Hissink, Rotterdam, The Netherlands, Dr. Kaden, Berlin, Federal Republic of Germany, Mr. Marsalek, Hamilton, Ontario, Canada, Dr. Massing, Düsseldorf, Federal Republic of Germany, and Dr. Thomas, Perth, Australia. I am also indebted to Professor Herricks, presently guest Professor at the University of Essen, for his assistance in editing this paper.

Special thanks are due to the Emschergenossenschaft, Essen, which allowed to use their material, especially the results of work contracted to the University of Essen for the renewal of the Emscher system. Last but not least, the open and foresighted support of Professor Dr. Ganser, Director of IBA Emscher-Park, Gelsenkirchen, is greatly appreciated when he encouraged to go beyond my own work and to convey the IBA Emscher-Park spirit.

NATO ASI Series, Partnership Sub-Series, 2. Environment – Vol. 45
Environmentally Devastated Areas in River Basins in Eastern Europe
Edited by Alfons Georges Buekens and Vasily Victorovich Dragalov

1. Introduction

The renewal of old industrial areas is a growing concern in industrialized countries. The short-sighted use of resources, especially water resources, has changed the natural environment and now threatens further use of old industrial areas.

There are no standard methods, rules or fixed procedures for renewing old industralized areas. Further, there are no standard objectives that specify use, expected quality, and long term management for devastated areas. Although the lack of specific guidance may slow some projects, there may actually be benefits in the freedom to do what is right rather than what was required in the past. When renewing old industrial sites, it must be recognized and accepted that each project will be "special" reflecting specific climatological, geographic, ethnological and cultural background, and socio-economic conditions in an area.

Although no standard methods exist and objectives may be poorly defined there is a recognition that general requirements must be met for the renewal of old industrial areas. Several general requirements follow:

- Human well-being must be secured. This includes that every person should have a dwelling that provides protection from changing weather and climatic variability, particularly from natural disasters such as flooding. An economic system should allow the head of the household to meet family including children and the aged needs for food and education. This sense of human well-being is much more important than material well-being.

- The environment, including air, soil, vegetation, surface water, and groundwater should be in a condition that it supports human well-being ensuring especially human health and maintains all living organisms in a sustainable state of high quality.

- The environment as a whole, especially water resources, should only be used to an extent that protects resources for future generations. A critical issue is that future generations should not be forced to pay for the mistakes and environmental degradation of earlier generations. Clearly water resources must be used wisely and protected to assure future sustainability.

However, we know that no generation on earth has fully met these requirements, and we are just beginning to grasp the full implications of past industralization. Renewing old industrial areas is one of the first steps we can take to address needs of society, economy and the environment.

In considering renewal priorities, water availability and water quality are obviously strongly linked with human well-being and the socio-economic condition of man. Consequently, renewal programs can benefit from an integrated approach to water management. Unfortunately, when we examine old industrial areas we find that integrated water management did not exist in the past, and water resources management only met selective goals.

As an example of the general situation, and renewal opportunities possible with integrated water management, experiences gained in the renewal of the old industrial sites covering the Emscher River Basin are provided. In the following text the examples are given in italics.

The background of the Emscher area is as follows. The Emscher area is the core of the Northrhine-Westfalian industrial region usually called the Ruhr region (Figure 1). With approximately 5.3 million inhabitants the Ruhr region is one of the oldest and largest industrial areas in the world. The reason for its industrial development lies in its geographic location within in north-west European bituminous coal belt, which extends from Silesia to England through the Ruhr area, Belgium and northern France. The carbon layers penetrate the surface of the earth on both sides of the Ruhr valley. Consequently, it was already possible in the middle ages to extract coal with simple technical means.

The major industrialization in the Ruhr region occurred within a few decades. While 130 years ago the area was still largely agricultural, it has changed completely into an industrial area dominated by coal and steel. While in 1820 274.204 inhabitants were counted, in November 1989 there were 5.337.118 people (KVR, 1990). On the basis of coal and steel, the area became the most industrialized and energy producing part of Germany. The Emscher area with approximately 865 km^2 *and seventeen cities, which corresponds to the Emscher river basin, forms the large urbanized and industrialized centre of the Ruhr region. Although the importance of heavy industries was decreasing in the last decades, still 2.5 million people are living today in the Emscher area.*

From its nature the structure of the Emscher river is not suited for larger settlements. In the middle ages it has been a flat and partially swampy area with only small water courses. Around 1900 Middeldorf (1904) reported a low flow of 3 m^3/s*. This flow consisted already at that time of 1.5* m^3/s *mining water, 1.1* m^3/s *industrial and communal wastewater and 0.4* m^3/s *natural flow. Due to coal mining the land surface subsided up to 20 m at some points opposing natural drainage. In order to serve the pressing drainage problems for the industrial region in 1904 communities, mining and heavy industries founded a waterboard, the Emschergenossenschaft, with the aim to secure drainage and to re-establish hygienic living conditions.*

The extremely difficult drainage conditions could only be solved by a number of measures, which today present the Emscher system. A major component of this technical system is the reshaping and straightening of the natural surface waters to receive and to carry the waste flows produced by population and industries. However, it has not been possible to place this collection system underground as the land subsided constantly due to coal mining. Further, due to land subsidence the Emscher river and its tributaries in some parts had to be diked and bridges had to be adapted to the continuous land subsidence with increasing annual costs. Most tributaries of the Emscher river obtained concrete inlays to increase their hydraulic capacity and to allow for the constant maintenance necessary in the subsiding areas. Pumping stations and flood retention basins were installed to prevent flooding (Figure 2). With this solution a hygienic and least dangerous drainage was possible.

Still today all domestic and industrial waste flows are collected by this open system. Originally, waste flows only were treated mechanically, which with increasing pollution due to the industrial inputs, was no longer feasible at some stage. Biological treatment of all flows then was provided in 1956 to protect the Rhine river. To allow for free surface drainage in spite of land subsidence the Emscher mouth to the Rhine river 1974 was located downstream to Dinslaken. Today, biological treatment takes place at the Emscher mouth treatment plant located at Duisburg-Dinslaken and at the treatment plant Duisburg-Kleine Emscher In the Emscher area itself there remained six mechanical facilities for pretreatment. The mechanical treatment facilities together with oxygen concentration aimed at avoiding odour emission of the waste flows as well as extensive sedimentation in the river beds. Furthermore, at the coking plants there are dephenolization plants that pretreat the waste flows before they are discharged into the river.

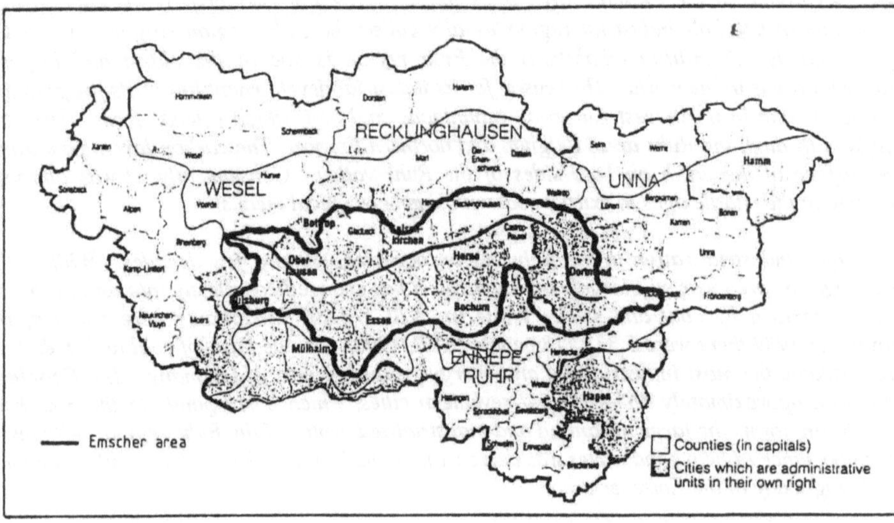

Figure 1. Location of the Emscher area within the Ruhr region (after KVR, 1990)

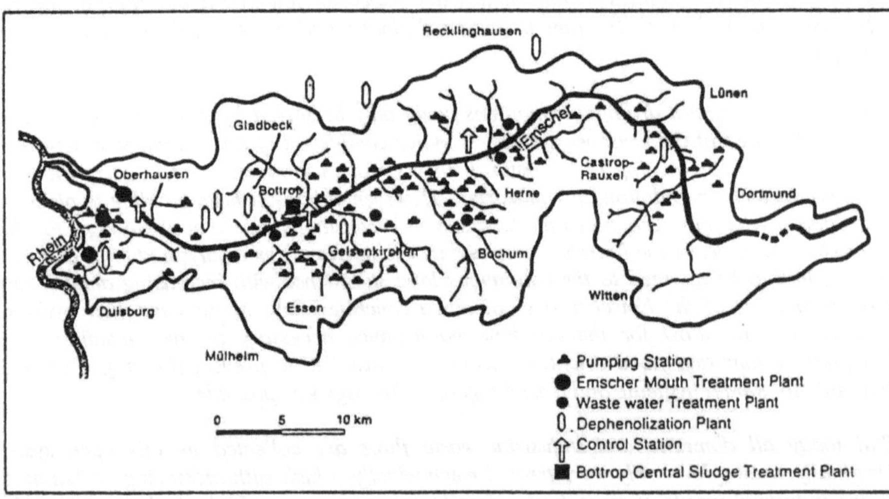

Figure 2. The Emscher river and its tributaries showing also existing pumping stations (Emschergenossenschaft, 1983)

2. Problem Identification

The problems associated with old industrial areas are many. Old industrial sites are typically plagued with ineffective or inadequate wastewater treatment and stormwater drainage. Past industrial activity has also often left environmentally damaging residuals on surfaces and in the ground. Industrial sites have large impervious areas and mostly inadequate runoff control and treatment for stormwater. To address the problems posed by the renewal of old industrial areas, a vision is needed by managers, engineers and scientists, that is also shared by the public.

2.1 Sanitation and Industrial Wastewater Treatment

A major problem recognized in many fast growing cities and industrial areas, in the past and today, is a development of sanitary provisions that often does not keep pace with the rapid urbanization and industrialization. Domestic wastewater and human faeces in unsewered residential areas are discharged directly and untreated into canals and other open water courses, or are infiltrated into the soil, with that contaminating the various sources of water. This practice not only leads to unhygienic conditions and low human perception values of open waterways, it also exposes the people to the risk of epidemic diseases, such as cholera and typhus. The increasing contamination of wastewater streams with industrial by-products, which are often toxic, eventually reaches the point where the natural recuperative capacity of the receiving watercourses is exceeded. They may become unfit for further human use.

Unless measures are taken to secure and/or restore the original situation, these degraded water courses may produce human health problems and limit further economic growth. Many now developed countries were already confronted with this serious problem and have by now institutionalized and implemented Master Plans for the majority of their cities to neutralize such risks. Where man fails to consider water resources degradation, permanent damage may result. If the ecosystem deteriorates and water resources are destroyed, the basis for urbanization and industrialization is withdrawn.

The need for safe drainage and wastewater disposal has been recognized early in the Emscher area. In consequence the Emschergenossenschaft has pioneered many developments of new techniques in wastewater treatment. Today 98% of the sewered areas in the Emscher river catchment have combined systems. All together at more than 3300 points wastewater and storm run-off are drained into the Emscher river and its tributaries. The conditions of the local underground combined sewer systems is not known in all parts of the area. But it is assessed at present. It is anticipated that due to land subsidence most of the sewers are heavily damaged. Therefore, in future along with the renewal of the receiving water system, also the local drainage systems must be redone.

As mentioned in Chapter 1, the Emscher and its tributaries are all open drainage channels transporting wastewater and natural flows to the treatment plants. As water for public and industrial water supply to a large extent are imported from the neighbouring Ruhr and Lippe river catchments, due to the extremely high waste flows, the flows in the surface waters in the Emscher system are much higher than they would be under natural conditions. The total mean flow in the Emscher river under dry weather conditions of 10 m^3/s today consists of 8 m^3/s from communal and industrial wastewater, 1 m^3/s of mining water and approximately 1 m^3/s of natural flows. Furthermore the high impregnation of surfaces increases storm runoff and increases the danger of flood flows.

In consequence, under dry weather flow the concentrations of the major water quality parameters in the Emscher river approximately correspond with the concentrations measured in wastewater. For BOD5, for instance, 1988 the concentrations fluctuated between 110 and 140 mg/l, for COD between 300 and 430 mg/l and for ammonium between 18 and 32 mg/l. The Emscher to its largest part therefore is classified as water class IV, which indicates a very extensive pollution.

Although the flow of the Emscher river at its mouth does not even reach 1% of the flow of the Rhine river, the high BOD, COD and ammonia values originally had influenced the water quality of the Rhine river significantly. Therefore, at the Emscher river mouth a treatment plant including biological treatment was built. This treatment facility significantly improved the quality of the Rhine river downstream.

2.2 Stormwater Drainage and Flood Flows

If it rains on vegetated surfaces plants and soil surfaces are wettened. Depending on the storage capacity of the soil, water will be absorbed. Water that cannot infiltrate into the ground will be collected in surface depressions and finally runs off. At the same time water evaporates directly from the surfaces and evapotranspirates from plants. If enough water is stored in the upper soil layers, it finally percolates into groundwater. Normally runoff from vegetated natural surfaces is significantly delayed if stormwater is not fully absorbed by the soil. Also on impervious areas surfaces are wetted and depressions are filled. However, contrary to natural surfaces, evaporation is small and depressions have only small storage capacities. This leads to immediate runoff. Runoff quantity and peaks are higher if impervious surfaces are large and connected.

While on pervious and vegetated surfaces 0% to approximately 20% of the rainfall runs off, for roofs, asphalt or concrete surfaces the percentage of runoff amounts to 90% to 100%. From mineral surfaces still more than 80% of the rainfall may run off, dependent on the degree of the condensation of the surfaces. Figure 3 compares the water balance of natural vegetated and more impervious surfaces. The dynamic of rainfall runoff depends on the slope and condition of the surfaces. The form of runoff hydrographs is also influenced by the structure of gutters and culverts. If those collection systems are narrow and straight, runoff is accelerated in addition.

An excellent example for the effects of increasing urbanization and industrialization on flood flows is the Emscher area. Figure 4 provides the increase of imperviousness of the urban surfaces since 1950. It is astonishing that within the last 40 years twice as many urban surfaces have been impregnated than up to the year 1950 (Londong, 1984). Figure 5 accordingly presents the increase of flood flows and flood frequencies within the same time span. While between 1950 and 1959 only one flood was registered and from 1960 to 1969 eleven floods with lower flood peaks were registered, already during the 80's six floods with high flood peaks have been observed. In 1954 a peak flood flow of 143 m^3/s was registered while the flood peak in 1981 was 243 m^3/s.

Over the last decades in the Emscher area millions of DM have been invested into flood protection, especially for raising the dikes and building large retention basins. In spite of this investment, still local flooding occurs. As flooding to a large extent influences the socio-economic condition in the area, flood protection measures is considered as a major part of integrated water management in the Emscher area.

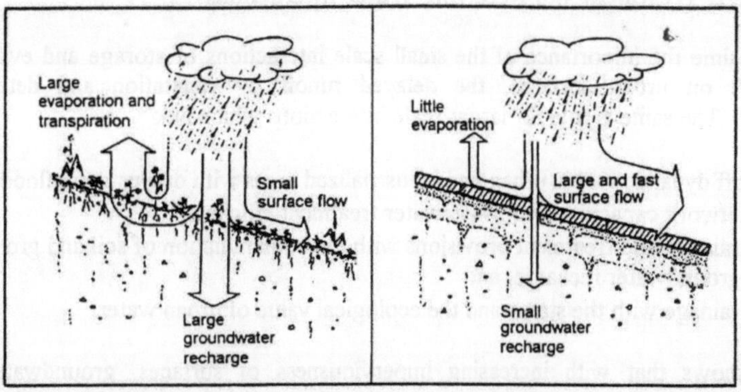

Figure 3: Water balance of vegetated and more natural and of impervious surfaces (Emschergenossenschaft and IBA Emscher-Park, 1993)

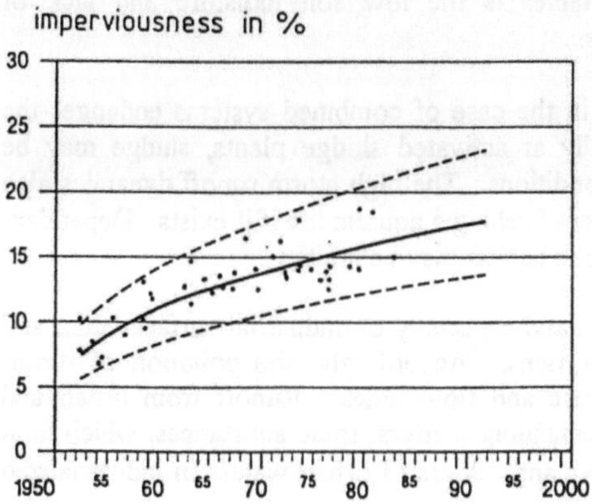

Figure 4: Increase of impervious areas within the Emscher catchment (Emschergenossenschaft, 1991)

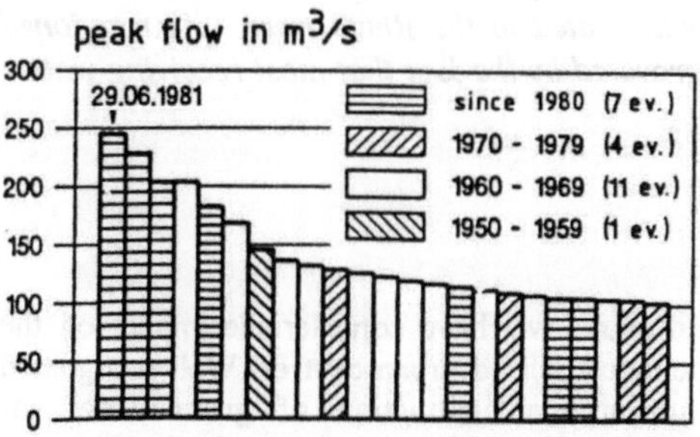

Figure 5: Increase of flood flows and frequencies since 1950 (Emschergenossenschaft, 1991)

2.3 Drainage, Treatment and Receiving Water Interactions

For a long time the importance of the small scale interactions of storage and evaporation of stormwaters on urban surfaces, the delayed runoff by vegetation and detention, was overlooked. The same is true for larger scale interactions which link

- the runoff dynamic within urban and industrialized areas with downstream flooding,
- sewer network capacities with wastewater treatment efficiency,
- storm drainage and treatment provisions with the contamination of soil and groundwater as well as groundwater recharge, and
- storm drainage with the status and the ecological value of urban water.

Figure 6 shows that with increasing imperviousness of surfaces, groundwater recharge deteriorates. While flood conditions increase in urban and industrialized areas, base flows of urban waters decrease. The movement of groundwater to surface water depends on the water level conditions. If the groundwater level is high, water supports base flows in surface waters. On the other hand, surface water is infiltrated into to the ground if groundwater levels are low. Especially in urban and industrialized areas there is a danger that small urban waters will dry out if groundwater levels are lowered by reduced groundwater recharge due to the imperviousness of surfaces. Often for an area typical plants and animals can no longer exist. Another consequence of low groundwater tables is the low soil moisture and lack of vegetation, which may change regional climates.

On the other hand, the high runoff dynamics in the case of combined systems endanger the efficiency of wastewater treatment. Especially at activated sludge plants, sludge may be moved to receiving waters under heavy rain conditions. The high storm runoff dynamics also cause intermittent shock loads to receiving waters flushing a aquatic life if it exists. Dependent on the frequency of loading reestablishment of the habitat may be limited.

Runoff quality affects river ecology. On urban and especially on industrial surfaces dust and dirt accumulates and are washed off when it rains. Accordingly, the pollution of runoff changes in wide ranges, depending on land use and flow times. Runoff from urban and industrialized surfaces often contain oxygen demanding matters, toxic substances, which limit the use of urban water significantly. The looks and odours of urban waters in industrialized areas are often unaesthetic.

In the Emscher area most of the mentioned interactions were observed. While Figures 4 and 5 showed the increase of flood flows with ongoing impregnation of surfaces, the biological treatment plant at he mouth of the Emscher river was protected against high wet weather inflows by bypassing all flows above 30 m^3/s untreated to the Rhine river. As mentioned above the receiving water interactions are aggravated by the fact that most receiving waters are lined and carry wastewater.

2.4 Water Related Environmental Impact

It may be clear that developments as described above will have considerable impact on the environment and ecology within and around the industrialized urban centres. Well recognized are land subsidence and erosion, due to unwarranted extraction of groundwater. or uncontrolled discharge of stormwater. Another effect of increasing urbanization and industrialization is the contamination of the natural aquifers. A serious concern is often the already existing pollution of groundwater by nitrate, resulting from unsewered residential sanitation or leaky sewer systems. Elevated levels of organic compounds and heavy metals and other toxic matters from the many industrialized areas enhance the problem.

An indirect effect is contaminated bottom sludge and sediment in drains and other water courses. Some contaminants in surface waters, like chemicals and heavy metals or spilled oil, tend to adhere to the bottom sludge and sediment. Once the waterways have to be dredged, the removed material cannot be stored in depots without protective measures, because otherwise the underlying soil will be impregnated with the contaminants accumulated within the spoil washed out during rainfall, which in turn may result in groundwater pollution. Consequently, once the bottom sludge and sediments in the waterways are contaminated, costly provisions are needed at the storage depots, such as bottom covering and circumference drainage provisions.

To wind up this discussion, the acidification of atmosphere and soil from industrial or traffic exhaust-fumes and from top dressing of agricultural land should be contemplated. Both, fumes and fertilizers often contain an abundance of chemical components, like chlorides, nitrates and phosphates. As far as such components enter into the atmosphere, they partly dissolve in the water vapour at hand, which results in what is known as acid rain. Where such components enter into the soil, they dissolve in the groundwater. This causes what is known as acid soil. And while both rain and groundwater spread out over the region, large areas are affected by this form of originally local contamination, sometimes with devastating results.

In many cases also rivers have been interrupted by constructing reservoirs for different purposes. This interrupts flowing waters, limits self-purification and the migration of fish. The alternation of flowing and standing waters does not allow for a typical and unique aquatic habitat.

The beginning of industrialization in the Emscher area was also the start of a conflict of interest between economic growth and environmental conditions. Only today it has become obvious that at the same time the living conditions were destroyed.

In the Emscher river itself, there is no dissolved oxygen. Only in the effluent of the river mouth treatment plant the oxygen concentrations range between 2 and 4 mg/l, averaging, for instance, in the year 1988 at 2.8 mg/l O_2. According to the industries of the area, the heavy metal contents in the Emscher river are significantly high. Due to mining inflows, chloride often exceeds values of 2000 mg/l Cl^-. Also many of the tributaries through the Emscher river function as wastewater collection channels. Only in the upstream regions of the tributaries and of the Emscher river itself, where no wastewater is discharged into the river system, the water reaches good water quality.

Bottom sludges of the Emscher river and its tributaries pose a tremendous problem for the renaturalization of the rivers. It is still under discussion if the bottom sludges must be removed completely and be treated as hazardous waste or if they can be concealed and incorporated into landscaping.

2.5 Socio-Economic Issues

Environmental problems and specifically water shortage and water pollution yield negative socio-economic consequences. An external indication for this are the declining population figures that are caused by emigration of people from the area and by the death rates exceeding birth rates. Usually, the average age of people cannot be taken as an indicator because medical developments still increases the average age of man to a larger extent than health effects generally would decrease it in industrial areas.

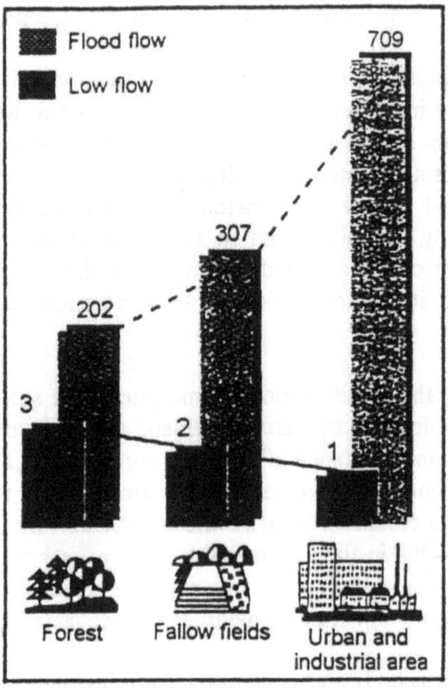

Figure 6. Interaction of base flows of urban waters and flood conditions (Emschergenossenschaft and IBA Emscher-Park, 1993)

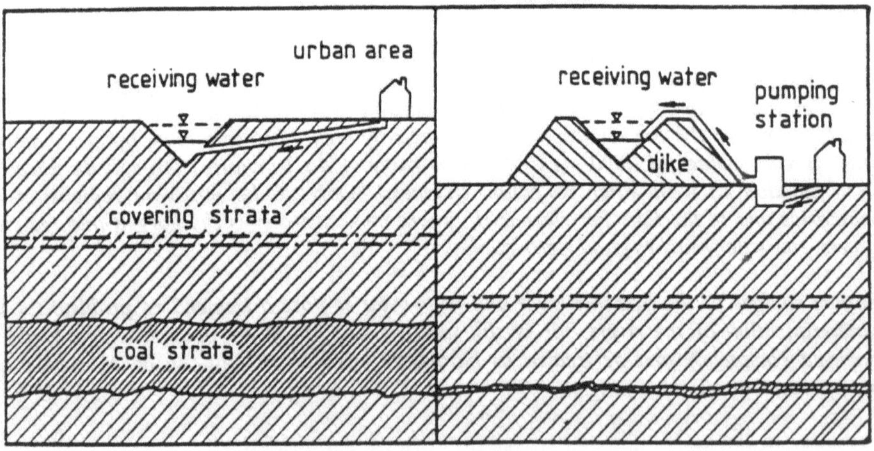

Figure 7: Schematic representation of dike construction in subsided areas (after Emschergenossenschaft and Lippeverband, 1983).

Another socio-economic consequence is the rate of unemployment and the increase of crime rates in old industrial and devastated areas. New and clean industries, which are necessary for economic renewal, will only settle in old industrial areas, if the environmental conditions are appropriate. There is obviously an internal cycle of ecological and economic conditions. On the other hand, old industrial areas need new economic investment in order to take revival measures.

A typical indicator for the socio-economic consequences of industrialization and especially of the decline of coal mining and steel industries in the Emscher area is the development of the age structure. While in 1960 22% of the population in the Ruhr region was under the age of 15, today it is only 14%. On the other hand, the percentage of people over 65 has risen from 9.5% to 15.6% (KVR, 1990). The unemployment rate especially in the Emscher area is rather high. While in 1957 there were still 607.000 coal miners, today there are only about 90.000. While in the steel industry in 1967 300.000 workmen were employed, today this figure is only about 80.000. In spite of all attempts it has not been possible to compensate for this loss of jobs. The Ruhr area has the highest unemployment rate in the old German states.

Nevertheless, some changes have occurred in industrial employment. Since 1970 the white-collar workers in both the public and private sectors have increased from 30.4% to 50.3% of the total work force, whereas the number of blue-collar workers has decreased from 57.3% to 42.3% of the total work force. Already today the Ruhr region not only depends on coal and steel industries, which today only contributes 16% to the regions net product, but still employ approximately 33% of all labour (KVR, 1990).

The open transport of wastewater in the straightened creeks have become a burden and increasingly are found disturbing by the people. Their bad image and ecological conditions hinders a new economic development of the area. More and more a nature-like condition of the rivers and creeks is demanded.

As already mentioned, when land subsided, the Emscher river and its tributaries were diked in order to maintain free surface flow in the receiving water system. In some places the dikes are higher than the houses in the vicinity. This also is felt as being an artificial barrier separating the cities into parts. It is one of the major concerns to reduce or to remove the artificial barriers. Figure 7 provides a schematic representation of the development of the dikes.

Since 1989 the Emscher area has been undergoing a substantial structural change by initiation of a variety of revival measures. Along with this change the demands of the population for living quality constantly grow. This indicates that there is a direct and fast interaction between economy and social conditions.

3. Principles of Integrated Water Management

The general goal of integrated water management is a sustainable utilization of water resources respecting the social, economic and environmental interests. The linkages between society, economy and environment are shown in Figure 8. Considering the close interrelationship between society and economy, the first two groups are usually aggregated into socio-economic issues. It should be recognized that the goals and objectives of integrated water management are formulated at various spatial scales, involving all three components.

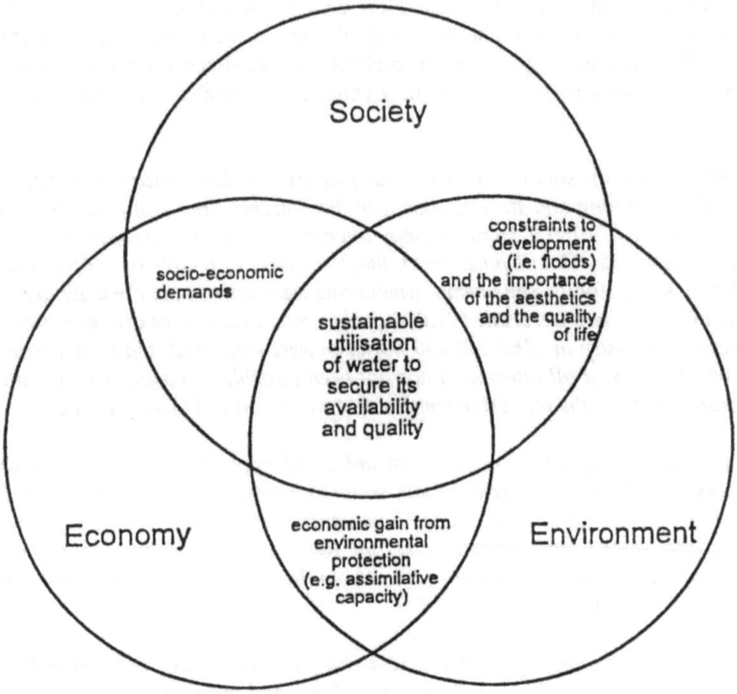

Figure 8: Linkage of society, economy and ecology

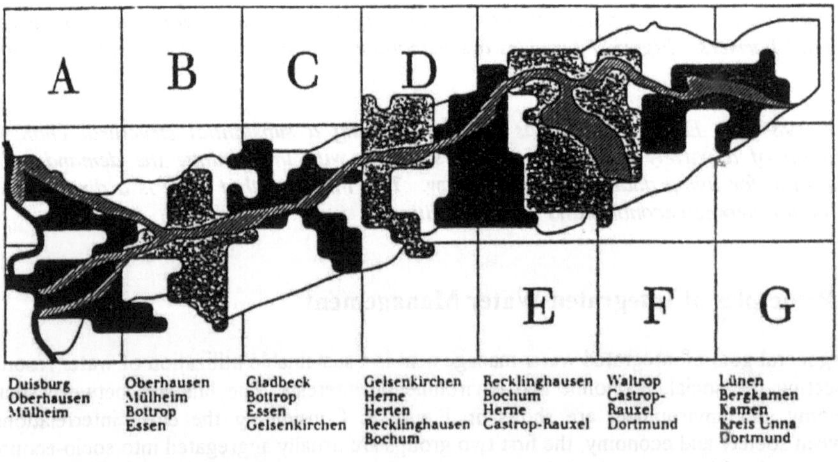

Figure 9: Emscher Landscape Park (IBA Emscher-Park, 1993)

3.1 The Holistic Approach

The fundamental qualities of integrated water management are its holistic nature, which recognizes the system complexity and interconnectivity of its elements, demonstrated by exchange of information, energy and matter, and the style of planning actions. The holistic approach also equally involves local and regional authorities, engineers and natural scientists, environmentalists and decision makers, politicians of all parties, governing and in opposition, as well as the people affected. The holistic approach sometimes is called ecosystem approach. Sustainable water management ensures that no matters are accumulated or energy is lost by recovery and reuse techniques. This approach requires novel, environmentally sound technologies.

When proposing technological solutions to water management problems, environmental costs must be considered. In this connection, the use of an environmental cost/benefit ratio has been proposed, where the cost represents the economic effort and the benefit is the environmental effect defined as the positive effect of the implemented measure (e.g. removal of a certain mass of pollutants) minus the environmental damage caused by the implementation (i.e. the use of natural resources and energy consumption). However, the equal presentation of social, environmental and economic issues in the comparison of alternatives is very difficult as the evaluation of non-monetary values is rather subjective.

Water management is cross-sectorial by its nature, and the division of the society and its institutions into various sectors becomes an obstacle in achieving the goals of integrated water management. The creation of inter-sectorial links, supporting cross-sectorial cooperation and integrated multidisciplinary actions in water management, represents the greatest challenge in implementation of integrated water management. This task is also made more difficult by the current sectorial educational systems which are not designed to impart the broad integrated knowledge necessary to promote a multidisciplinary, holistic approach to resource management problems. In consequence, integrated water management also needs novel administrative approaches and an holistic education.

The Emschergenossenschaft recognized at an early stage that with fading out of land subsidence in large parts of the system the necessity for open wastewater drainage is no longer necessary and that there is an extensive potential for the revival of the system. Furthermore, water laws were tightened and no longer provide for wastewater transport in rivers. Therefore, already at the end of the 70's the first planning for a renewal of the system began. Originally the concept mainly involved the construction of underground interceptors, which were mostly located parallel to the creeks in order to collect all waste flows and to allow for later renaturalization of the river beds.

In 1988 representatives of local governments, mining and industrial companies decided on a joint solution to the problems of the open and combined systems of natural and wastewater drainage. All activities were incorporated into a plan of the government that aimed at an ecological renewal of the Emscher area. This includes:

- *the abolishment and compensation of the economic disadvantages, which were left over from the old industrial structure*
- *the improvement of the environment in the densely populated area and the activation of the ecological potential in the surrounding areas*
- *the improvement of the built environment to gain higher living quality and better living, housing and working conditions.*

To achieve these goals the government installed in 1988 the International Building Exhibition (IBA) Emscher-Park. Building exhibitions are institutions with a long history in German architecture. They always aimed to bring together urban planners and politicians to address new challenges. With this objective in mind, the government of Northrhine-Westfalia staged the International Building Exhibition Emscher-Park for a period of ten years. At present, local authorities, private investors, individual citizens and the state of Northrhine-Westfalia are involved in more than 80 projects, which are divided into five main groups:

- *the ecological revival of the 350 km of the Emscher river system*
- *the modernization of workers' settlements and the construction of new housing*
- *the development of fallow land to areas following the principle of "working in the park"*
- *the preservation and the conversion to new uses of old industrial heritage sites*
- *the protection and reclamation of open spaces on the regional project "Emscher Landscape Park".*

Due to their scope, the projects are sometimes planned for a duration of 10 to 20 years. One of these long-term projects is the revival of the Emscher system. The anticipated Emscher Landscape Park indicated in Figure 9 has a potential size of 300 km^2. The revitalization of devastated industrial sites of up to 100 hectares each and large scale housing projects are planned to be realized within 5 to 7 years. Small and innovative housing projects involving the construction of up to 30 residential units usually only require one or two years. More details on the total effort can be found in the brochure "The Emscher Park International Building Exhibition" provided by IBA in various languages. Due to the topic of the paper, further discussions concentrate on water management aspects only.

Although in the Emscher area a holistic or ecosystem approach was not planned at the beginning of the revival efforts the projects initiated and done so far more and more draw the picture of an ecosystem approach. The example may also demonstrate that an ecosystem approach probably cannot be planned precisely but must grow.

3.2 Assessment of Water-Related Integration Issues

Water is a precious asset, as it is a necessity of life for both human beings and all other life. Key indicators for water management therefore are:

- the availability of water and
- the quality of the water.

Availability can be attributed both positive and negative aspects. Any human needs a certain daily amount of good quality water for its various purposes. As long as daily demand can be met from the existing water sources, the availability of water can be said to be positive. However, when it can not be met the ensuing shortage is felt as a negative value, forcing man to set priorities to his use of the available water. An excess of water, i.e. floods is unwanted and thus can be likewise considered a negative value. Quality of water is of similar complexity. Good quality usually only bears positive, bad quality negative aspects.

General values may be ascribed especially in densely urbanized and industrialized areas. These other values include:

- economic conditions,
- socio-economic status,

- **environmental and ecological values,**
- **aesthetic values, and**
- **human perception values.**

The economic value of water, i.e. potable water supply, water for industries, for irrigation, for fishing or for navigation is evident and widely recognized throughout the world, although each location may require different approaches to the water management necessary to preserve the economic condition.

In contrast, the availability of public facilities like a well functioning potable water distribution system within the urbanized area, available to all citizens, can be considered a matter of socio-economic status.

Less evident and consequently often less recognized are ecological, aesthetic and human perception values of the various urban water systems. Due to low standard of living and lack of money, the preservation of the natural assets has often no high priority and (local) political preferences select options that often do not include the ecological soundness of water systems that would be precious to the general appearance of the city and the welfare of its inhabitants. It may be said that the degree of development of a given area commonly reflects in the degree of preservation of the latter three values.

With the support of the Ministry of Environment of the State of Northrhine-Westfalia and the cooperation of the Emschergenossenschaft and the IBA Emscher Park in 1989 and 1990 a small number of interdisciplinary expert reports were established:

- ***for the decentralization of wastewater treatment and the development of new urban drainage concepts***
- ***for urban planning, ecological revival and integration of rivers and creeks into the urban landscape***
- ***on the possibilities of reduction of peak flows in the Emscher river***
- ***on the possibility of a new water quality concept for the Emscher river system.***

The principle has been that parallel two teams were contracted, one basically from universities, the other from companies. The completion between the more theoretical and academic thoughts and practical procedures proved to be very powerful in developing the final concept for the revival of the Emscher area.

Further expert opinions dealt with the more specific problems of low base flows and with the reduction of imperviousness of the area as well as with infiltration of stormwater.

3.3 Environmental and Ecological Issues

The most important environmental/ecological goal is the protection and enhancement of ecological integrity which requires the development of a new perspective on management activities. First, the scalar issues involved in the management of water resources in urban and surrounding areas must be considered. For example, ecosystems are dependent on influences that extend beyond the jurisdictional boundaries of most water management agencies. Birds that migrate between continents may be an important feature of a local ecosystem and require consideration of both local and distant factors when considering preservation and management.

Time scales are also important. The species present at any time in managed areas reflect a history of evolutionary development. Thus any species or condition is dependent on what has happened in earlier times in the managed area. Furthermore, short term management strategies do not reflect the full range of lifetimes represented in the species present. For example, some trees may have a life span of thousands of years. Animals may have a life span of hours to tens of year. Thus the management strategy cannot focus on present conditions, but must consider past history, and the current characteristics of the organisms present.

The ecosystem characteristics that are critical for system maintenance have been defined by Herricks and Schaeffer (1987) and are listed in Table 1. These characteristics can be used as a basis for identification of critical ecosystem integrity attributes, which include elemental dynamics, energy dynamics (physical), food web (trophic dynamics), biodiversity, critical species, genetic diversity, dispersal and migration, natural disturbance, and ecosystem development.

For many professionals, the prospect of dealing with such a complex system is daunting. Although the system is complex, this complexity is not an excuse for lack of management activity. It is possible to focus on specific elements of the ecosystem while maintaining a more holistic view of the total system process. It is also possible to make decisions on what we do know, recognizing that those decisions may change as we learn more about the ecosystems we manage. This integration of what engineers know, as the fundamental basis of "practice", is an important change in present management approaches that is produced by the ecosystem approach. There must be a willingness to fail, but failures can not be catastrophic. There is a clear need to monitor progress and the ecosystem conditions to determine if decisions actually meet management goals and objectives.

Table 1. Factors necessary to maintain a healthy, self-sustaining ecosystem (Herricks and Schaeffer, 1987).

1. Habitat for desired diversity and reproduction of organisms.
2. Phenotypic and genotypic diversity among the organisms.
3. A robust food chain supporting the desired biota.
4. An adequate nutrient pool for desired organisms.
5. Adequate nutrient cycling to perpetuate the ecosystem.
6. Adequate energy flux for maintaining the trophic structure.
7. Feedback mechanisms for damping undesirable oscillations.
8. The capacity to temper toxic effects, including the capacity to decompose, transfer, chelate or bind anthropogenic inputs to a degree that they are no longer toxic within the system.

Environmental and ecological issues demand sustainable water resources management and managing cycles of various types of matter or materials. In general, environmental protection goals could be listed as follows:

- apply source controls wherever possible
- reduce impregnation of surfaces
- infiltrate stormwater on-site into the ground
- stop pollution at the source
- prevent accumulation of non-degradable toxic substances
- practice reuse of water and other materials

- prevent transport of contaminants
- sustain or enhance the existing surface and groundwater uses
- manage energy use
- develop and use non-polluting energy sources
- minimize land desertification
- prevent water-related diseases

Source controls of various types, especially of pollution, generally belong to the most cost-effective remedial measures, particularly in the case of diffuse sources of pollution. Once the contaminants are released into the environment and dispersed to low concentrations, controls or treatment are often impossible. Some of the released contaminants may temporarily accumulate in certain parts of the ecosystem (e.g. in sediments) only to be released at later times. Some can be prevented by recycling wastes and preventing their transport though the environment.

The protection and enhancement of ecological integrity require a new approach to water management. This new approach integrates the available information from a variety of sources and focuses attention on critical ecosystem values.

The separation of clean and polluted water is a fundamental prerequisite for achieving the ecological basis of the new Emscher system. To link the sewage flows underground and reshaping the formally straightened river beds, the natural impression of the rivers can be reclaimed and integrated into the surrounding countryside. Efficient decentralized sewage treatment will release the pressure from the biological central treatment plant at the confluence of the Emscher and the Rhine river. The decentralization of treatment, the water quality in the Emscher river itself can be improved significantly. Furthermore, decentralized treatment will also provide clean water to increase base flows of the rivers in the area. Together with this effort as much stormwater should be infiltrated into the ground also to increase base flows of the tributaries by natural seepage.

In this context three significant pilot projects should be mentioned:

- *The biggest construction site in the Emscher region at present is a new sewage treatment plant at Bottrop.*

- *At Lepkes Mühlenbach at the boundary between Oberhausen and Essen is one of the first pilot projects for the revival of tributaries to the Emscher. It is an impressive example of the ecological and aesthetic improvement that can be achieved by reshaping the old straight drainage channels.*

- *The Schüngelberg residential area in Gelsenkirchen is an example of the decentral storage and infiltration of stormwater, which again helps to increase base flows in the nearby Lanferbach.*

A revival of the Emscher river system is also linked with the planned East-West green corridor, which is the basis of the future Emscher countryside park. Actually, this Emscher Landscape Park is a late further development of a plan to create 7 regional green corridors, which was first elaborated in the 20's by the Siedlungsverband Ruhrkohlenbezirk. However, this early plan has never been properly realized. The originally planned corridors numbered A to G are shown in Figure 9. The extent of the new Emscher Landscape Park is marked in this figure. In future, the revived river system and the Emscher Landscape Park represent a linked habitat which for an industrial area to a high degree provides ecological integrity.

3.4. Socio-Economic Issues

Socio-economic issues express the needs of various interest groups in urban areas. Usually these issues are highly interrelated and partly contradictory or competitive. They originate from activities and perceptions of both, the individual and the institutional sector.

In defining socio-economic objectives one has to consider the complexity of the system, including the cultural, geographical, economic and political conditions. Socio-economic objectives of water management in urban and industrial areas are not at all separate from or in addition to other more specific technical, environmental or ecological objectives, such as water supply needs, providing protection against flooding, and preserving natural environmental conditions. Essentially all water management objectives have socio-economic dimensions. Without question, sustaining and protecting human life is the most important socio-economic objective of water management.

Maintaining diversity of the ecosystem in at least a part of urban areas is an important socio-economic goal of water management. While recognizing scientific arguments for maintaining diversity of flora and fauna, e.g. a more stable, biologically robust environment able to withstand external impacts, there are clearly socio-economic values associated with this goal. Many developed areas have obliterated the natural systems and now must incur great costs to take remedial action. To quote an old saying, "an ounce of prevention is better than a pound of cure".

Provision of open space, particularly space that includes vegetation and water and is known as blue-green space, is another crucial socio-economic goal for urban and urbanizing areas. The planning and design of that space must go well beyond determining the total quantity of space. Important characteristics of blue-green space include continuity, functionality, access, safety, and the continuous presence of clean water, assuming that such presence is consistent with the natural climate and physical geography. One benefit of establishing such space is support of recreation and related business that serve those who use open space. Careful provision of space may ultimately affect the micro-climate of a growing urban area by influencing humidity levels and the frequency of fog. Finally, blue-green space can control odours, provide places for active recreation activity, and will have an aesthetic value for a large fraction of human population.

Conscientious, compatible use of the usually-limited water resources is another socio-economic goal. Water should be managed in a way consistent with the local climate and physical geography. To do otherwise is ultimately in conflict with the natural environment and wasteful of financial and natural resources.

Finally the inability to quantify, in monetary terms, a socio-economic dimension of an urban water management project does not determine its importance. Analysts should differentiate between price and value in situations where price is a current, transient factor and not an accurate measure of ultimate value. The conventionally used monetary costs should be supplemented by environmental costs, which consider positive environmental effects as well as environmental damage caused by implementation of environmental measures. This is needed for attaining the sustainability of water resource management.

Although for some old industrial areas it would be desirable to have a coordinated drainage concept as it already exists in the Emscher area, there its status is unsatisfactory. An open wastewater system does not represent the conditions necessary to attract new and clean industries, nor does it fulfil today's living standards. In the Emscher area it was understood that the desired socio-economic change can only be realized by complying with ecological and environmental demands. The Emscher Landscape Park introduced in the previous chapter and shown in Figure 9 is an excellent example for the socio-economic dimension of the revival of the Emscher area.

Incidentally in adjacent areas to river stretches where wastewater has been separated from the rivers and the old straight channels were removed and water like river paths were landscaped, property prices and rents went up with the consequence that for low income families new problems were inadvertently created.

3.5 Water Management Objectives

The objectives of water management are often expressed in terms of unimpeded or enhanced water uses. If such a water management plan were developed for pristine conditions, it would be possible to formulate the associated objectives in terms of a set of water uses. For example, Jermar's water use classification (1987) can be expanded in the following way:

- Natural functions: supply of soil moisture, biological/ecological functions (including fisheries both commercial and sport; wildlife habitat), transport of various elements, climate regulation, aesthetic enjoyment, and general environmental functions including self-purification and educational resource.

- In-situ uses: groundwater and soil moisture regulation, shipping and navigation (require maintenance of adequate water depths by regulation and dredging - both impacts on water quality), land drainage, waste transport and disposal (including industrial and municipal effluents, combined sewer overflows and stormwater discharges from separate storm sewers), hydropower generation and water-based recreation.

- Withdrawals of water: water supply for domestic, industrial and agricultural purposes. For domestic supply, emphasis is put on safe supplies, for industrial and agricultural supplies, the availability of water not requiring excessive treatment is important.

In the context of urban and industrial water resources, the most pertinent water uses are water supply (safe, reliable and equitable), drainage and flood protection (affordable), sanitation with maximum reuse, recreation (protecting public health), aesthetic and cultural values, and ecosystem health.

The above objectives do not receive the same priorities. At different stages of societal development, different objectives receive more importance. In the pre-industrial society, emphasis was placed on drinking water supply, transportation and water supply for irrigation. In the industrial society, generation of hydropower and waste disposal and transport are prioritized. Finally, in the post-industrial society, high emphasis is placed on aesthetics and ecology. Figure 10 sketches the interaction of industrialization reflecting priorities set for water management and the environmental condition.

While the existence of changing priorities must be recognized, at the same time lower priority uses cannot be neglected over a long term, because of interdependency of various uses. High mobility of water and physical connectivity of water resources, natural or man-made, contribute to interdependency of actions of water uses. This becomes particularly obvious under the conditions of intense use which provides an impetus for integrated water management with reduced conflicts among users and the spreading of benefits arising from management actions.

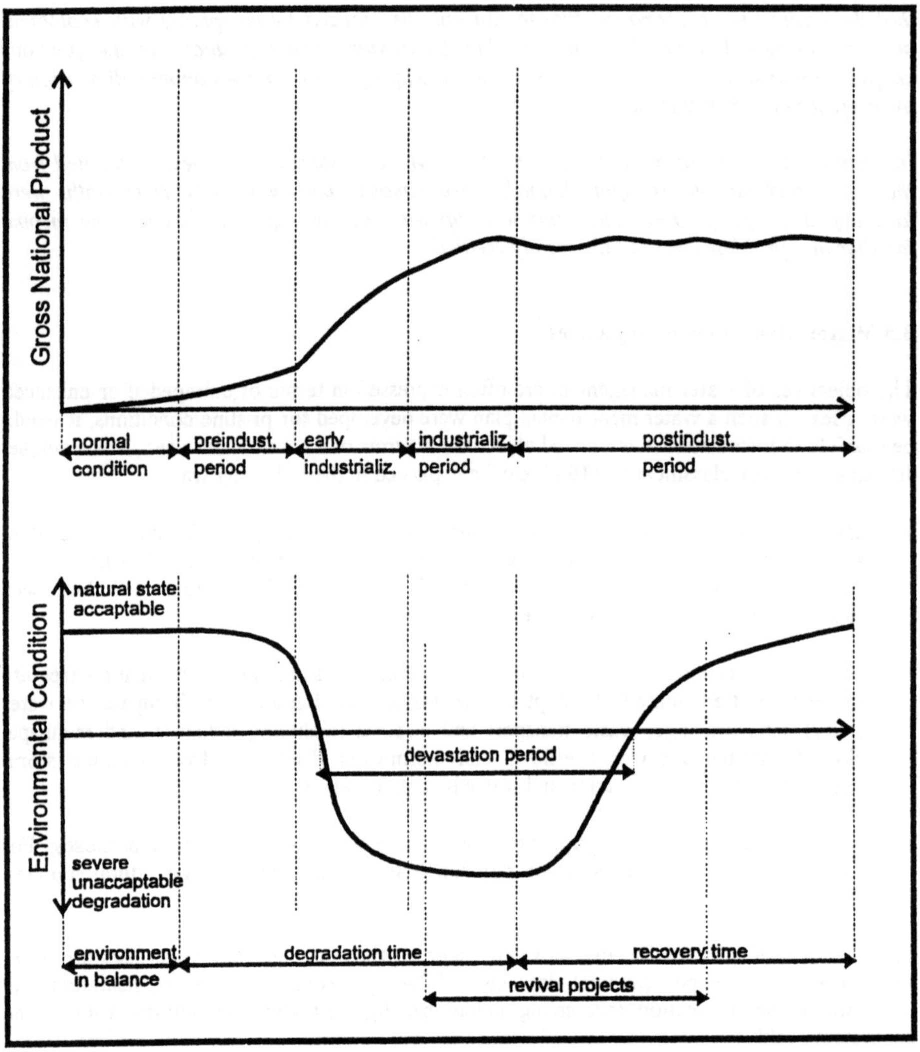

Figure 10: Interaction of industrialization,economics and environmental condition

Safe, reliable and equitable water supply is one of the most important objectives of urban water management. Conventional solutions, involving water import from distant areas, are no longer acceptable, because of great environmental costs incurred in the area of water withdrawals and along the transport route. Similarly, processes involving high energy consumption, such as desalination operated with energy from thermal plants, should be avoided because of adverse environmental impacts. Thus, in the ecosystemic approach to water supply, it is important to stress the management of demands, use of local water resources (e.g. through a building code requesting the installation of water cisterns in new housing), water reuse, and equitable allocation of water supplies withdrawn with low energy consumption.

The protection of population and material property against harmful effects of water comprises a number of specific objectives - prevention of human life losses during flooding, prevention of damages of material property by flooding or groundwater inundation and/or wave forces, and prevention of disruption of economic activities and accompanying losses by inundation (both on the surface and underground). In urban areas, the concentrations of population, properties and economic business activities are particularly high and the potential losses caused by flooding or inundation are extremely high. Consequently, the objectives of protection against harmful water effects, and particularly against the loss of human life, receive the highest priority. In protection of property and business activities, much greater risks are acceptable and return periods ranging from 2 to 100 years are typically used, depending on potential damages.

Sanitation and protection of surface water quality are very important objectives not only from the point of view of environmental protection, but also from public health point of view. In selection of control measures, emphasis must be placed on appropriate technologies matching the local conditions in terms of local climate, resources (funds and human skills), and cultural traditions. It is of interest to note that various biotechnological measures, used for many centuries by indigenous people, may provide low cost, low energy demand solutions providing local employment and a very effective environmental protection. Examples of such solutions include wetlands and fish ponds, i.e. the areas located east of Calcutta, used to treat municipal sewage.

Ecological objectives and their close relation to the socio-economic goals, have been discussed in detail. It should be noted that there are opportunities to include ecological objectives in practically all projects dealing with urban water. One example is the rehabilitation of drainage systems - rehabilitation programs offer excellent opportunities to incorporate ecological objectives into the older system, which were generally built during an era of environmental neglect and abuse.

Water based recreation is widely practised in industrial and post-industrial societies. To protect public health, various guidelines for quality of recreational waters have been promulgated. Particularly important are recommendations for microbiological water quality, typically based on permissible concentrations of indicator bacteria (e.g. faecal coliform, or Escherichia coli), implying faecal pollution and the presence of pathogens.

Aesthetic and cultural values of water can be incorporated into design. Specific features include open drainage of rainwater, runoff conveyance in cascading open channels, treatment in ornamental ponds, and recharge of groundwater aquifers by surface water of relatively good quality. Incorporation of water into the urban design brings about economic benefits in the form of higher property values and general attractiveness for modern industries, particularly when dealing with redevelopment of environmentally damaged areas. Successful designs require an early incorporation of water aspects into the planning process, retaining a multidisciplinary team in planning, and well organized public information/involvement programs.

Water management in industrial areas strives to meet the demands of economic growth by optimal development and utilization of water resources and to create an optimum living environment, which would protect the society against harmful effect of water. These objectives emphasize two groups of interests - society and economics. Only after introducing the third group of interests, environment (including air, land, water, and biota, excluding people), a truly integrated approach can be achieved. For a holisitic approach, all objectives must be developed on a river basin scale by a stakeholder group broadly representing social, economic and environmental interests. Table 2 compares various water management approaches.

Table 2. Comparison of various water management approaches.

Technical Objective	Approach		
	Piecemeal	Environmental	Holistic/Ecosystemic
Water supply	Transport water over great distance (disregarding impacts at the source)	Provide advanced treatment, no management of demands	Manage water demands by water conservation, recycling of industrial water, water re-use; provide treatment
Wastewater transport and disposal	Develop treatment regulations for various sources and individual constituents (more or less independently)	No management of generation of wastewater; develop regulations, plans for local receiving water body considering all pollution sources and constituents, implement treatment	Reduce generation of wastewater through conservation, recycling; enhance self-purification of natural water bodies (by restoration of biota); provide some treatment
Protection against local flooding	Zero-runoff increase regulation - for individual developments	Develop master drainage plans, build drainage channels, pipes to convey flows	Reduce runoff generation by source controls, view stormwater as resource - use in irrigation, water supply; for limited runoff flows - provide treatment in ponds and wetlands

In the Emscher area the water management objectives ideally can be followed up, as the responsibility for objectives except water supply lies with the Emschergenossenschaft. As mentioned earlier, the Emschergenossenschaft is a board comprising cities, industries and other stakeholders representing the public. This is an ideal condition to follow an ecosystems approach by integrated water management. However, economic constraints sometimes restrict the implementation of the desired measures or even worse influence setting priorities.

4. Integrative Planning Procedures and Technologies

Although various forms of single or multipurpose water management have been practised in many areas, a comprehensive approach to water management in industrial and urban areas is still uncommon. Therefore, this chapter tries to elaborate on procedures for integrative planning and technologies for sustainable water management.

4.1 Sustainable Technologies

Technologies for sustainable water management combine natural solutions with technical skills. As it is impossible within this paper to discuss all possible techniques for water supply, urban drainage and wastewater treatment in equal depth, possibilities of sustainable drainage in densely populated and industrialized areas are exemplified.

4.1.1 Principles for Storm Drainage

In the past, abatement of storm runoff and combined flows was usually implemented at the "end of the pipe". Flows were conveyed as fast as possible out of populated and industrial areas whereby consequent flooding of downstream areas was countermeasured by large retention facilities. Wastewater often was treated only mechanically or even released untreated to receiving waters. The accumulation of nutrients in water bodies such as the North Sea and the increasing awareness of the ecological value of urban waters in the FRG for instance has led to tighter requirements that are difficult to achieve by the inherited "end of pipe" approach.

New principles for stormwater management are:

- avoid impervious surfaces and infiltrate stormwater at the point of origin
- if necessary, detain and treat stormwater prior to infiltration
- use stormwater as much as possible for household, commercial and industrial purposes
- if decentral infiltration is impossible, collect stormwater, treat it if necessary and infiltrate it in central infiltration facilities
- if the underground does not permit stormwater infiltration, detain stormwater and lead it into the nearest water course
- collect only polluted stormwater and treat it in central facilities
- make use of the sludges produced in treatment for construction material.

By following these principles, runoff peaks and volumed are reduced significantly. Consequently, mechanical, biological and bio-chemical treatment methods also become effective for stormwater treatment.

4.1.2 Source control of stromwater runoff

Figure 11 summarizes the possibilities for technical infiltration of storm runoff. Aerial infiltration is especially suitable for backyards, parkways and sport fields. Here, the permeability of the soil must be sufficient to absorb the design rainfall. For the trough percolation facilities stormwater may be stored if the permeability of the soil is not sufficient. In any case, stormwater is still infiltrated through the top soil, allowing micro-organisms the opportunity to reduce pollutants in the stormwater. However, when shaft percolation (infiltration pits) are used, stormwater reaches the soil sublayers directly, not using micro-organism treatment benefits. The same applies for infiltration trenches, which are usually filled with gravel. A distribution of stormwater within an infiltration trench may be accelerated by using perforated pipes. The storage volume of pipe and trench infiltration is usually larger than at other facilities. Design elements such as aerial infiltration, and various infiltration pits can be combined depending on local conditions, as shown in Figure 11. There are a number of possibilities to pre-treat stormwater prior to infiltration using sedimentation, filtration, separation, and biological degradation.

Since vegetation and water improve the urban climate by thermal equalization by evaporation, maintaining vegetation in urban areas is important. For example, the evaporation rate in built-up areas is approximately 300 mm per year to 500 mm per year, from deciduous trees and conifers about 600 mm per year, at damp meadows up to 900 mm per year, at open water surfaces depending on the turbulence of the water 500 mm per year up to 1000 mm per year, and finally from reeds even up to 1800 mm per year. Evaporation occurs by resorption of heat, which is taken from the surroundings.

When new developments are planned, it is possible to create green surfaces by limiting traffic and parking areas to a minimum, creating green strips between roads, bicycle ways and pedestrian ways and even by the greening of roofs.

An investigation in the Emscher area yielded that in some cases impervious surfaces could be reduced by nearly 50%. However, on an average, in the already urbanized areas only a reduction by 5% is more typical.

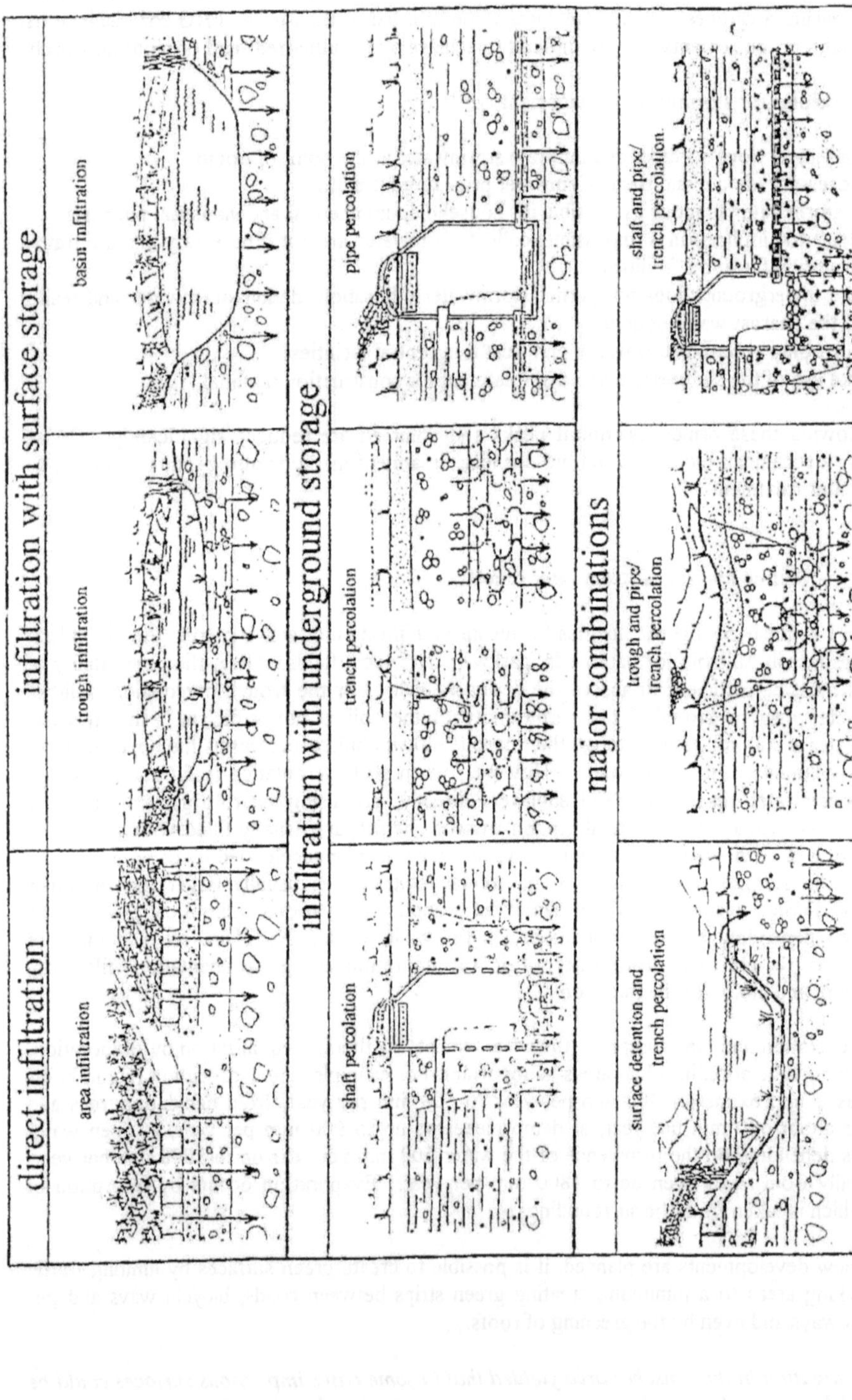

Figure 11: Possibilities for technical infiltration of stormwater

For the Emscher river catchment with a ratio of imperviousness of 20% it was estimated that the overall ratio of imperviousness only may be reduced by about 1%, if 5% of the impervious areas are disconnected from the sewer system. Furthermore it was assumed that in addition per hectare approximately 50 m³ of rainfall may be distracted for secondary water uses such as sprinkling of lawns. Such measures are important to support base flows in the small tributaries of the Emscher river. However, their contribution of reduce flood peaks is quite limited. The effectiveness of both measures on the flood-peak in the Emscher river may be seen in Figure 12, 2nd row. Under loading conditions with a return interval of 2 years the peak-flow may be reduced by 10% while for a return period of 200 years the flood-peak only is reduced by 2% at the most.

4.1.3 Flood Protection Measures

Except for source control in urbanized and industrialized areas for flood protection only the creation of large flood retention basins helps. In any case it must be avoided that by increasing the hydraulic capacity of receiving waters within urbanized and industrialized areas the flood problem is shifted downstream.

From Figure 12 it can be concluded that to control floods source controls and small detention basins will not solve the problem. The only possibility to abate flooding in an area like the Emscher river catchment is to provide for large scale flood retention basins. These basins show a significant effect on flood-peak reduction. 25 existing and 31 additional retention facilities were investigated as well as 4 additional storage volumes along the Emscher river itself. The retention volume of all basins is 6.5 million m³ in total which corresponds to 400 m³/ha impervious area. Flood flow retention reduces flood-peaks by 30% under flood conditions with a 200 year return interval (Figure 12, 4th row).

4.1.4 Combined Sewer Overflow and Stormwater Treatment

In storm sewers of separate systems the variation of flows is high, so that treatment of storm runoff becomes rather difficult. In principle, systems for treatment of stormwater basically are storage and sedimentation. However, a better solution is to modify separate systems according to Figure 14. This modification limits the dynamics of flows and improves the efficiency of treatment facilities. If space is available for sustainability stormwater should be treated in wetlands linked to the natural drainage system of rivers.

Almost all drainage systems in the Emscher area are combined.

In conventional combined systems, where storm runoff from all impervious surfaces is collected together with wastewater, under storm conditions not all combined flows can be treated due to the high dynamic loading. If combined systems are modified according to Figure 14, the dynamics decrease and consequently the facilities are easier to operate. For combined flow treatment storage and sedimentation are used. Again, sustainability wetland type of treatment should be chosen, if space is available.

Advanced combined sewer overflow treatment is sometimes requested when receiving waters have to be protected for ecological reasons or for water supply purposes. The following are considered as advanced combined sewer overflow treatment measures:

- increase of the detention level of combined flows providing that treatment can cope with additional wet weather flows carried to the treatment plant

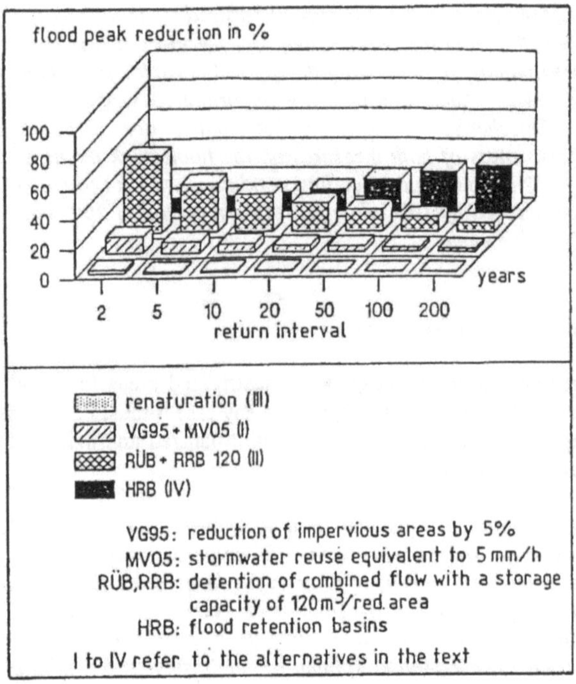

Figure 12: Reduction of peak-flows by different improvement measures (Emschergenossenschaft, 1991)

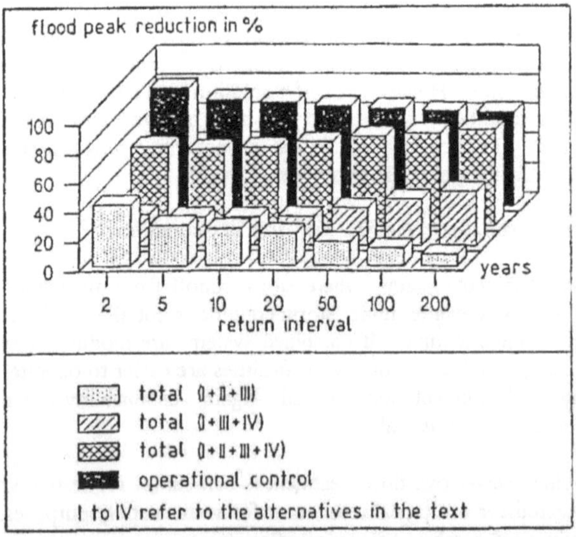

Figure 13: Reduction of peak-flows for different combined measures (Emschergenossenschaft, 1991)

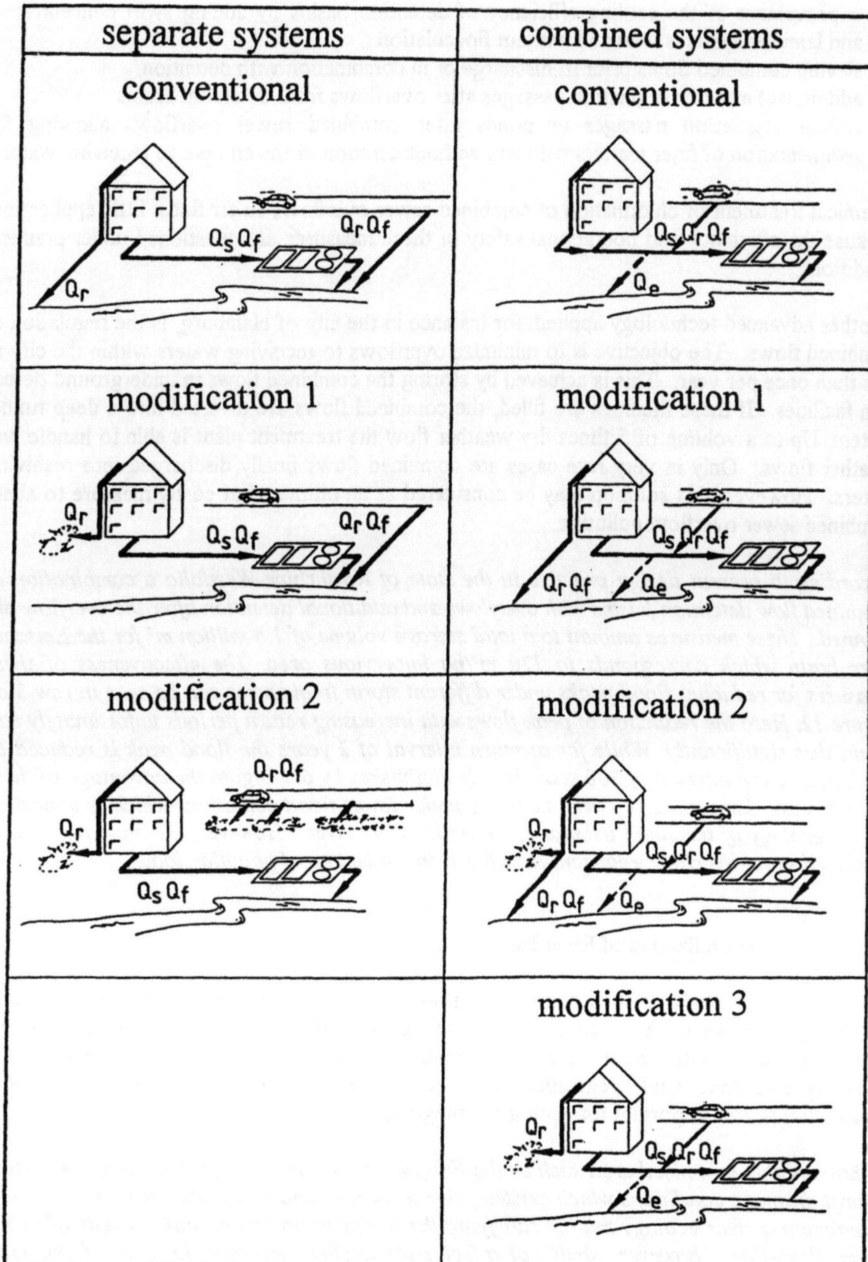

Figure 14: Modifications of conventional separate and combined systems

- improvement of the settling efficiency of detention basins by adding swirl concentrators and lamellar systems with and without flocculation
- sieving combined flows prior to discharge or in combination with detention
- adding wet and dry vegetation passages after overflows from detention basins
- adding vegetation passages or ponds after combined sewer overflows allowing for sedimentation of finer matters with and without aeration of the effluent to receiving waters

Chemical treatment or chlorination of combined sewer overflows so far finds little application, because the efficiency and operational safety of these measures are questioned under practical conditions.

Another advanced technology applied, for instance in the city of Hamburg, is the regulation of combined flows. The objective is to minimize overflows to receiving waters within the city to less than once per year. This is achieved by storing the combined flows in underground detention facilities. If these facilities are filled, the combined flows are diverted into a deep tunnel system. Up to a volume of 5 times dry weather flow the treatment plant is able to handle wet weather flows. Only in very rare cases are combined flows finally discharged into receiving waters. However, this solution may be considered as an ultimate but costly measure to abate combined sewer overflow pollution.

According to present design practice in the state of Northrhine-Westfalia a combination of combined flow detention basins with overflows and additional detention after the overflow are planned. These measures amount to a total storage volume of 1.8 million m³ for the Emscher river basin which corresponds to 120 m³/ha impervious area. The effectiveness of these measures for reducing flood peaks under different storm frequencies can be seen in row 3 on Figure 12. Here the reduction of peak-flows with increasing return periods unfortunately still diminishes significantly. While for a return interval of 2 years the flood peak is reduced by 50%, at a return interval of 200 years it only diminishes by 6%. Again the advantage of these measures are dampening of overflow-peaks under lower flood conditions which is important for the ecology of the small tributaries of the Emscher river. However, the main purpose of combined sewer overflow treatment facilities is the reduction of pollution loads.

4.1.5 Renaturalization of River Beds

For the renaturalization of river beds in devastated areas it is advisable to start with old maps showing the location of the original rivers. If space allows the old location of the rivers should be established. Toxic sediments must be removed, the beds must be stabilized wherever necessary and river bank vegetation must be allowed. River bank vegetation without interruption is very important for ecological integrity.

Within a highly urbanised area such as the Emscher river basin it is impossible to reproduce natural drainage conditions which existed prior to urbanisation. The goal is to produce and to maintain a river ecology mainly satisfying the aesthetic and recreational needs of urban areas. Peak-flows, however, shall not affect river ecology irreversibly. One of the most important criteria for this is the shear stress on the river bottom. In an expert opinion it was found that a loading as frequent as once in two years will satisfy urban ecological needs (Arbeitsgruppe Planquadrat, 1991). The low shear stresses necessary are obtained by widening the flood plain width of the rivers and by appropriate vegetation.

The concept for the renaturation of the rivers (Emschergenossenschaft, 1992) demands, that the present triangular profile of the rivers is replaced by wider crossections stepped for low and average flow conditions. Again, on the basis of the real potential for widening the river sections the effectiveness of this measure on the flood-peak was estimated (row 1 in Figure 12). Herewith only minor reductions of 3% to 5% of the flood-peak under different return intervals can be achieved. The relatively large retention volume, which is provided by widening the river beds only has little effects on flood-peaks. However, the advantage of this measure mainly is seen in the possibility to create river side vegetation and to lower the bottom shear stresses.

4.1.6 Treatment of Domestic and Industrial Wastewaters

For treatment of domestic and industrial wastewaters in old industrial areas, sustainable technologies are very difficult to apply. A sustainable technology for instance is the wetlands east of Calcutta, which together with fish ponds treat the waste flows. However, this type of treatment requires space and relies on the right topography. If space is confined the many technical concepts available should be applied. However, in selecting the treatment processes, emphasis should be placed on maintenance cost and operational safety.

To achieve wastewater treatment in the Emscher area, approximataely 350 km of transport sewers are necessary along the tributaries of the Emscher river as well as 40 km of collectors parallel to the Emscher river. Part of this system are combined sewer overflow treatment facilities as explained earlier.

Today's central treatment of combined flows at the mouth of the Emscher river treatment is decentralized to six locations. A major problem is to find appropriate locations for wastewater treatment facilities within the densely populated area. Therefore a compromise of decentralization and central treatment had to be found. On the other hand, today's treatment standards require a certain treatment plant size considering that large amounts of the waste flows are industrial. Six locations for decentral treatment have been chosen whereby four locations already existed, i.e. the treatment location at the mouth of the Emscher river, the locations at Bottrop, Bochum and in Dortmund-North. Additional locations will be treatment facilities in Gelsenkirchen and Castrop-Rauxel.

All new treatment plants will be equipped according to the newest technologies for biological treatment in order to ensure compliance with the tight treatment requirements existing in the Federal Republic of Germany. Especially the efficient reduction of carbon, nitrogen and phosphor are intended. Also for the decentral treatment facilities, the common treatment of communal and industrial waste flows will be maintained, whereby dangerous industrial waste flows are requested to be pretreated by the individual industries prior to discharging the waste flows into the collecting system.

The resulting treatment sludges pose a significant problem. However, it is intended to mix treatment sludges with coal dust to improve dewatering as well as to prepare the sludges for later thermal uses. Sludge handling is anticipated to be centralized at two locations in Bottrop and in Dortmund-Nord. At the latter location an energy concept for the use of other wastes and the sludge is planned.

For the disposal of the mining flows with high chloride contents new concepts are searched for, as the chloride contents cannot be reduced in existing treatment processes. Today in the Emscher area each year 44 million m^3 mining water have to be drained. This corresponds with an average flow of 1.4 m^3/second and a chloride load of 23.5 kg per second. However, it is estimated that with time the mining flows will slowly decrease. The goal is to keep also the mining flows away from the tributaries of the Emscher river. There the chloride concentration shall not exceed 2g Cl/l. However, this value is still under discussion.

For the transport sewers to the treatment plants it was recommended to provide them with a capacity of four to six times dry weather flow although today only two times dry weather flow is handled by the treatment plants. The reason is to allow for future development of technologies to treat combined flows on the treatment plants. Furthermore it was suggested to plan the individual facilities also stagewise in a way, that at a first step existing minimum requirements are achieved. With ongoing experience the facilities should be extended to meet also growing ecological requirements and needs.

It has been suggested for the operation of the intercepting-storage-transportation system for waste flows that at a later point operation may be managed by drawing stored flows according to their pollutional contents and in view of treatment capacities to achieve best treatment. Furthermore it was suggested to use weather radar data and rainfall forecasts in order to optimize use of available flood flow storages.

Also a variety of combinations of above measures were investigated, which are summarised in Figure 13. One can see that combining the cases I to III (1st row) the reduction of flood-peak diminishes with increasing return intervals. Only the combination I to IV (3rd row) shows the desired effect. At a return interval of 200 years the reduction in flood-peak is 50%. If in addition run-off control is applied the reduction of peak-flows is even larger.

4.2 Consecutive Steps to Integrated Planning

Problem identification is a first step for integrative water management and includes:

- investigation of all water resources and a specification of their capacities
- detailed identification of water pollution levels in groundwater and surface waters
- specification of the effects of deteriorated water quality on man and fish
- specification of aquatic life in all parts of the surface water system
- evaluation of the water budget specifying dry, wet and mean conditions
- mapping of all discharge points of stormwater, industrial, commercial and household wastewaters
- specification of existing treatment measures (mechanical, biological, chemical) and treatment efficiencies
- survey of the potential of commercial wastewater discharges into communal sewer systems (cleaners, dentists, laundries, galvanic outfits)
- mapping of the ecological potential in still existing green areas
- specification of land uses throughout the area
- specifications of dust fall-out and rainfall contamination
- mapping of flooding areas according to their occurrence
- specification of soil contamination and mapping of contaminated areas
- definition of groundwater movement

Planning should be staged as follows:

- river basin wide management plans
- operational plans for watersheds or parts of it
- city wide master plans
- planning and design of individual drainage structures
- plans for operation of drainage systems or individual drainage structures

Criteria must exist for different subsystems within the total system in order to coordinate and plan individual measures. It is also necessary that the interfaces between the different subsystems are clearly defined. In today's planning due to different responsibilities subsystems are often planned parallel. The key of integrative planning is to interact between individual planning levels. A possibility for such interaction is shown in Figure 15, where integrative planning is subdivided into three levels:

- Level A: determines development of quality concepts that consider the ecological function of the system, possible water uses, defines water quality criteria and specifies user-dependent quality types as well as characterizes the specific water system.

- Level B: represents the actual planning of individual measures which in the first place must fulfil minimum requirements and secondly must be adaptable to fulfil advanced requirements. These measures refer to the conception of treatment plant facilities, the design of drainage networks including facilities for treatment of storm runoff and combined overflows, the definition of measures in receiving waters. This level also addresses in detail the necessary costs and financing.

- Level C: presents the control of the effects of implemented measures, locates monitoring stations, defines analyses programmes and necessary maintenance as well as derives from the findings new developing tendencies.

The key to operating such a system is that Level C must be given a chance to improve measures planned under Level B. The findings of Levels B and C may well lead to an adaptation of the water quality goals originally anticipated under Level A. This fact must be provided in the planning process.

Water quality criteria must relate to the objectives and to water uses at a very early stage of the planning process, whereby the type of pollution and its effect, the type of the pollution source as well as the type of the receiving water system should be taken into consideration.

In order to recognize deficiencies in procedures and knowledge as early as possible, projects must be as manifold as possible and should reach from the stabilization of still natural parts to solution for devastated urban and industrial areas. The projects should be planned in such a way that technical instalments can be adapted to possible better knowledge and increased needs of the future. It is nearly impossible to try to achieve with today's planning a definite final condition. In many cases it is useful to plan step-wise and to allow further improvements of later planning steps in the future.

According to the development stage and the need the methods necessary for integrative planning should be built up over time. The benefit of a growing system is that the responsible authorities will finally identify themselves with the methods applied. It is not advisable to import complete systems developed on other cases. However, most of the methods needed are available free of charge or at low costs through UNESCO, WMO, WHO and also from national authorities. For some state authorities in the Federal Republic of Germany, for instance, it is common practice that planning methods are provided at no charge to other agencies if disclosure agreements are signed.

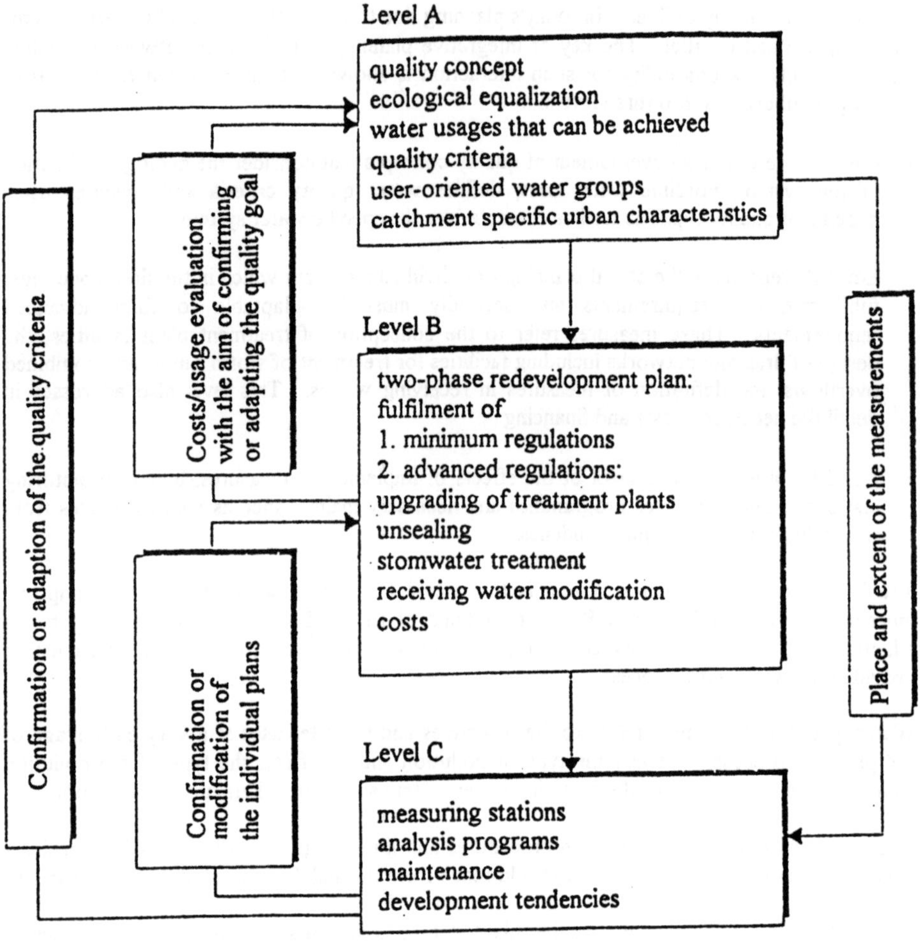

Figure 15: Interaction of planning levels

4.3. Implementation Strategy

The implementation of an integrated management approach requires a number of steps, which are summarized below and then explained in some detail.

1. Establish a common cause of all interest groups
2. Establish a stakeholders group
3. Develop a plan of action - by selecting the best alternative
4. Develop a funding mechanism
5. Engage in political lobbying to gain political support
6. Involve politicians in promoting the plan of action
7. Revise institutional arrangements to support the plan implementation
8. Implement the plan
9. Verify the plan success by post-audit

To be successful in water management, all interest groups must feel that they are a part of the ecosystem in the studied area and that their interests are served by the proposed course of action. This common cause was also described as enlightened self-interest (Christie et al., 1986). Since all citizens in the basin face a common problem, they must be committed, in their own interest, to find a solution. The definition of this common cause is often provided by voluntary associations, or political parties (e.g. the green parties) or government agencies.

In every river basin, there is a number of agencies, organizations, associations and programs, involved in various aspects of river basin management. Such institutions are often depicted in a two-dimensional (i.e. vertical and lateral) schematic diagram reflecting responsibilities and jurisdictions of these groups. The challenge of the integrated plan is to rearrange these parties into a circular arrangement, in which all parties carry the same weight and communicate with all the others (Hartig and Vallentyne, 1989). This is generally achieved under newly created bodies referred to as stakeholders groups or committees. It is essential that such a committee involves concerned citizens and voluntary associations, because their involvement will facilitate the financing and implementation of the plan. Other participants include research establishments (university, private and government laboratories), industry representatives, government representatives (all levels) and engineering companies.

The next step is to develop a plan of action. This is often done in several stages, where the first stage may represent just problem definition of objectives, and the second stage would be the actual plan of action. In this process, it is essential that the plan is based on ecosystem science describing the basin, its water resources, environmental conditions and socio-economic aspects in full comprehensiveness reflecting the interconnectivity of various actions. Where scientific data are missing, the collection of such data may be a part of this plan, particularly in the problem definition stage. The plan must be specific in terms of objectives, correction of the existing problems, the listing of remedial options, identification of preferred options, and definition of the final state of the basin with all the desirable water uses and functions restored. The plan should also include the means of verification that the objectives have been achieved and can be sustained. It is common to develop a number of planning alternatives, subject these alternatives to public and professional scrutiny, and then select the "best" planning alternative. An important part of the plan are the estimates of costs of various remedial options, reflecting both initial (capital) costs and operating and maintenance costs.

Plan implementation is an important stage which involves extensive activities often dispersed throughout the basin studied. These activities are conducted by various contractors retained by the basin committee and often it will be also necessary to retain professional managers for these activities. Important elements include co-ordination of remedial activities so that they proceed in a timely manner and cost overruns are avoided.

Following the plan implementation, a post-audit is conducted. The essential steps in post-audit comprise the comparison of the predicted and actual results and evaluation of the planning methodology. Remedial measures have to be designed using data containing large uncertainties and that may affect the performance in relation to the plan goals. Consequently, the design should be flexible to allow for corrections where design objectives are not met. It has been asserted that in spite of proliferation of the literature on systems analysis of water resources, the measurable outcome of such activities is marginal and there are very few published post-audits (de Donnea, 1982). To resolve this controversy and to further advance the integrated water management and planning, more attention and publicity should be paid to the post-audit of integrated planning studies.

Renewed drainage concepts and the revival of the receiving waters are closely linked. Therefore the measures for the renewal of the systems must be carried out parallel and require a careful coordination. The upstream catchments, especially the drainage area to the treatment plants Dortmund-Nord and Bottrop obtain a high priority. At first the treatment facilities must be built before wastewater can finally be separated from natural flows. Planning must involve at the same time the construction of new transport sewers, storm and combined flow treatment as well as the renaturalization of the urban waters. Only if the development of these measures is integrated, can the costs be minimized and the systems utilized at the earliest possible time. With this also the inhabitants can see the success of the entire effort. As the Emschergenossenschaft is financed by all communities and industries in the area, it is also important that all parts are equally affected by the revitalisation.

However, for carrying out the measures a number of conditions exist:

- *land subsidence must be mostly finished,*
- *the revival of surface waters must start upstream and be carried out step-wise moving downstream,*
- *wastewater and natural flows shall be separated in the tributary areas sequenced according to the construction of the new treatment facilities,*
- *the necessary grounds must be available for construction of facilities and the renaturalization of river stretches,*
- *prior to planning flood protection and removal of land subsidence damages must obtain higher priorities,*
- *for the revival of the surface waters special attention must be paid to parts with high ecological potential, and*
- *measures where a high ecological and landscape effect can be expected as compared to the time and finances involved should be given first priority.*

Just to give an impression of the timing of planning and construction of new treatment facilities, an early estimate done in 1990 is given in Figure 16. Figure 17 supplements this estimate by the planning of the renewal of individual river stretches and for the separation of waste and natural flows.

The first projects for the revival of the Dellwiger Bach in Dortmund and Lepkes Mühlenbach in Essen and Oberhausen, as well as for the Vorthbach in Bottrop and the Emscher itself within the Westfalen Park in Dortmund yielded good and encouraging experiences. The projects have been accepted by all agencies involved as well as by the people living in the areas.

Even under very favourable conditions it is expected that the ecological and urban renewal of the Emscher area will take 30 years. This assumes that the necessary administrative procedures will be carried out extremely quickly. This might be difficult for some parts of the step-wise planning anticipated. However, step-wise planning for the reasons mentioned is of utmost importance for carrying out the large number and time-consuming revitalization measures.

5. Organizational Requirements and Constraints

It is clear from the problems identified in the previous section that the water resource problems of old industrial areas can only be tackled by efficient water resources management bodies having sufficient political support. In most nations such organizations are found, either at provincial or municipal level. Recognized problems are taken up and tackled sooner or later. The general organization and management structures present in some situations may be

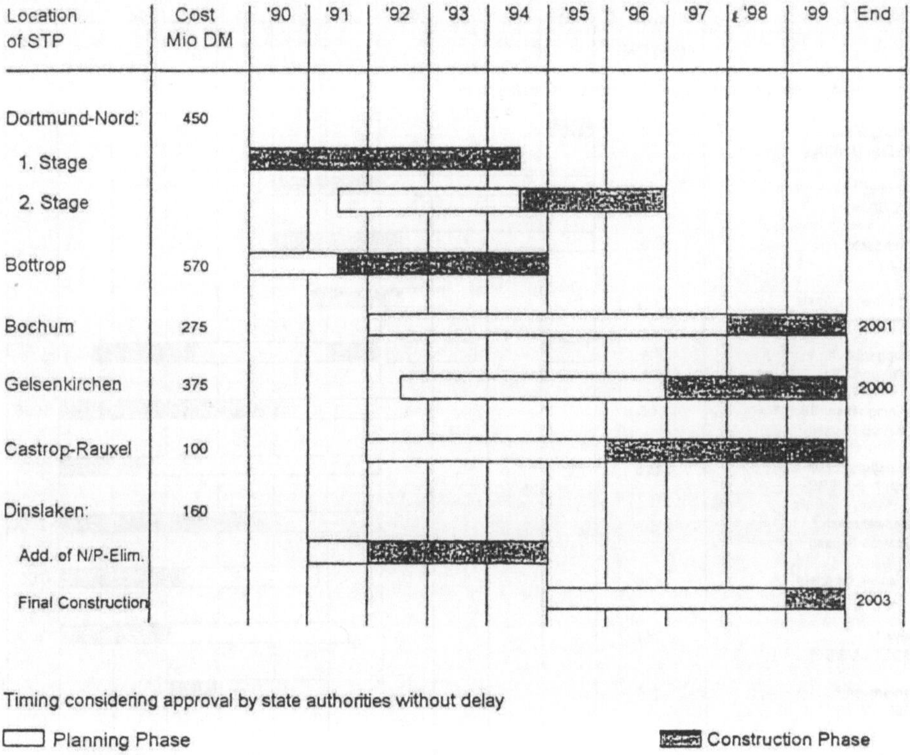

Figure 16: Anticipated planning and construction periods for the new treatment facilities (Emschergenossenschaft, 1991)

inappropriate for problem identification and solution. Such organizations or management structures fail to bring politicians, technicians and the rising 'green movements' together to resolve their fundamental philosophical differences with regard to the way in which a recognized problem should be solved.

5.1 Managerial Conditions

There exists a significant difference in the follow-up of the projects brought to life. Although knowledge is available and money for investments can be obtained, there is often a lack of maintenance. This problem, which is usually considered solely technical or financial, is but more often organizational and institutional.

Another constraint is deficient management during disasters, which may result in great damage, both to the welfare of the citizens and to the socio-economic life. In general disasters do result from two causes:

- natural circumstances, like heavy storms;
- human faults and technical disasters.

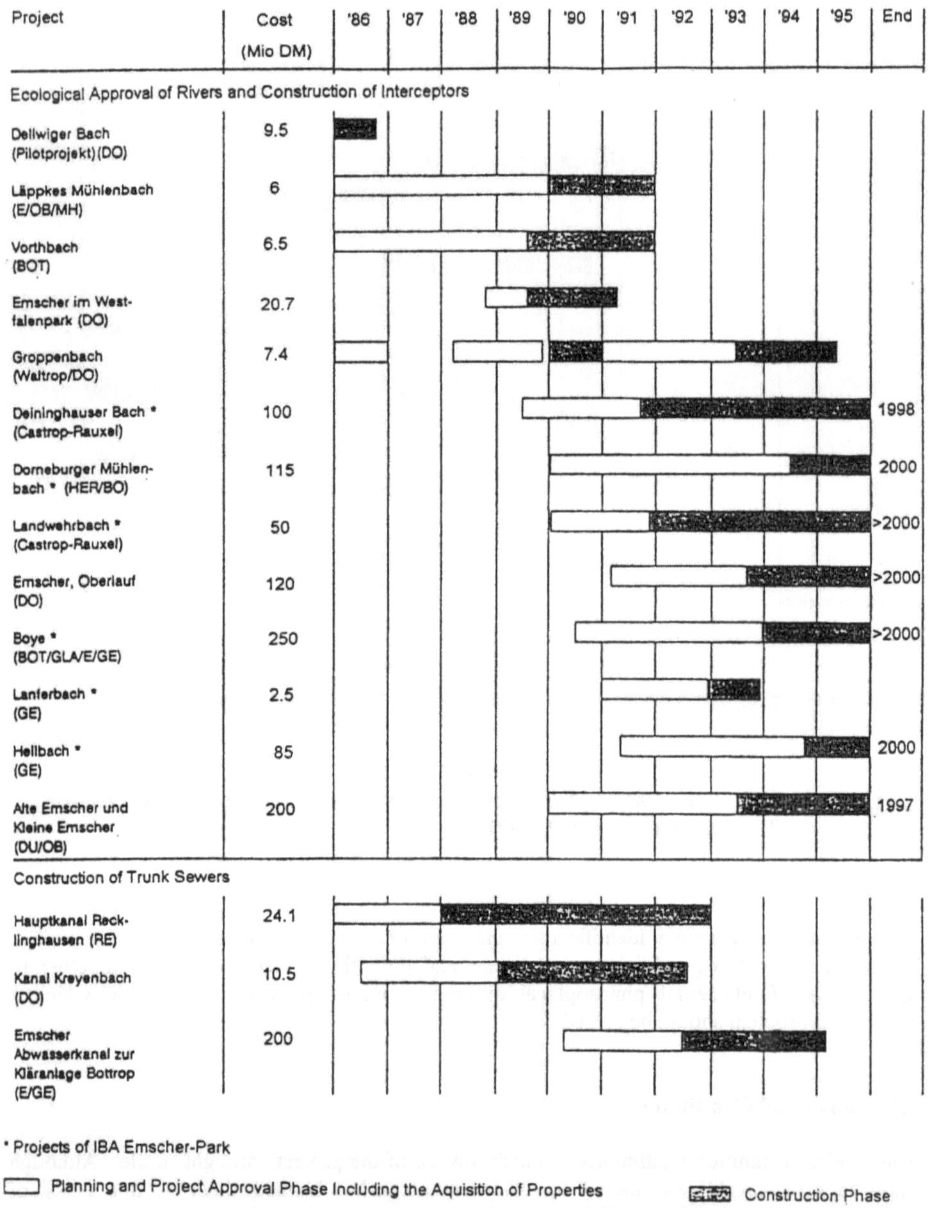

Figure 17: Anticipated planning and construction periods for the renaturalization of rivers (Emschergenossenschaft, 1991)

An important question, related to such events, is the degree of acceptance (from the public) of such occurrences. In most developed countries this degree of acceptance will be low, because the existing management organization is simply expected to foresee such events, have the money and other resources and to be able to cope with the problem.

The existence of the Emschergenossenschaft provides an excellent background to cope with the renewal of the Emscher system. The reason is that the responsibility for planning, construction and operation of the whole water system lies in the hands of that one authority. If the responsibilities would be split, as is usually the case in many countries, the renewal of the Emscher system due to coordination needs obviously would have had much less progress. The momentum of the renewal only can be maintained if the responsibility is kept on a local or regional level. As soon as the responsibilities would be shifted to higher levels the identification in connection to the area would be lost and again the progress would suffer.

5.2 Planning Constraints

An organizational issue is the lack and inappropriate use of data. Water related projects in urban areas require reliable and accurate data for both design and management purposes. The time and space resolution needed may not always be available from standard sources. As an example, correct storm drainage design requires rainfall data of a resolution in between 1 to 5 minutes, whereas the meteorological services rarely produce these data at a resolution lower than 5 minutes. And even if the required data would be obtainable, the existing historical data have not been processed accordingly, or even not at all.

The availability of appropriate management information, like up-to-date maintenance data, long-term process data and reliable financial data, is of at least equal importance. And what is more: when available, they must be used in a correct manner. Unfortunately, the necessity of such management information is sometimes disputed, as some decision makers, even nowadays, are still under the impression that they can do without.

In the composite process of planning of water resources management, there is often a discrepancy between financial and technical responsibilities. The final decision often lies with political decision makers, who may have little technical background. When selecting a technical solution to an environmental or ecological problem, a purely technical approach unconnected with financial realities, can create solutions that are unrealistic.

There is a general tendency to deal ineffectively with risks. Unknown features of the planning process tend to paralyse decision makers, limit the innovative potential and result in stagnant technology and inappropriate solutions. New methods of risk assessment and a public discussion of the risks involved in a particular approach are needed.

The political context under which planning processes must progress, often block effectively an efficient planning procedure. In general the planning process tends to be reactive rather than proactive. Because of the obstacles listed above, science, technology, engineering design, ecological analysis, impact assessment and a range of other factors all operate as reactive systems.

The actual planning process is also restricted by a number of constraints - the separation of fiscal responsibility (rests with politicians) and technical responsibility (rests with technocrats).

One of the major planning constraints in the Emscher region is the lack of a conclusive documentation of water uses, quality conditions and the formulation of immission goals for the improval of the system. Most of the projects done so far follow the idea "learning by doing". However, for reaching the overall goal of the renewal of the Emscher system a quality management plan is needed.

5.3 Obstacles to the Implementation of Integrated Water Management Plans

In spite of generally acknowledged attractiveness of integrated water management, there are many obstacles and constraints encountered in integrated management planning. A personal perspective of obstacles - threats to livelihood, fear of job loss, hopelessness resulting from perceived loss of control; mistrust of governments and industry; confusion caused by issue-by-issue reporting; and, differing perspectives resulting from diverse background and specializations (Christie et al., 1986). An industry perspective of obstacles - the desire for demophoric growth (the growth of energy consumption); competition, mistrust of competitors and government, and suspicion of emotional influence that environmental organizations are thought to exert on government (Christie et al., 1986). A voluntary association perspective - inadequate representation in the decision making process, organizations and finances that are weak, and ineffectiveness of adversarial organizations in responding to common causes (Christie et al., 1986). A government perceived perspective - lack of legislation incorporating the integrated management concept; lack of public support for integrated policies; and lack, of trans-institutional networking reflected in poor inter-agency planning (Christie et al., 1986). Common obstacles - lack of holistic perspective - our knowledge of interrelationships is limited, instead piecemeal action is more typical; predominance of self-interests in utilization of water resources; and, lack of preventative approach - much of our response is retroactive - reacting to crises caused by past mistakes, rather than pro-active - trying to anticipate and avoid future crises (Christie et al., 1986).

More objective are various constraints on integrated water management, with the major ones arising from lack of various resources needed for implementation of specific technical objectives. Such resources include human resources, technological resources, economic resources, natural resources, and their acquisition is further affected by political will and cultural influences. The availability of various resources will be different in countries at various stages of development. Other constraints frequently encountered include differences between the technical perspective emphasizing engineered systems as opposed to natural systems, lack of generally accepted valuation systems for environmental benefits, professional conservatism, relying on proven practice and blocking technological innovation, lack of understanding of the nature of and control opportunities for diffuse sources of pollution, and preoccupation with transport processes detracting from the assessment of impacts on receiving waters.

Having identified some of the typical obstacles, it is now desirable to look for strategies that would overcome these obstacles and achieve the selected objective - the implementation of integrated water management.

There are some deficiencies if it comes to the approval of individual projects. One example is putting infiltration facilities to reality. The inherited drainage design requests for the connection all impervious areas to drainage networks. In some cases it proved to be rather difficult to introduce the new methods of infiltration of stormwater and to obtain approval for this. However, at present the responsible water authorities are alerted and hopefully the attitudes will change quickly.

6. Public Involvement and Awareness

In the recent past, major public projects, like renewal of old industrial sites were conceived by planners, designed by architects, and implemented by engineers with the only public involvement occurring when funding was approved through the political system. In the United States, that system came to the end in the early 1970's with passage of the National Environmental Policy Act. A major requirement of the "Environmental Impact Statement Act" as some call it was the involvement of the public in the decision making process at all stages of development. In other countries in the world similar requirements for public involvement have been established. No longer is it possible to design and build small or large publicly funded project without involving the public in the integrated planning effort.

Bringing the public into the decision making process can include either public involvement or public participation. Public involvement, is by far, the older method of bringing the public into the decision making process. Public involvement takes the form of public review and comment on well developed planning. In the typical public involvement program notice will be published of an anticipated action, plans will be available for review, and the public is encouraged to comment. Comment may be written, or given verbally in public hearings.

Public participation is a much newer procedure. In public participation the public is informed of a planned action at the earliest stages of the planning process. Members of the public, usually called stakeholders, technicians, engineers, and managers meet together to plan, develop, and implement a project. The major difference between involvement and participation is that in participation the public is intimately involved at all stages of the project rather than being limited to providing comment when certain stages of planning or development are reached.

In the renewal of old industrial sites, particularly when an integrated water resources planning and management is required, there is a clear need, if not demand, for full public participation in projects. Public participation is not a more efficient and speeder method of bringing the public into the decision making process. In fact, full public participation effectively changes the entire planning and design process producing, what many feel, is a better design that can move through the construction or implementation phases very quickly. Figures 18 and 19 illustrate the differences between the more traditional public involvement and the public participation process. Note that the public involvement model weights engineering development too heavily and is not focused on policy and goal setting at the beginning of the planning process. The public involvement approach is fundamentally "top-down" only involving the public late in the planning process. The consequence of this approach is that key policies and environmental priorities are not dealt with until near the end of the project when change is more difficult.

The public participation model is a "bottom-up" planning process that addresses public concerns and reviews policies and environmental priorities from the earliest stages of the planning process. A characteristic of the participation model is that the public, often with no technical or professional training, are asked to consider and act on technical as well as non-technical issues. This changes the role of technical specialists from independent and somewhat isolated participants to resources who are dependent on effective interaction with the group. The public also is changed. Stakeholders are educated during the participation process, building their expertise, understanding, and a recognition of the need for compromise and trade-off to achieve a common goal. The primary advantage of the participation process is the early identificaton of critical issues and the development and evaluation of engineering alternatives early in the planning process. As indicated in Figure 19, it can be expected that early stages of a project will be confused by the number of issues defined, but it is possible then to address each issue and develop an implementable plan that meets the varied interests of the stakeholders.

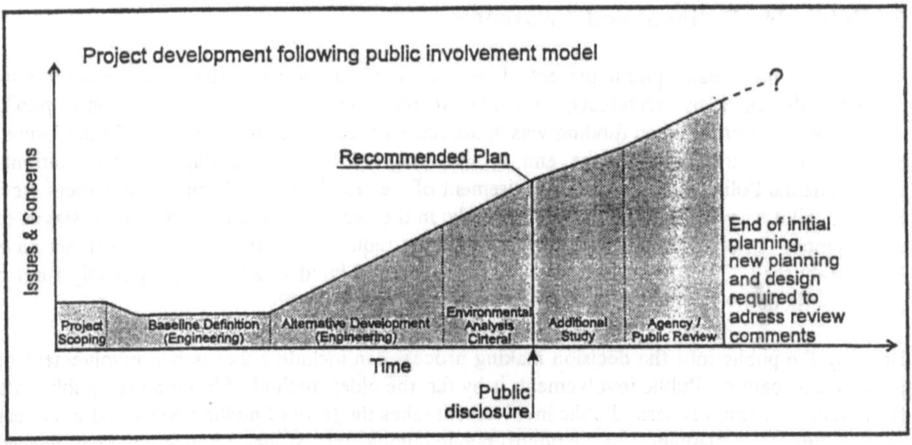

Figure 18: Project implementation under public involvement

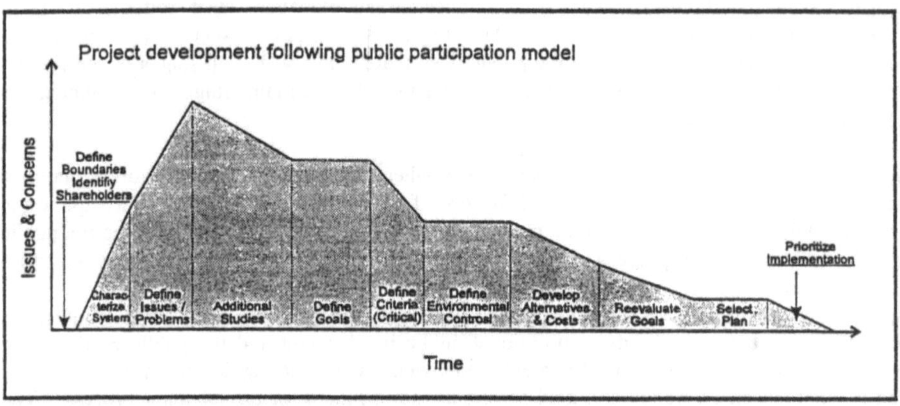

Figure 19: Project implementation with public participation

The reasons for public involvement in integrated water resources planning should be clear. First, we are all the public. Second, support for projects, particularly projects involving water resources development or protection require long-term public support. That long term support can best be achieved by keeping the public informed and participating throughout the project. Finally, water resources projects will compete for funding resources with many projects that interest the public. Involving the public at early stages of any project allows consideration of funding priorities early, and the confidence of the public that the funds allocated will be well spent.

When considering public involvement/participation in integrated water resources projects, the model used and the final approach taken must consider the history and culture of the public participating in the process. The actual model used should be adapted to different political and social/societal settings.

Originally the concept for the revival of the Emscher system was developed by staff of the Emschergenossenschaft and other state authorities. Action was spurred by people living and working in the Emscher basin who had complained for many years about industrial pollution and the smell sewage transported in open channels. However, the public long accustomed to the condition of the Emscher did not demand an effort with the scope of the Emscher renewal. It is only now, with completion, and success, of the initial renewal projects that the public is becoming aware of the full scope of the renewal efforts. The public is now making new demands for revival.

At the beginning of the renewal project, authorities in the Emscher basin provided information sheets on the technical measures used in the revival that were designed for water management professionals. When the IBA Emscher-Park started to coordinate projects for the revival of the area they put special emphasis on public awareness and made special efforts to integrate the public in projects. This public awareness and participation campaign included:

- *initiating public competition for the solution of problems*
- *initiating projects that involved women and children*
- *inviting local residents to opening ceremonies of restored buildings and industrial sites*
- *using restored sites to provide theatres and playgrounds and holding festivals on restored sites throughout the Ruhr region*
- *involving residents in housing renewal to keep the costs low*
- *involving local universities such as University of Essen into all phases of the renewal projects to provide scientific and technical support while depending on the universities for the generation of new ideas*
- *sponsoring design competitions at universities to encourage student participation in the renewal efforts*
- *providing information boards throughout the area reminding people on the individual*
- *coordinating and providing public relation leaflets on all ongoing projects*
- *actively using the press to report on projects and inform the public of events*
- *publishing a newspaper devoted to the revival of the Emscher system*

By combining public awareness and public participation activities, the developing renewal program gained tremendous momentum and public support. The end result is and increased interest in the revival of the Emscher system. The public also starts to realize that their financial support is needed to continue the revival effort. The public has also come to recognize the role they can play in the revival effort. For example, stormwater control programs involve individual property owners using measures to re-use or infiltrate stormwater on their property. The public awareness efforts in the Emscher renewal have provided a better understanding of the infrastructure required for a program on the scale of the Emscher renewal. The IBA Emscher-Park informs the public with a number of folders on individual projects as well as with brochures on specific topics. Furthermore, an information sheet is published regularly. All information is available from IBA Emscher-Park in Gelsenkirchen.

7. Administrative Background for Integrative Planning

New methods of integrated water resources management require appropriate legislative and administrative support. In general there are no single typical policies which would fit all specific conditions. Normally the parliament has to pass laws, which also include water laws determining the rules for the use of water and the constitution of corresponding administrative bodies. The duty of these bodies is to execute the rules and to supervise the compliance with the rules.

At some stage the environmental and ecological impact of water use must be taken into consideration as well as town and country planning tendencies. These interests require an integration of the water laws with laws regulating urban and landscape planning as well as land uses.

The experience shows that it is desirable to separate water authorities on national, regional and communal levels. In any case, the responsibilities should be chosen in such a way that administrative boundaries are not artificial or political but rather related to river basin catchments or water related defined parts of it. The responsible authority for all water related planning within a catchment must be linked to national or even international planning levels.

If it comes to integrated water management, the major responsibility should match the river basin wide responsibility. In any case, the water authorities have the duty regularly assess water quality and quantity to issue permits setting conditions for the use of the water which at an integrated approach must include the preservation of environmental and ecological values. They must also be responsible for controlling and maintaining the set conditions. In the Federal Republic of Germany it has proven to be helpful if violators of set regulations are punished and must pay fines to the state. It is very important to have a strong force for implementation of the laws and regulations. There must not be any exceptions.

Examining the specific legislations in various countries it seems that integrated water management can be successfully implemented without the legislative creation of special bodies for it. Usually structures to cope with integrated planning do exist. It is only necessary to convey the appropriate rights and orders to the existing authorities.

In recognition of the geographical, cultural and social differences in different regions, the legislative structure must accommodate for these differences. Water responsibilities may be divided as follows:

- controlled by central government, the establishment of special authorities may preferably be delegated to regionally oriented bodies
- controlled regionally, stake holder committees and corporation among individual agencies may be more effective

Comparing both approaches the first one may be more susceptible to autogradic water management, which largely is driven by experts and managers. The second alternative is more inviting involvement of a large spectrum of stake holders who try to formulate the management objective, retain experts to develop a scenario, participate in the selection of best options, implement these options and participate in the post-audit evaluation of these remedies.

In order to aid the coordination and the presentation of the projects in the Emscher-Park International Building Exhibition, the government of Northrhine-Westfalia established a planning company operating as a private enterprise, the IBA Emscher-Park Pty. Ltd. The IBA

Emscher-Park was equipped with a budget of 35 million DM and has its headquarters in Gelsenkirchen. At the present they operate with a staff of approximately 30. Its task is to prepare and to stage the architectural and planning competitions, to implement the results together with the project organizers and to act as an intermediary throughout the planning process.

Decisions as to which projects are to be included into the IBA Emscher-Park activities are made by a steering committee chaired by the Minister of Transportation and Urban Development, Mr. Franz-Joseph Kniula. Members of this committee are representatives of state government departments, towns, industry, trade unions, nature conservation organisations as well as planners and architect associations. The board of trustees is chaired by the Minister president of Northrhine-Westfalia, Mr. Johannes Rau.

8. Financing

Remedial actions can be quite expensive and their funding may be beyond the means of stakeholders. Consequently, it is required to develop a funding mechanism that would support the financing of the remedial plan over some long-term period. Differences in political systems, institutions and general perceptions will prevent establishment of general principles in this field. However, it is of interest to discuss here some of the possible options. In general, the principle "user or polluter pays" should be applied. In this sense, financing through general taxation is unpopular, because it ignores the above principles. Some financing through this means is however unavoidable and would involve contributions from municipal, territorial (state or provincial) and national taxes.

Other funding mechanisms, based on the polluter/user pays principle, are more appropriate. Those include taxes on consumptive resource uses (e.g. water withdrawals for cooling, irrigation), developing full cost pricing by internalizing pollution costs and full-cost pricing for new development, and developing options for charges and tax incentives that shift more of environmental protection costs from the taxpayer to the consumer. Again, while many of these options seem attractive, political realities may render them unfeasible in various political systems and situations.

The implementation of the plan, with large funds and responsibilities involved, may call for revisions of the existing institutional arrangements to support the plan implementation. Many examples of such institutions can be found in the literature, again reflecting the local political arrangement. The existing institutions are generally characterized by large fragmentation of responsibilities and jurisdictions. Specific institutional arrangements are rarely transferable to other jurisdictions, because they generally reflect the local socio-economic conditions (Lord, 1984). The issues of legislative and administrative support were discussed earlier in Chapter 7. In many countries, the political and legislative environments are not satisfactory, with reported cases of political interference, transitional changes in policies and institutions, conflicting and fragmented legislation, and competition between local and central interests.

At present the IBA Emscher-Park coordinates approximately 80 projects. The responsibility for the individual projects lies with local authorities, private companies and different organizations in the region.

The projects are frequently financed jointly by the state, the cities and private companies. In some cases, for example the Landscape Emscher Park, all costs are met in principle by public funds. Many projects also receive support from the central government of the Federal

Republic of Germany and the European Community. In summer 1993 the total current expenditure on projects amounted to 2.5 billion DM, 1.7 billion DM of which came from public funds and approximately 800 million DM from private sources.

The revival of the Emscher system is designed so that the costs are still acceptable to society and that the living and economic conditions in the area are not endangered. In 1990 the following costs were estimated in detail:

- *construction of interceptors and connecting sewers as well as of storm outfall and combined overflow treatment facilities* — *3 billion DM*
- *construction of the decentral treatment plants* — *2 billion DM*
- *revitalization of the surface waters and flood protection measures* — *2 billion DM*
- *unexpected matters* — *1 billion DM*
- *total* — *8 billion DM*

These figures do not include all measures all measures in the drainage catchments itself which are taken to reduce storm flows. These costs were estimated at an additional 2 billion DM.

It must also be recognized that the effect of the measures will not only be obtained at the very end of finalization of all projects.. Each treatment plant, each interceptor and in each river stretch renaturalized is a step towards the improvement of the ecological renewal of the area. It will be inevitable that some projects will be delayed also due to financial restrictions. However, the momentum already now seems to be so big that the total effort is not in danger. For the construction of the treatment plants within the next 15 to 20 years and the intercepting sewers as well as for the renaturalization of many parts of the river system, an annual finance volume of approximately 320 million DM is estimated.

For the entire duration of the revitalization of the Emscher system it is expected that approximately one third of all costs are subsidized by the state. The remaining finances including also the funds necessary for operation and maintenance of the new system must come from by the wastewater fee collected from inhabitants and industries. It was calculated that the wastewater fee of presently 0.7 DM per m^3 will have to be raised to 2.10 DM per m^3, in order to provide the necessary finances. These cost calculations show that the area decided not to rely on international funds. In addition to the wastewater fees mentioned, the wastewater costs for the individual cities of the amount of 2 DM per m^3 must be added. As a result, at present the wastewater fee of 2.7 DM is collected and in future this amount is estimated to be 4.70 DM per m^3. This amount seems to be reasonable and compares to wastewater fees collected from the inhabitants in other parts of the Federal Republic of Germany.

9. Conclusions

The following is concluded from the evaluation of different case studies presented at the two UNESCO workshops mentioned at the beginning, and from the experiences gained with projects in the context of the revival of the old industrial Emscher area:

- **Environmental revival of devastated areas can only be achieved by economic development through clean industries.**

- **It must be the people living in the area who must request the change, carry it and live it.**

- The right balance between no or low cost initiatives in the area, state funding, private investment and international help must be found. The basis for the success is always local initiatives.

- Small low cost pilot projects carried by local authorities are often more effective than large scale national or international funding programmes. Therefore, it is advisable not to start with large scale programmes but to start with local activities in order the create the necessary momentum.

- Established long-term programmes are of little use. However, there must be a long range framework that is filled step-wise with projects. Each project must be evaluated for its success. The experiences and findings must immediately be translated to approve or to adapt long-term goals.

- To revive a devastated system takes as much time as it took to establish the bad condition. It took 60 to 90 years for the Emscher system to develop into today's status. Therefore estimates that the system can be revived within 30 years seem to be too short.

The many projects aiming for the revival of the old industrial zone of the Emscher ara today present a holistic approach involving integrated water management in most activities. The spirit existing today after five years of intensive planning and implementation proves that revival of old industrial areas can be achieved also in times of financial constraints. The vision of the Emscher area developing into the greenest industrial zone of Europe can be achieved. However, one must not overlook that

- *the area will always remain densely populated and will have an industrial background*
- *the area will always have small rivers and creeks and is located away from large receiving waters*
- *the morphological conditions with subsided areas having no natural drainage will always require technical drainage provisions to keep the area flood-free*
- *the area will always have to cope with the relicts of the old coal mining activities by constantly disposing mining waters*
- *the area will always draw its water supply from the adjacent river basins and therefore the waste flows in the area will exceed the amount of natural flows in the rivers and creeks within the area.*

References

Bayerisches Staatsministerium für Landesentwicklung und Umweltfragen 1994: Wasserwirtschaftliche Rahmenplan Main. Munich.

de Donnea, F.X. 1982: The Role and Limits of Normative Models in Water Resources Policy and Management. In: Water Resources and Land-Use Planning: A Systems Approach. Laconte, P.; Y.Y. Haimes (eds.), Martinus Nijhoff Publishers, The Hague, pp.7-19.

Christie, W.J., Becker, M, Cowden, J.W., Vallentyne, J.R. 1986: Managing the Great Lakes Basin as a Home. J. Great Lakes Res. 12(1): 2-17.

Emschergenossenschaft 1989: Möglichkeiten der Umgestaltung von Wasserläufen im Emschergebiet. Essen, May.

Emschergenossenschaft 1991: Rahmenkonzept zum Ökologischen Umbau des Emscher-Systems. Materialien zum Umbau des Emscher-Systems. Heft 1, Essen, July.

Emschergenossenschaft 1991: Konzept zur Verminderung und Behandlung des Regenwasserabflusses. Materialien zum Umbau des Emscher-Systems. Heft 2, Essen, November.

Emschergenossenschaft 1991: Konzept zur Verminderung der Hochwasserabflüsse. Materialien zum Umbau des Emscher-Systems. Heft 3, December.

Emschergenossenschaft 1992: Konzept zur Beurteilung und Stärkung der Niedrigwasserabflüsse. Materialien zum Umbau des Emscher-Systems. Heft 5, Essen, August.

Emschergenossenschaft 1992: Konzept zur Umgestaltung der Wasserläufe. Materialien zum Umbau des Emscher-Systems. Heft 6, Essen, September.

Emschergenossenschaft 1993: Wohin mit dem Regenwasser? - Arbeitshilfe für einen ökologisch-ausgerichteten Umgang mit Regenwasser in Baugebieten. Materialien zum Umbau des Emscher-Systems. Heft 7, Essen, January.

Geiger W.F. 1991: Reduzierung von Hochwasserabflüssen in den Wasserläufen des Emschergebietes. Gutachten für die Emschergenossenschaft Essen.

Geiger W.F.; Hvitved-Jakobssen, T.H. 1992: Gutachten zur Wassergüte im Emscher-System - Anforderung, Kenntnisstand und Handlungsmöglichkeiten. Commissioned by the Emschergenossenschaft, Essen.

Hartig, J.H.; Vallentyne, J.R. 1989: Use of An Ecosystem Approach to Restore Degraded Areas of the Great Lakes. Ambio, XVIII(8): 423-428.

Herricks, E.E.; Schaeffer, D.J. 1987: Selection of Test Systems for Ecological Analysis. Water Sci. Tech. 19, 47-54.

IBA Emscher-Park 1991: Wassertechnische Gutachten zur Umgestaltung des Emscher-Systems. Kurzgutachten und Bewertung. Commissioned by the Emschergenossenschaft, Essen and the Internationale Bauaustellung Emscher-Park. Published by the Internationale Bauaustellung Emscher-Park GmbH and the Emschergenossenschaft, Gelsenkirchen.

IBA Emscher-Park 1993: Dimensionen der ökologischen Erneuerung. Ministerium für Umwelt, Raumordnung und Landwirtschaft des Landes Nordrhein-Westfalen, IBA Emscher-Park, Gelsenkirchen.

IBA Emscher-Park 1993: The Emscher-Park International Building Exhibition - An Institution of the State of North-Rhine-Westfalia. Edited by Marion Zerresson und Sabine Radomski, IBA Emscher-Park, October.

Jermar, M.K. 1987: Water Resources and Water Management. Elsevier, Amsterdam.

KVR 1990: The Ruhr. Kommunalverband Ruhrgebiet, Essen, August.

Londong, D.; Geiger, W.F. 1992: Stormwater Management Strategies and Their Effects on Peak-Flows in the Highly Urbanized Emscher Region. In: NOVATECH, Lyon, November 3-5.

Niemczynowicz, J. 1993. New Aspects of Sewerage and Water Technology. Ambio, 22(7): 449-455.

Arbeitsgruppe Planquadrat 1991: Umgestaltung der Wasserläufe im Emschergebiet, Rahmengutachten zur städtebaulichen, ökologischen und landschaftlichen Integration der Wasserläufe. In: Emscher Park-Planungsgrundlagen, Vol. 2/1 of the IBA Emscher Park, Gelsenkirchen and the Emschergenossenschaft, Essen

Tjallingii, S.P. 1990: Strategies in Urban Water Design. In: Massing, H., Packman, J.; Zuidema, F.C. (Eds.) Hydrological Processes and Water Management in Urban Areas. IAHS Publication No. 198, Wallingford, UK, pp. 323-329.

Pollution Potentials of Storm Runoff from Urban and Industrial Areas

Čedo Maksimović
Ana Tomanović
Dubravka Pokrajac
IRTCUD, Institute of Hydraulic Engineering
Faculty of Civil Engineering, University of Belgrade
YUGOSLAVIA

1.1. INTRODUCTION

In the projects of pollution control and integrated rehabilitation of polluted and devastated areas, it is common to opt for mechanical (primary), biological (secondary) and even physical-chemical (tertial) treatment of domestic waste waters. However, in the sites heavily polluted by industrial or agricultural activities, evenly dispersed over large areas or concentrated in point sources (dumps of industrial wastes, etc.), pollution from these locations can generate harms for both surface and ground waters. The pollution of storm-water runoff is analysed in what follows.

The paper presents some preliminary results of a comparison of field measurements with numerical simulation of generation and migration of pollutants in urban and industrial areas. Since most of the non-point pollution generated during storm events an emphasis is being placed both on pollution by urban storm runoff and on percolation of pollutants into underground aquifers from industrial wastes disposal sites, as a consequence of storm events.

In most cases the pollutants being lifted and washed from the ground (roofs, streets etc.) are transported in gutters, sewers and channels by overland flow, or partly infiltrate before reaching a receiving water body. On its way to the receiving water or the underground aquifers body the polluted water interacts with the elements of the urban infrastructure systems as shown in Fig. 1. Local conditions, however, differ considerably, with respect to the duration and intensity of rainfall, the extent of impervious surfaces, the inclination of surface and sewers, the natural and manmade impediments to flow, etc.

NATO ASI Series, Partnership Sub-Series, 2. Environment – Vol. 45
Environmentally Devastated Areas in River Basins in Eastern Europe
Edited by Alfons Georges Buekens and Vasily Victorovich Dragalov

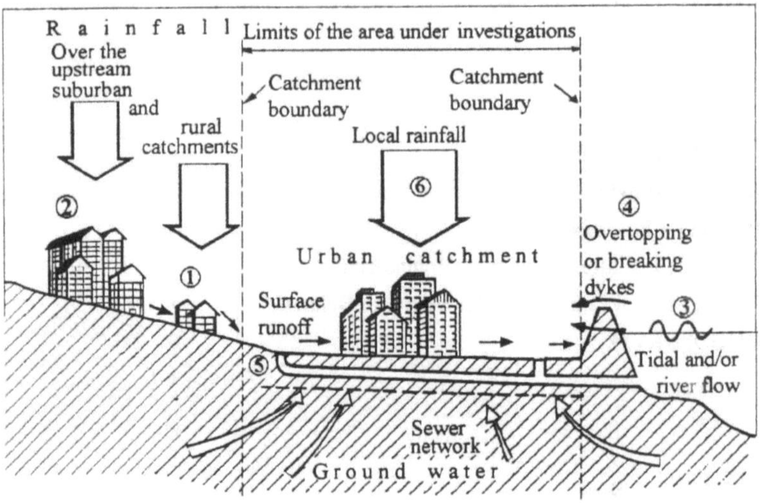

Figure 1. Interaction of urban waters with infrastructure systems

In the heavily polluted and environmentally devastated areas, health hazardous situations occur when the heavily polluted waters reach the intakes of drinking water (Fig 2).

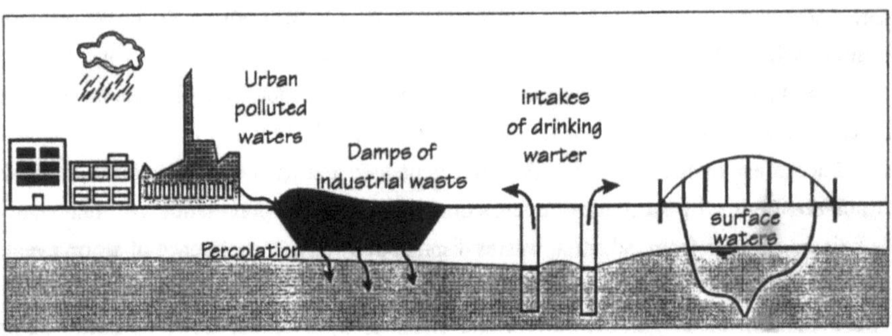

Figure 2. Interaction of polluted storm runoff waters with intakes of drinking waters

For the analysis of existing systems and for the design of new infrastructure it is a prerequisite to have a reliable assessment of the quantity (spatial and temporal distribution of water) and quality (spatial and temporal distribution of concentration of pollutants).

Table 1. The range of mean values of certain water quality characteristics of urban water discharge taken fron various sources)

Water characteristic	Range of event mean value	Water characteristic	Range of event mean value
Suspended solids (Comb. sys.) [mg/l]	341-674	Suspended Solids(Sep. sys.) [mg/l]	190-496
Electric conductivity [μS/cm]	08-470	pH	6.47-6.78
BOD [mg/l]	7.3-15	COD [mg/l]	47-146
TOC [mg/l]	26-28.3	DOC [mg/l]	3.1-5.1
NH4 [mg/l]	1.92-2.75	P [mg/l]	1.6-2.95
Pb [μg/l]	60-525	Cd [μg/l]	2.8-6.4
Zn [μg/l]	320-2000	Cu [μg/l]	23-184
Ni [μg/l]	35-57	Colif. [100ml]	2.2-5.6x10^6

As a general conclusion one can state that one of the main shortcomings of all these models is the lack of good quality data obtained by field measurements with sufficient time resolution. Data on water quality obtained by sampling and off-line analysis (with typical time resolution of 5 minutes) are insufficient because the water quality characteristics during storm event change much faster, especially during the first minutes of rainfall
(Figure 4.).

Aiming at improving the quality and reliability of modelling, the authors have carried out an extended field monitoring program including measurements of several quality characteristics at a time resolution of 10 seconds. Besides the rainfall intensity and surface runoff quantity, the following water quality characteristics have been monitored: suspended solids, pH, conductivity and temperature. These characteristics have been selected because they are easy and not too costly to measure by electrochemical methods, with possibility to record data with a fine time resolution. Similar measurements have been performed at two sites (surface storm water runs into one gallipot): in Lund, Sweden (Spongberg and Niemczynowicz, 1993) and in Belgrade, Yugoslavia (Tomanovic, 1994). The program of measurements and data collection has been arranged as a part of a bilateral project between the University of Lund and the University of belgrade. Two sets of equipment for measurements and data recording have been financed by the Swidish governmental agencies. The

To obtain these information there are at least two major possible approaches:

- measurements in which only a statistical mean value of various pollutants are determined;
- measurements in which the concentration of pollutants and discharge of water are measured continuously, so that they can be used for the development and calibration of dynamic models.

The authors have been involved in an experimental program in which the second approach has been applied. It served as a basis for development of a model for a dynamic simulation of pollution loads from urban areas during storms (wash-off of pollutions). The development of this model and its application are shown at the first part of the paper. The second part deals with the assessment of hydrodynamic properties and quality of water that percolate from the waste dumping sites. The results obtained by the application of these two models could be either combined with some other analysis applied directly in solving concrete problems (for example sizing the required retention basins, selecting and sizing pumps, or other elements of treatment plants, etc.).

1.2. QUALITY OF SURFACE RUNOFF

The measurement of mean values can provide us with general guidelines of non-point source of pollution generated by storm urban runoff. The ranges of mean values of the most relevant water quality characteristics of these waters are presented in the Table 1.

In many cases, however, more information is needed. Modelling of runoff quality in addition to quantity, in recent years has become possible. Most of the advanced software packages used in developed countries (SWMM, STORM, MOSQITO, FLUPOL, etc.) offer this option. The reliability of these models depends on the quaity and quantity of data used for their development and calibration, but also on the experience of their users.

The models differ by their basic assumptions and means of handling the input parameters and data. There are also differences in handling build-up and washoff of pollutants from a surface. In the present model it is provided that the modelling of build-up is less sensitive to the initial conditions provided that the total yield has been modelled correctly.

experimental set-up for the catchment in Belgrade is shown in Fig 3. In the initial phase of measurements several problems with the setup have been experienced such as: clogging of sensors and of discharge measuring weir, inadequate initial preset range, etc. However with required cleaning and maintenance the system performed satisfactorily.

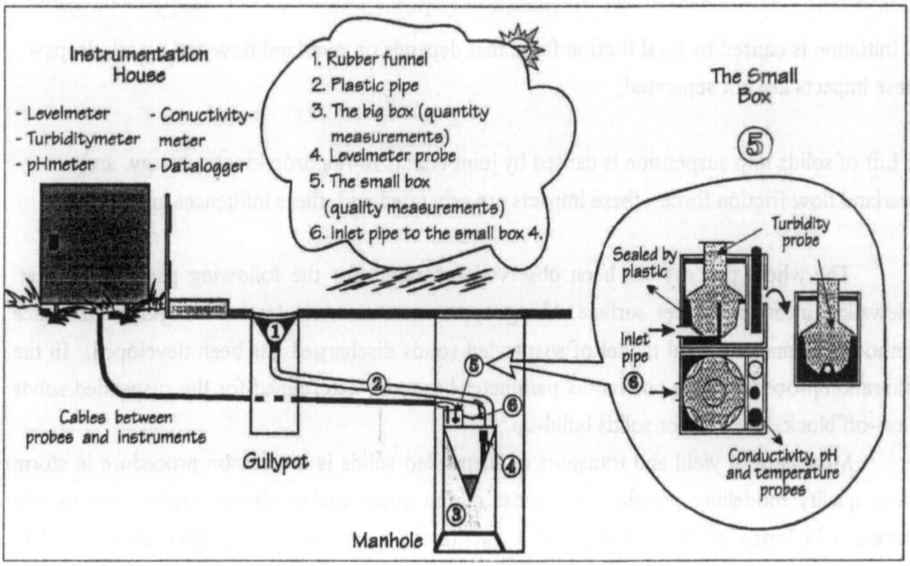

Figure 3. Experimental facility in Belgrade

The program of measurements has been carried out during 1993 (in Lund during 1991). It resulted in a set of 52 events, 29 of which have been discarded because of various irregularities or because of too small total precipitation rate. These data, together with data collected in Lund were used for distinctive statistical analysis, which revealed relevant factors of the surface water pollution generation. Based on these findings a mathematical model of suspended solids runoff from non-pervious areas has been developed. For calibration of this model one half of the recorded events has been used, and the other half served for its verification.

The model contains two main blocks (Tomanovic, 1994): model of solids build-up on the surface, and model of solids wash-off from the surface. Since the accumulation of solid particles at the surface has not been directly measured in this study, the exponential equation proposed by Sartor and Boyd in 1972. The distribution of solid particles across the street surface, based on available results published in the literature, is incorporated in the model. The solids wash-off is based on the deterministic description of the following processes: (1) surface runoff - a kinematic wave theory is applied; (2) initiation of suspension; (3) transport of suspended solids by surface runoff - transport equation of solid particles is used.

Two different methods for description of the initiation of suspension have been applied, and compared:

1. Initiation is caused by total friction force that depends on overland flow and on raindrops - these impacts are not separated;

2. Lift of solids into suspension is caused by joint effects of raindrop kinetic energy and overland flow friction force - these impacts are separated and there influences are summed.

The whole process has been observed separately for the following parts of a street: sidewalk, gutter and street surface. Using appropriate numerical schemes (finite difference method), the mathematical model of suspended solids discharged has been developed. In the calibration procedure only one model parameter has to be determined for the suspended solids wash-off block, and two for solids build-up.

Modelling of yield and transport of suspended solids is a common procedure in storm water quality modelling practice. The most of the other quality characteristics are usually expressed in terms of concentration of a fraction of suspension solids after an appropriate calibration by the local samples. An example of the results of measurements and modellings of water and suspended solid runoff is shown in Fig 4. Thje results of measurements of other quantities (pH, conductivity) are available.

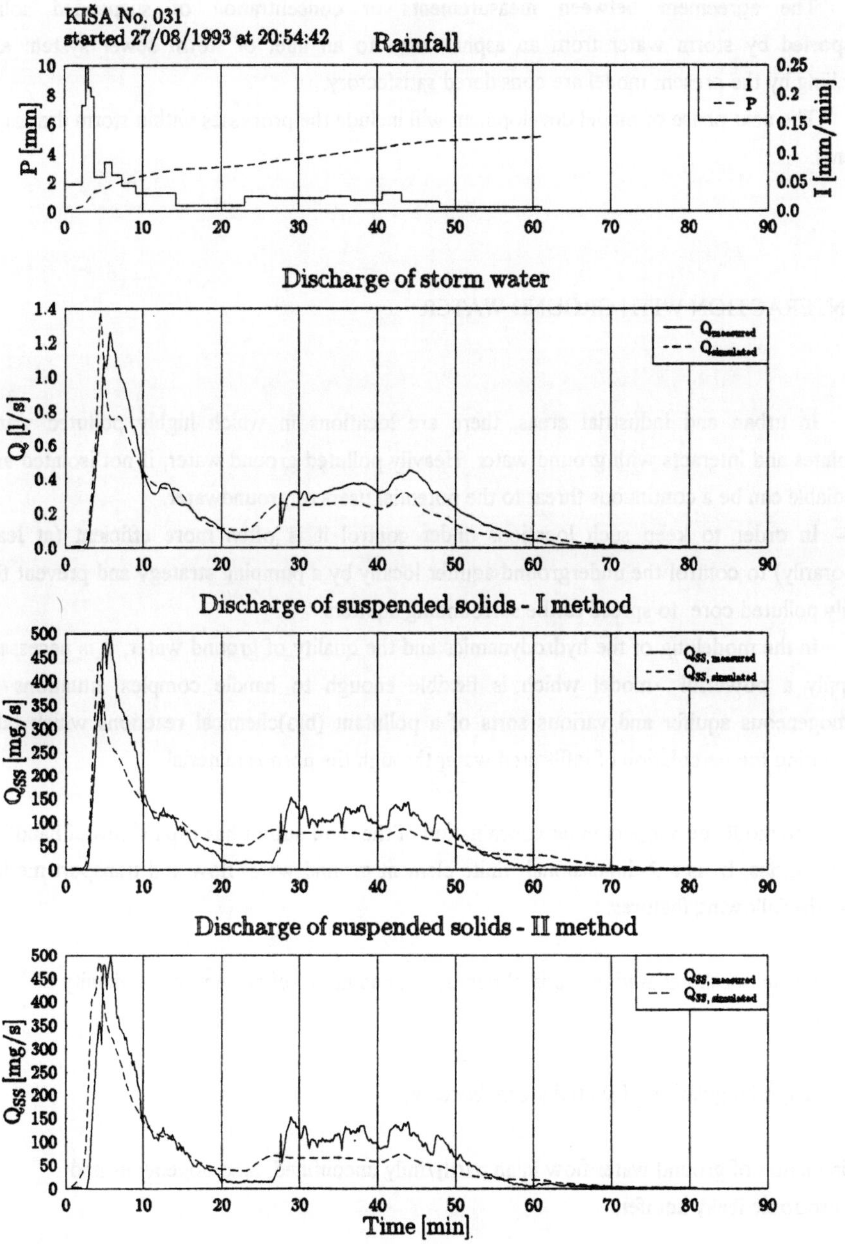

Figure 4. Measured and modelled discharge of water and suspended sediment for one storm event - experimental catchment Miljakovac, Belgrade

The agreement between measurements of concentration of suspended solids transported by storm water from an asphalt area to an inlet of storm sewer system and modelling by the present model are considered satisfactory.

The next phase of model development will include the processes within storm drainage systems.

1.3.INTERACTION WITH GROUND WATER

In urban and industrial areas, there are locations in which highly polluted water percolates and interacts with ground water. Heavily polluted ground water, if not isolated and remediable can be a continuous threat to the potential users of groundwater.

In order to keep such locations under control it is often more efficient (at least temporarily) to control the underground aquifer locally by a pumping strategy and prevent the heavily polluted core to spread to the surrounding aquifers.

In the modelling of the hydrodynamics and the quality of ground water, it is necessary to apply a powerful model which is flexible enough to handle complex situations of inhomogeneous aquifer and various sorts of a pollutant (bio)chemical reactions which take place during the percolation of infiltrated water through the porous material.

The model developed under coordination of the third author has capabilities of handling such situations. It is a 2-dimensional finite element ground water flow and transport model, having the following features:

- automatic mesh generation, optimal numbering and mesh refinement in the vicinity of wells,

- Delunay triangulation of spatially variable data,

- simulation of ground water flow in an partly/fully unconfined, inhomogeneous and anisotropic leaky aquifer,

- simulation of convective and dispersive pollutant transport as well as various sorption and decay type reactions.

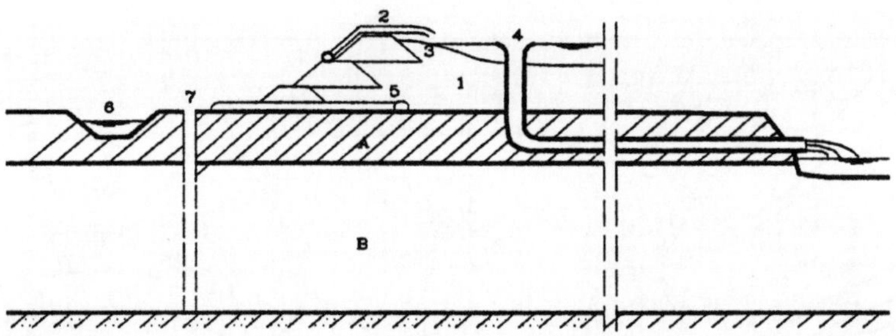

Legend:

1. Deposited ash
2. Spigots
3. Beach
4. Evacuator
5. Drains
6. Channel
7. Wells

A. Cover layer
B. Aquifer

Figure 5. Flow in the vicinity of the ash impoundment

In the testing and development phases the model has been used for the analysis of ground water protection from the pollution by fly ash impoundment of the thermal power plant "Nikola Tesla B", upstream of Belgrade water supply system. The cross section of the impoundment is given in Fig 5.

Ash transport is hydraulic as a low-concentration suspension. Therefore, the significant part of the impoundment volume is saturated with water. Ash is discharged through the spigots (2) onto the beach (3) and into the lake. By the time the water reaches the evacuator (4) ash is settled on the impoundment surface. The stability of the surrounding dikes is assured by the horizontal drains (5) which discharge drainage water into the surrounding channel (6).

Water is entering the underlying aquifer (B) through the layer of low permeability (A), carrying some pollutants washed off some the ash. Major pollutants for the analysed ash are sulfates because of their large amounts, and arsenic, due to its toxicity. The impoundment is surrounded by a system of wells (7) which control the amount of water infiltrating into the aquifer, and prevent polluted water to escape outside the controlled zone. Series of computations were made in order to determine the optimum well operation during seasonal changes in natural ground water regime.

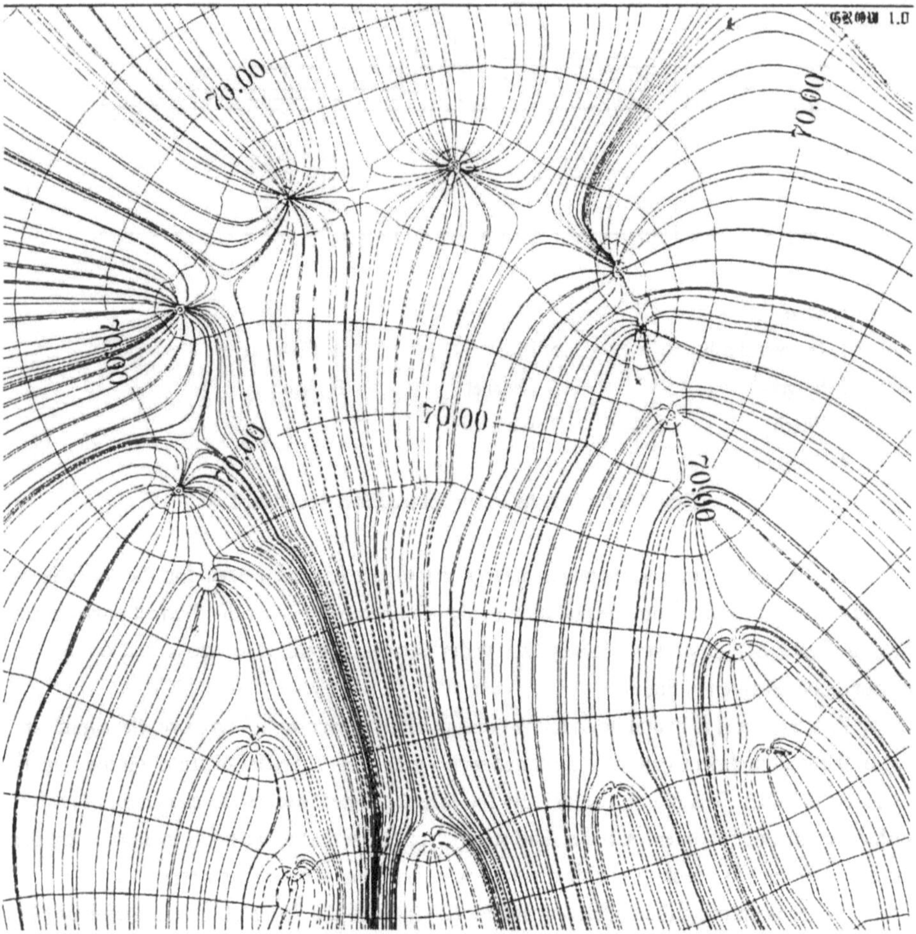

Figure 6. Flow in the vicinity of the ash impoundment
(izo-piezometric lines and pathlines towards the pumping wells)

Some of the results of a typical hydrodynamic simulation are shown in Fig 6. They are presented as izo-piezometric lines and pathlines for the optimum situation and proper operation of protective wells. All the water infiltrating from the impoundment is captured by the wells, and treated before discharged into the river Sava.

1.4. CONCLUSIONS

Integrated rehabilitation of heavily polluted or devastated areas is possible if each of the causes of pollution is well studied and a proper cure applied, as a part of the integrated solution. The present paper presents an example of the analysis of two components of superficial and underground hydrodynamic and water quality issues.

1.5. BIBLIOGRAPHY

Bertrand-Krajewski J.L., Scrivener O., Briat P. (1993) Sewer Sediment Production and Transport Modelling: A Literature Review, Journal of Hydraulic Research, Vol 31, No 4.

Cordery I. (1977) Quality Characteristics of Urban Storm Water in Sydney, Australia, Water Resource Research, Vol. 13, No 1.

Xanthopoulos C., Hahn H. (1993) Anthropogenic Pollutants Wash-off from Street Surface, Proc. of the Sixth International Conference on Urban Storm Drainage, Niagara Falls, Canada, Septembar 1993.

Sartor J.D, Boyd G.B. (1972) Water Pollution Aspects of Street Surface Contaminants, U.S.A. Environ. Prot. Agen. Rep., EPA-R2-72-081.

Spongberg A., Niemczynovwicz J. (1993) Measurements of Wash-off from an Asphalt Surface, Sixth International Conference on Urban Storm Drainage, Septembar 1993.

Tomanovic A. (1994) Modelling of Surface Runoff Water Quality from the Asphalt Areas Based on the Field Measurements, Msc. Thesis, Faculty of Civil Engineering, University of Belgrade.

ASSESSMENT OF POLLUTION HAZARDS IN THE ESTARREJA AQUIFER (PORTUGAL). REMEDIATION STRATEGIES.

Teresa E. Leitão,
J.P. Lobo-Ferreira
Laboratório Nacional de Engenharia Civil
Avenida do Brasil, 101
1799 Lisboa Codex - Portugal

Dulce A. Pássaro
Ministério do Ambiente e Recursos Naturais
Direcção-Geral do Ambiente
Av. Gago Coutinho, 30
1000 Lisboa - Portugal

ABSTRACT: A groundwater contamination case-study including a proposal for its reclamation is presented in this paper. A fifty years old chemical industrial area is located in the case-study region of Estarreja, Portugal. The industrial wastes coming from the factories have been and are still kept in three different waste disposal sites. The liquid effluents, i.e. the industrial waste waters, have been directly rejected to the soil into unlined ditches. In this study, the groundwater pollution extent caused by the industrial activities is assessed for the unconfined superficial aquifer composed by a quaternary sandy formation. An hydraulic pumping system for the reclamation of the polluted groundwater is proposed.

The study, that is briefly presented in this paper, was divided in the following items: (1) regional assessment of the groundwater and soil contamination; (2) review of existing geochemical data and development a preliminary groundwater quality assessment; (3) problem identification and hydrogeological analysis, including mathematical groundwater flow and transport modelling; (4) development of a specific 43 wells monitoring network, sampling and laboratorial analysis for more than 20 different chemical elements; (5) groundwater risk assessment; (6) evaluation of potential remedial actions and (7) modelling and design of an hydraulic pumping system for the contaminated groundwater recovery in the case-study area.

1 INTRODUCTION

In Portugal, groundwater is the major water source for the supply of domestic, industrial and agricultural water needs. According to Lobo-Ferreira and Oliveira (1993) the total water supply needs of mainland Portugal are aprox. 70% satisfied with groundwater resources. Groundwater pollution problems in some regions of this country raised the need towards an assessment of the state-of-the-art in today pollution problems, both in space and time. This assessment aims future reclamation projects, where necessary.

In this paper a synthesis of the achievements of a study carried out during a three years period are presented (cf. Leitão et al. 1992 and Leitão et al. 1994). The study addresses the impact caused by the rejections of effluents of the industrial complex located 10 *km* north of

NATO ASI Series, Partnership Sub-Series, 2. Environment – Vol. 45
Environmentally Devastated Areas in River Basins in Eastern Europe
Edited by Alfons Georges Buekens and Vasily Victorovich Dragalov

Estarreja, in Portugal (Figure 1). The case-area is 25 km^2 large. The aquifer studied is a quaternary sandy aquifer, with a maximum depth of 15 *m*.

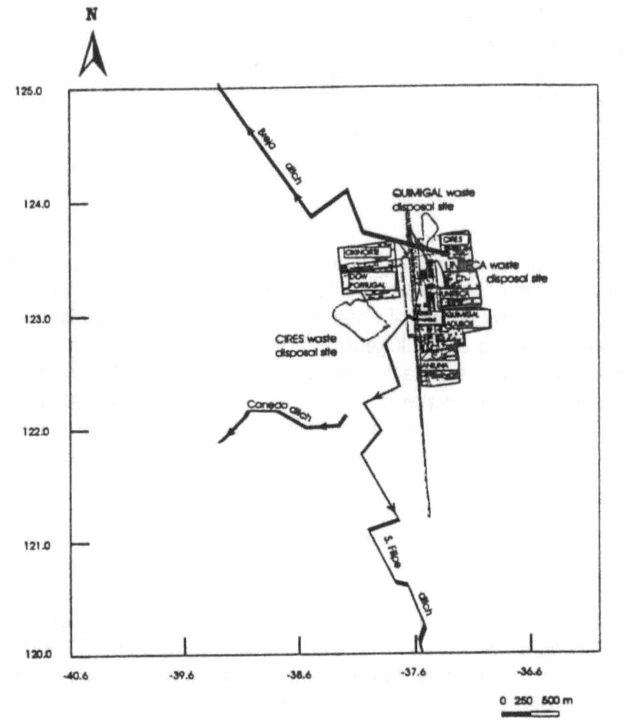

Figure 1 - Schematic representation of the area studied and its main industrial units

This article focus the following general items: (1) broad chemical characterisation of the liquid effluents and the solid wastes, and identification of the existing potential pollutants; (2) hydrogeologic characterisation and conceptual modelling, based on the data collected in 43 new monitoring wells drilled in the area; (3) analysis of the groundwater quality results using both the monitoring carried out through a multiparametric probe and the chemical analysis of the water sampling; (4) use of mathematical models for flow and transport analysis and (5) evaluation of possible reclaiming strategies for the remediation of the area.

2 GENERAL CHARACTERISATION OF THE AREA AND THE EXISTING POLLUTANTS

The region of Estarreja was subjected to a strong industrial development during the last 50 years, the industrial complex being the main local industrial activity, composed of the following units: Quimigal, União Têxtil e Química (Uniteca), Companhia Industrial de Resinas Sintéticas (Cires), Isopor (Dow Portugal) and Oxinorte.

This industrial units, schematically represented in figures 1 and 2, are recognised as being a general threat to the environment, either caused by the solid wastes and the liquid effluents produced or for the toxicity of some of the material or by-products used in the industrial processes. This situation is aggravated by the fact that this factories are located on very permeable soil, with a piezometric level very close to the surface; the flow pattern also plays an important role in the pollutant spreading as a result of its piezometric geometry (Figure 3), in an area of also intensive agricultural activity, highly dependent on groundwater.

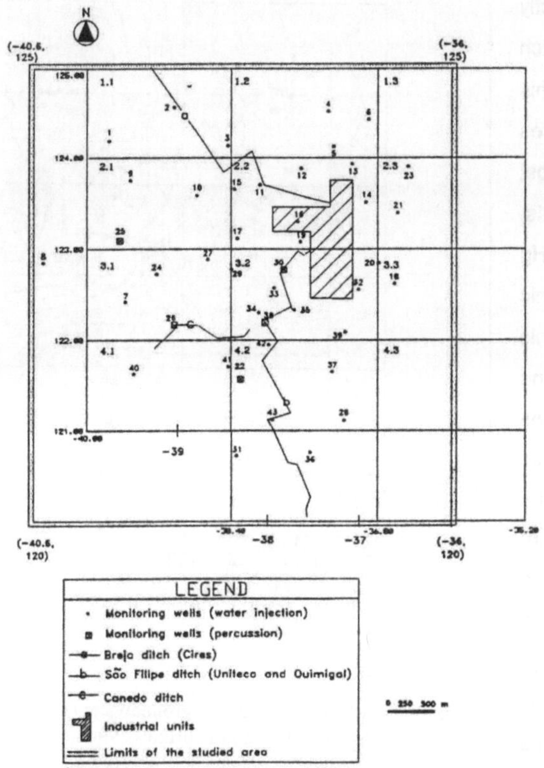

Figure 2 - Location of the monitoring wells, the liquid effluent ditches, and the industrial units

The industries reject high quantities of solid wastes directly to the waste disposal sites not lined nor drained. It is estimated that the Uniteca waste disposal site has 60,000 tons of industrial residues, the Cires waste disposal site has 30,000 tons, and the one of Quimigal has 80,000 tons of solid wastes (this amount is diminishing by the present transfer of the wastes to other disposal sites out of the study area). This wastes were settled without any proper lining procedure to avoid percolation to the soil and groundwater. The solid wastes coming from Dow Portugal (4,000 tons) are kept on an evaporation tank, waiting for future treatment.

A detailed characterisation of all the wastes coming from each industry is not presented here; further details may be found in Leitão et al. (1992). In a general way one can say that the more threatening pollutants existing in the solid wastes are (1) heavy metals, such as Hg, As, Zn; (2) the organic compounds like benzene, phenols, mononitrobenzene and (3) some anions like chlorides, nitrates and sulphates.

Since the beginning of the industrial activity the liquid effluents produced by Quimigal and by Uniteca are directly rejected in São Filipe ditch (excavated in the soil). This unlined ditch is several kilometres long causing the leaching of most elements present in the effluents; some examples of that are As, Hg and other metals, inorganic compounds and also some organic pollutants. In the winter-time, and also when the effluent discharges are higher, it is very usual to notice the flooding of this ditch. The flooded agricultural areas are subject to pollutant spread causing a high concentration of pollution in the soils. Only recently a drainage system has been built to carry this effluents; nevertheless, some rejections in the old ditch have been observed along the field work, proving that a part of the effluents is still being rejected there. Also, the Breja ditch is responsible for the drainage of the liquid effluents coming out of Cires.

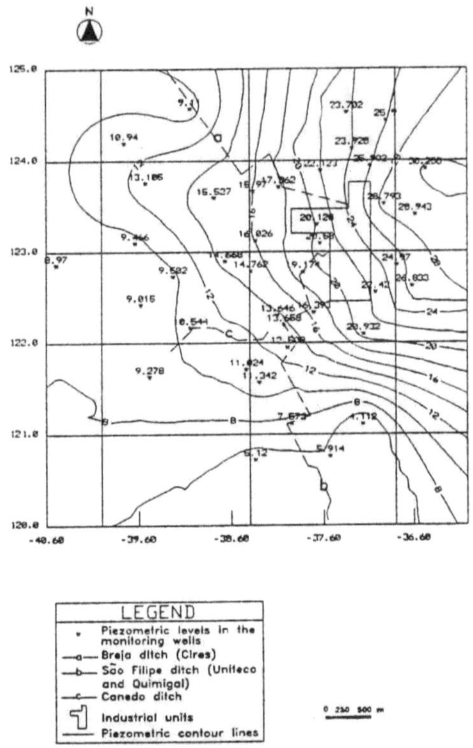

Figure 3 - Mean values for the groundwater head between October 1993 and February 1994

Making the situation even worst, the atmospheric pollution is also a problem, causing the deposition of some pollutants in the soil. One example of that is the SO_2 emissions from Uniteca and Quimigal. The consequences of that can be detected by a progressive general acidification of the soils, and consequently an increase in the mobility of the metals present in the soil.

The negative effects of this problem have been detected in Ria de Aveiro, located some 20 *km* SW of Estarreja. Very high content in Hg and other elements have already been analysed in its sediments.

3 AQUIFER MONITORING

3.1 Monitoring aim and practices

The monitoring network of Estarreja's quaternary aquifer consists of 43 monitoring wells, drilled in the area in the framework of this study. Their aim was to obtain a qualitative and quantitative analysis of the groundwater and a general characterisation of the hydrogeology of the aquifer.

The main purpose for the quaternary sandy aquifer monitoring was to obtain a global picture of the horizontal and vertical groundwater pollution for the study period, using *in situ* measurements and also the chemical analysis of the water samples collected in those wells.

Next task was to associate the pollutants detected in specific areas with their potential sources of pollution. For that purpose, a more dense network was settled around the areas located in the vicinity of the pollution focal points.

The monitoring wells were drilled using two different techniques: percussion and water injection. The selection of the drilling methods intended to fulfil the following technical requests: (1) the methods should not disturb significantly the chemical and hydrogeological properties of the aquifer; (2) they should be screened allowing a full contact with the water in all the saturated zone depth of the aquifer and (3) they should allow the extraction of representative water samples. The first method referred to is very appropriated for this type of studies because it uses very few quantity of equipment in the drilling process and it also allows soil extraction during drilling. The percussion method is also indicated for more accurate pumping tests. This technique has the drawback of being very expensive, specially when considering the installation of 43 wells! Consequently, only five wells were drilled using this technique. The other 38 wells used a water injection technique, suitable only for sandy soils. In this technique the drilling is achieved through a very simple process, well known by local people, that uses the injection of water as a tool to induce water pressure in the soil formations, allowing the hole drilling. The water used in the drilling process induced no relevant changes in the quality of the water (Leitão et al., 1994).

The water sampling was carried out in order to get representative samples of the water quality for the sampling period. To achieve this results the following aspects were considered: (1) type of polluting sources (liquid effluents, waste disposal sites, atmosphere); (2) focal points of pollutants rejection; (3) groundwater flow properties and (4) chemical and physical properties of the pollutants. A detailed study of the behaviour of the pollutants within the aquifer was also carried out.

3.2 Monitoring without pumping

After the well installation and before sampling the groundwater for chemical analysis, some of the most important chemical and physical parameters were measured, *in situ*, avoiding to disturb the natural equilibrium. One week is usually considered to be the minimum lag of time between the well installation and this measurements, so that the natural condition can be re-established.

Aller et al. (1989) and Canter at al. (1987) suggest some of the general procedures for monitoring the main chemical and physical aquifer parameters, that can be summarise in the following way:

- piezometric level measurement, aiming the regional mapping of its variations in space and time;
- registering of the following parameters, for several depths in the well vertical profile: temperature, dissolved oxygen, resistivity, total dissolved solids, salinity and, whenever possible, the pH and Eh.

These parameters were measured using a multiparametric probe. The probe measured the parameters every 0.5 *m*, in the first 3.5 *m*, and every 1 *m* for depths greater than 3.5 *m*. This allowed a vertical profile of the parameters in all wells.

The measurements referred to were important to assess whether or not the aquifer suffered any kind of important changes in what concerns quality parameters. The results obtained contributed to a sounder decision on the choice of the sampling depths, and to assess if they could be representative of all vertical profile.

3.3 Monitoring in dynamic conditions

The sampling procedures were carried out after the conclusion of the monitoring program phases described in the previous section. All the information gathered during that process was carefully analysed and showed to be very useful, especially in detecting, when existing, different properties along the vertical profile of the wells. In the cases were different conductivity values were observed, the selected sample depth that was chosen was the one corresponding to the highest conductivity values.

When the aim of sampling is to get a representative water sample for a well, some procedures can be performed for certification. For that purpose, and before taking the water samples for chemical analysis, a continuous pumping in each well should be carried out; in the pumped water the values for conductivity, temperature and pH, should be continuously measured (Aller et al., 1989). A representative water sample should be extracted only when the values for the parameters mentioned have stabilised. Usually the parameters stabilise after the extraction of a volume correspondent to three times the saturated volume within the well column (Clark, 1993). When the equilibrium is achieved it is considered that the samples are representative of the water in the aquifer. At this moment the pumping rate should be gradually decreased for the adequate sampling rate (between 0.1 and 0.5 *l/min*). The preservative procedures for the water samples should be carried out immediately after its collection.

4 GROUNDWATER QUALITY CHARACTERISATION

4.1 General remarks

The general characterisation of the aquifer was carried out throughout different steps that proceeded the water sampling - i.e. before the final stage of the assessment. It can be divided in the following phases:

- Analysis of previous assessment studies for the area (1992);
- Hydrogeologic characterisation (April 1992 - March 1994);
- Groundwater sampling and chemical analysis for the existing infra-structures (April 1992);
- Drilling of 43 monitoring wells (June - October 1993);
- Groundwater monitoring (October 1993 - November 1994);
- Groundwater sampling at different depths (November 1993 and September 1994);
- Analysis of the groundwater quality results (October 1993 - November 1994);
- Flow and transport modelling (December 1993 - March 1994).

A plan for the aquifer monitoring was drawn. The following procedures were carried out during 2.5 years:

i. Piezometric level record every month;
ii. Bi-monthly record for conductivity, temperature and dissolved oxygen in the wells, with a multiparametric probe connected to a data-logger system;
iii. Groundwater sampling for chemical analyses: (1) sampling in the existing traditional large wells, for a preliminary characterisation; (2) sampling in the monitoring wells for a general characterisation, and (3) sampling water to analyse the pollutants that have occurred in higher levels.

4.2 Groundwater characterisation

Groundwater characterisation in the area under study was carried out in two different ways: (1) using the multiparametric probe for the general properties like temperature, conductivity and dissolved oxygen and (2) using the results of the chemical analysis of the water samples taken from the wells.

The first type of analysis was useful as a general indicator of the pollution "black spots", detected throughout the conductivity values (as the main pollutants are heavy metals). Other important role of this type of analysis was to allow the water sampling in a normal period, i.e. with mean conductivity values, not resulting of exceptional situation correspondent to any period of discharge.

Considering the fact that the main pollutants in the area are heavy metals, the conductivity measurement is a good parameter to measure the ionic activity. Usually, the bigger the ionic concentration in the water, the bigger are the conductivity values.

The bi-monthly record with a <u>multiparametric probe</u> for conductivity, temperature and dissolved oxygen in the wells allowed the following conclusions:

i. the conductivity values are generally very high (max. 64 400 μ*S/cm*);

ii. the higher conductivity values for groundwater are found nearer to the waste disposal sites and the liquid effluent ditches;

iii. in a general way, the conductivity increases with depth;

iv. in most of the wells a stratification in the water quality was observed.

The water samples were taken at the depths corresponding to the highest conductivity values. The results obtained in the <u>chemical analysis</u> can be summarised in the following terms:

i. the pollutants occurring in higher concentrations (higher than the maximum level allowed by Portuguese law) are: mercury, arsenic, zinc, chloride, nitrates, sulphates, benzene and nitrobenzene;

ii. the area with contaminant levels higher than the maximum allowed is about 25% of the case-study area;

iii. the contamination is more evident near the ditches and the waste disposal sites - the same result was found with multiparametric probe measurements. The pollution plume is spreading with the main groundwater movement, towards west.

Besides this general comments, the results obtained for each elements are hereinafter briefly presented, and projected in figures 4 to 6. This type of representation gives an idea of the general pattern of the pollutants and allows the interpolation of values in the areas with lack of data. It also allows the delimitation of the areas were the pollution concentration is above the maximum allowed in Portugal. These areas were considered in the reclaiming scenarios.

Nevertheless, this kind of representation (with interpolation) has the disadvantage of giving an idea that the pollutants variation along the area is smooth, but most of the times it shows

to have rapid transitions in concentration within the area. This is a fact especially observed in the area were the pH and Eh values change very rapidly, as a result from the influence of the waste disposal sites. These comments are important for map interpretation. Even with the disadvantages referred to, we think that the representation selected is appropriated for the assessment of the pollution, aiming its reclamation.

Heavy metals

The values observed for the dissolved and total content of mercury were extremely high. The maximum value observed corresponds to water sampled in well 19, as high as 745 *μg/l* (maximum allowed in drinking water is 1*μg/l*). The areas where the values showed to be higher are the ones located close to S. Filipe ditch and to Canedo ditch (Figure 4). Also a high concentration value for mercury was detected in the groundwater located upwards of the complex, but in a concentration smaller than the maximum allowed by Portuguese law.

Another interesting aspect observed in the water analysis was that the total concentration of Hg is always higher than the dissolved one. An important part of the Hg is adsorbed by suspended solids with a diameter smaller than 0.45 *μm* (mesh used to filtrate the water). A high correlation coefficient was observed for the dissolved Hg and total Hg. Adsorption processes observed for suspended solids in water indicate the possible extension of this phenomena toward the solid matrix as being an important process for the retention of this metal.

The high concentrations in chloride and sulphate in the groundwater suggests the possibility of complex formations between this two ions and the metals.

The results obtained may confirm the origin and the effects of this metal in the groundwater. The most relevant sources of pollution are related with the Uniteca waste disposal site which has an arsenic concentration of 100 to 500 *mg/kg*. Besides this important source, two other important focal areas contribute to the pollution: S. Filipe and Canedo ditches (Figure 4).

The concentrations in arsenic detected in the groundwater are also a matter of concern. The maximum value was registered in the water of well 30, with 6760 *μg/l*, being the values of the wells located near by of 3520 and 1360 *μg/l*, in wells 38 and 19, respectively.

The main polluting sources for arsenic are located in the Quimigal waste disposal site, where the mud concentrations are very high. Several steps are being performed at this moment for the wastes removal; 40% of the total amount of wastes was already removed. The remaining wastes however are still an important source of deterioration.

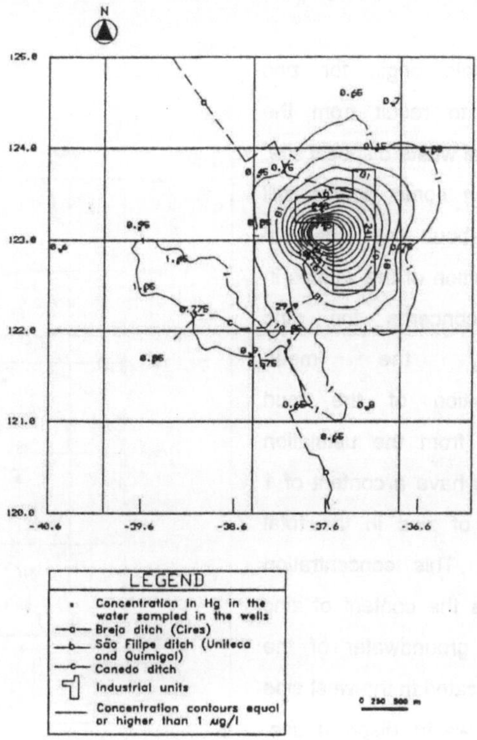

Figure 4 - Spatial distribution for the mercury concentration in the groundwater samples collected in the monitoring wells

Besides the waste disposal sites, S. Filipe ditch is also a source of pollution. The liquid effluent flowing in that ditch has a high content in arsenic. The liquid discharge from Quimigal was 100 m^3/h, in 1988. The samples taken from the liquid effluent from 1987 to 1991, showed a maximum value of 38000 µg/l (maximum value found in 1990, for a total of 13 samples); the mean value in arsenic content for those years was 8355 µg/l (in the Portuguese law, the maximum permissible value for waste waters coming from industries for arsenic is 1000 µg/l). In Leitão et al. (1992), the chemical analysis for the waste effluents are presented in more detail.

In what concerns the spatial distribution of zinc in the area, two peaks of concentration are observed: (1) to the NW of the Complex, i.e. an area located downwards from the Quimigal waste disposal site and (2) in the surroundings of Canedo ditch. Some other areas besides this two areas have zinc content in the water mainly for the wells located in the vicinity of S. Filipe ditch. The area with high zinc content in the groundwater is much more extensive than the one detected for the other two mentioned elements. This element shows a higher mobility for the existing hydrogeological conditions.

The main origin for zinc seems to result from the Quimigal waste disposal site. Although some doubts still exist about the chemical composition of the wastes in what concerns the zinc content, the mean composition of the mud coming from the ustulation process have a content of 1 to 4% of zinc in the total volume. This concentration explains the content of zinc in the groundwater of the wells located in the west side of the waste disposal site. The data obtained showed also that some zinc is coming from S. Filipe ditch.

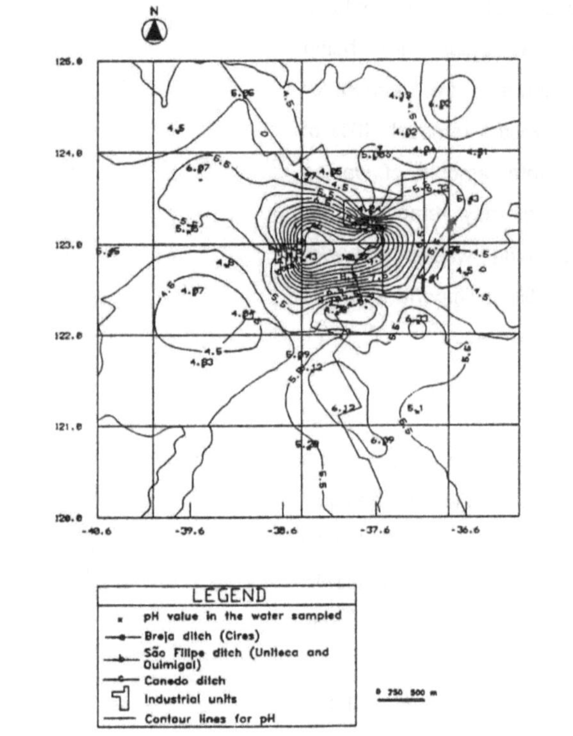

Figure 5 - pH mean values in the groundwater samples collected in the monitoring wells

The presence of iron in the groundwater can be seen in most of the wells located downwards of the chemical complex. The areas with the highest concentrations are similar to the ones detected for zinc.

The content in nickel, copper, lead, cadmium and manganese were also analysed. Its presence was analysed for five wells, as it is believed that its content in the groundwater is rather lower.

In three of the five wells the nickel content in the groundwater was higher than the recommended values; in two of them the values were higher than the maximum allowed.

For copper and lead the results from the chemical analysis showed its presence only in well 38; nevertheless, this value is lower than the maximum value allowed.

Cadmium was detected in the groundwater samples collected in wells 22 and 38, showing very high values in well 38.

The pollution by heavy metals, as a whole, is a matter of concern. The concentration values of these elements are very high. This cations are usually associated with high amounts of anions that also show very high contents in the sampled groundwater. The main anion pollutants are chlorides, nitrates and sulphates.

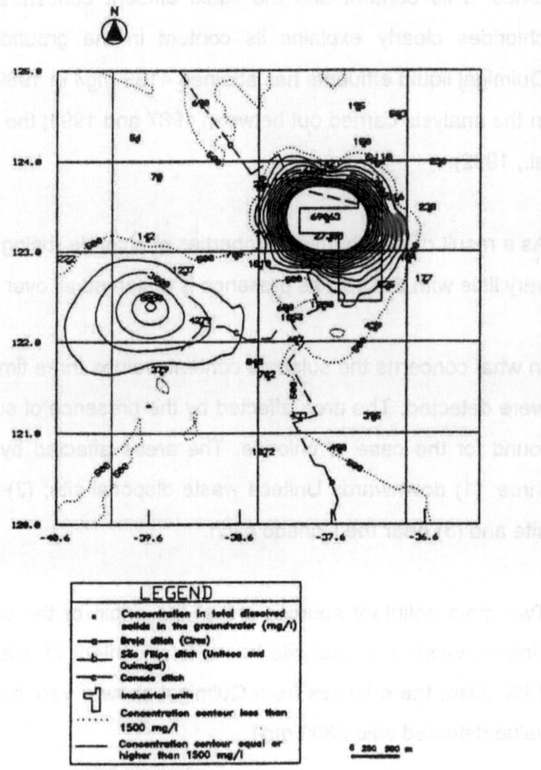

Figure 6 - Spatial distribution for the total dissolved solids content in the groundwater samples collected in the monitoring wells

Other elements

The results obtained for the chemical analysis show a very strong correlation between the total dissolved solids (TDS) and the chloride content. A correlation also exists for the conductivity (Figure 6) and the sulphates content, however being weaker.

For well 16, the chloride content in groundwater is extremely high. The case is not one of salt water intrusion because: (1) there are no high values in the wells located nearer the sea and (2) the concentration of this ion is clearly higher in the vicinity of the Uniteca waste disposal site were the chloride content is very high.

The wastes from Uniteca have a total content of chlorides between 7 and 11% in its 60000 tones. This content and the liquid effluent concentration of Quimigal (S. Filipe ditch) in chlorides clearly explains its content in the groundwater. The chloride content of the Quimigal liquid effluents has attained 4189 *mg/l* in 1989; this was the highest value obtained in the analysis carried out between 1987 and 1991; the mean value being 630 *mg/l* (Leitão et al., 1992).

As a result of the chemical properties of chloride (being a very conservative ion thus reacting very little with the soil) its presence is observed all over the study area.

In what concerns the sulphate content, values three times higher than the maximum allowed were detected. The area affected by the presence of sulphates is more restrict than the one found for the case of chloride. The areas affected by a high sulphate content are mainly three: (1) downwards Uniteca waste disposal site; (2) downwards Quimigal waste disposal site and (3) near the Canedo ditch.

Two main pollutant sources are at the origin of the sulphate content in groundwater. The Uniteca waste disposal site has a total content of calcium sulphate equivalent to 10 up to 14%. Also, the effluents from Quimigal show a very high content in sulphate; the maximum value detected was 2300 *mg/l*.

The chemical nature of sulphate allows the reaction with the porous formations; also the biochemical interactions restrict the propagation of the sulphates in the area. Possibly the existing sulphates occur as a complex formation that are afterwards precipitated. In reducing conditions, especially when organic matter exists, a part of the sulphates can also be reduced to sulphide. Well 26 is an example of this case, characterised by the high content in sulphides, with its typical smell, and also reduced organic matter. The drastic reduction of the Eh values from the S. Filipe to Canedo ditch should be responsible for the sulphate and organic matter reduction processes along with the gases release observed in this waters.

In what concerns the nitrate content, one can observe the presence of values exceeding the maximum content in the surroundings of the chemical complex, and also in the wells located near S. Filipe ditch.

Nitrate reduction is also a phenomena occurring near Canedo ditch. This is observed by the high content in ammonia (5.67 *mg/l*) instead of nitrate. The reduction of nitrates occurs generally in reducing conditions and low pH (Figure 5).

The nitrate content is originated mainly from the liquid effluent of S. Filipe ditch. Contamination due to fertiliser application is also supposed to exist, being difficult to distinguish from the ditch source. The chemical analysis of the liquid effluents carried out between 1987 and 1991 showed that the nitrate content reached 5116 *mg/l*, that is 100 times higher than the maximum value admitted by Portuguese law for liquid effluent discharges.

In what concerns the values for dissolved organic carbon, ten chemical analyses were made. The results obtained ranged from 1.2 to 12.0 *mg/l*.

Other analyses were made in five wells for potassium, calcium, sodium, phosphates and ammonium. Of these parameters the only one that showed acceptable values was phosphate. All the others showed, in some wells, values that exceed the maximum allowed. Examples of that were found for sodium, in wells 26 and 30, for potassium, in well 26, with 16.6 *mg/l* content (the maximum allowed is 12 *mg/l*). Also for well 26, located in an area of extremely reducing conditions, ammonium levels are ten times higher than the maximum allowed.

5 RECLAIMING STRATEGIES FOR THE GROUNDWATER POLLUTION PROBLEM

After a detailed characterisation of the groundwater pollution problem, in the previous sections, hereinafter, some reclaiming procedures will be assessed.

The complexity of this groundwater pollution make its reclamation process complex: (1) several types of pollutants show a high contaminant level in the groundwater; the pollutants causing major damages having different chemical behaviour, making very difficult an *in situ* chemical reclamation; (3) the flow pattern in the area spreads the pollutants towards NW and

SW and (4) the strong spatial variation of pH and Eh (mainly due to the Cires waste disposal containing calcium hydroxide $Ca(OH)_2$) changes drastically the chemical behaviour of the pollutants in surrounding areas.

The aspects just pointed out are responsible for very different physical, chemical and biologic pollutant behaviour in the aquifer. These characteristics control in a strong way the behaviour of the elements, especially the stability of the metals that are very depend of the pH, Eh and organic matter content. As a result of this considerations it was excluded, at this moment and based in the actual knowledge, an *in situ* reclamation process. A process in which the polluted groundwater is extracted for treatment seams to be the best approach.

Before starting a groundwater pollution reclamation programme, through a pumping and treatment system, a pollution control strategy to prevent further pollution must be achieved. This should consist in removing, isolating or treating the main pollutant sources (Quimigal and Uniteca waste disposal sites, the liquid effluents of S. Filipe ditch, Canedo and Breja ditches), in order to prevent further migration and a stronger leaching induced by the pumping process.

The effects of the waste disposal sites in the pollution of the groundwater must be controlled. In this sense, two kinds of actions must be carried out in a short term: (1) covering the waste disposal sites to avoid rain to infiltrate and percolate causing the leaching and percolation of existing pollutants, and (2) physical isolation of the waste disposal site throughout impervious barriers to be injected in the subsoil around the contamination area (the maximum aquifer thickness is 15 *m*) or stabilisation/solidification of the waste. The second action can also be performed with waste removal for treatment. Any of these actions can be carried out in short term and can improve significantly the present situation.

The liquid effluents in the ditches are also an important source of pollution. It is essential to avoid any further liquid emission from the factories, capable of contaminating the soils and groundwater. For that purpose, the effluents should be subjected to primary or secondary treatment before being rejected.

Besides these pollution control measures an important action plan to reclaim the present pollution must be implemented. In this paper, the extraction of groundwater with an optimised pumping system is presented.

The main purpose of using pumping wells for groundwater extraction is to control the contaminant plume and avoid further spreading of the pollutants. The pumped water must be afterwards treated. In order to achieve these objectives the optimum pumping rate capable of containing the plume, and also the minimum number of wells and their locations were calculated.

Furthermore, the travel time of a conservative tracer was computed, considering an emission point far from the pumping wells.

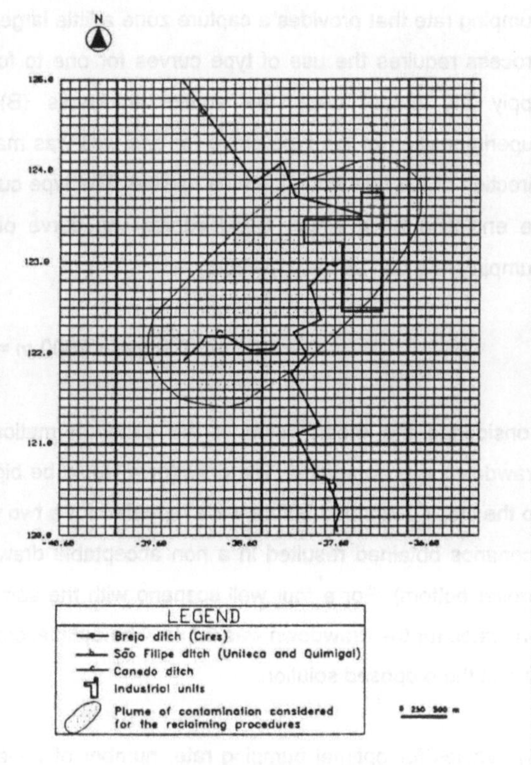

Figure 7 - Contaminated plume location

Mathematical modelling of flow in the quaternary aquifer of Estarreja was made using MODFLOW (cf. McDonald and Harbaugh, 1988). An analytical method (Javandel and Tsang, 1986) was used to calculate the optimum pumping rate for the extraction of the polluted water, the location of the wells and the number of wells. This method considers mean transmissivity values, and saturated thickness not taking into account differences within the aquifer. The analytical model was used only to obtain a first approach of the parameters (pumping rate, number and location of wells) for the extraction. These values were afterwards inputed in the MODFLOW model in order to validate and to calibrate the data.

To map the polluted area for which groundwater should be extracted through pumping, a superimposition of the areas with chemical concentration, greater than the maximum allowed by Portuguese law was performed (Figure 7). This plume was used in the modelling process described hereinafter.

As mentioned before, the analytical approach described by Javandel and Tsang (1986) to design contaminant recovery systems was applied to the Estarreja aquifer. An optimum pumping rate that provides a capture zone a little larger than the plume was estimated. This process requires the use of type curves for one to four wells. The parameters needed to apply the model were the aquifer thickness (B), and the Darcy velocity (U). A superimposition of the type curve for one well was made keeping the x-axis parallel to the direction of the flow and along the midline. The type curve is adjusted so that the plume can be enclosed by a single Q/BU curve. The curve obtained was 2500 *m*. The optimum pumping rate can be determined by:

$$Q = 7.91\ m \times 0.9\ m/d \times 2500\ m = 17\ 797.5\ m^3/d$$

Considering the mean value of the sandy formations porosity (in the area 20%), the drawdown was computed. The drawdown would be bigger than the maximum possible and so the steps described before were repeated for a two well curve. For two and tree wells the scenarios obtained resulted in a non acceptable drawdown (piezometric values below the aquifer bottom). For a four well scenario with the same estimated pumping rate, the value evaluated for the drawdown was 5.5 *m*. The optimal distance between each well was 230 *m*. This is the proposed solution.

The values for optimal pumping rate, number of wells required and optimum well spacing were imputed in the mathematical model (MODFLOW) in order to check if those values were capable of containing the plume, this time considering the local properties of the aquifer as the permeability, aquifer thickness, etc. The results obtained are shown in Figure 8.

To calculate the polluted groundwater volume the number of model cells that are inside the plume were counted (431 cells), corresponding to an area (A) of $4.31 \times 10^6\ m^2$, the average value for the aquifer thickness (B) being 7.91 *m* and the effective porosity (S) 20%. The volume can then be evaluated as follows:

$$\text{Vol.} = A.B.S = 7 \times 10^6\ m^3$$

The time that is required to pump this volume of water is:

$$T = Vol./Q = 400 \text{ days}$$

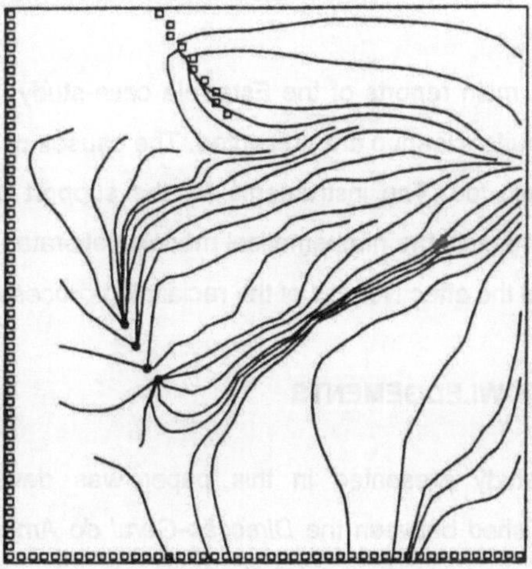

Figure 8 - Flow lines produced by four pumping wells, located to prevent any leakage from the space between them

Later in 1994, some actions started aiming the reclamation of existing pollution, considering the scenario of water extraction proposed and its treatment. A first step was done by the Estarreja Municipal Council (Câmara Municipal de Estarreja - CME) with the establishment of a new firm intending the co-ordination of all action plans to recover the area. A project has been presented by CME in order to treat the wastes coming from Uniteca and Quimigal together with the mud coming from Cires. This treatment would be done with the construction and use of an artificial wetland were the solid wastes would be deposit, treating also the water pumped out from the aquifer, through chemical and biological processes. These processes include the *on site* treatment of soils and groundwater.

6 CONCLUDING REMARKS

This paper briefly presents a synthesis of the assessment developed aiming the characterisation of the present state of pollution in the quaternary sandy aquifer of Estarreja. The pollution is caused by liquid effluents and solid waste disposals produced by industrial activity in the area.

Some pollution control scenarios are proposed in order to contain the existing sources of pollution and to avoid further leaching and spreading of contaminants. Without this pollution control it is impossible to apply any efficient measures to reclaim the aquifer.

A cleanup strategy is proposed using the design of an optimal pumping scheme for removing the groundwater contaminated. Mathematical modelling of flow and transport was used.

In the main reports of the Estarreja case-study, here summarised, the necessary tools for the aquifer cleanup are presented. The causes of pollution are known now, the contaminants dynamic too. The instruments for the support of the decontamination processes already exist, namely the mathematical models calibrated for the aquifer and the monitoring wells to control the effectiveness of the reclaiming process.

ACKNOWLEDGEMENTS

The study presented in this paper was developed under a co-operative agreement established between the *Direcção-Geral do Ambiente* (DGA) and the *Laboratório Nacional de Engenharia Civil* (LNEC), for the study of *Methodologies for the Remediation of Contaminated Groundwater and Soils*. The study was also supported by Junta Nacional de Investigação Científica, through STRIDE contract nº STRD/AMB/58/92.

We acknowledge the work of Mr. Piercarlo Ciabatti, Geologist of Hydrocontrol S.C. a r.l., Italy, Human Capital and Mobility research trainee of WEEL in the Groundwater Research Group of Hydraulics Department of the Laboratório Nacional de Engenharia Civil, Portugal, for his co-operation in the mathematical modelling of flow and transport and also for the pumping and treatment remediation modelling.

We acknowledge the contribution of Mr. Edward Zungailia and Mrs. Manuela Inácio, geologists, in the development of this study.

We would also like to acknowledge all the support of Câmara Municipal de Estarreja during the aquifer monitoring process and also for the work performed for wells location and their elevation measurements. To Mr. Duarte Esmeraldo, Chemical Engineer in the municipal council, we thank all the information and support given during this work.

We appreciate the support of Mrs. Maria José Henriques of LNEC drawing the figures of this paper, and also for her care and dedication during all the field work.

REFERENCES

Aller, L., Truman, W.B., Hackett, G., Petty, R., Lehr, J.H., Sedoris, H., Nielsen, D.M. and Denne, J.E., 1989 - *Handbook of Suggested Practices for the Design and Installation of Ground-Water Monitoring Wells.* US Environmental Protection Agency. EPA 600/4-89/034. Las Vegas, Nevada.

Canter, L.W., Knox, R.C. and Fairchild, D.M., 1987 - *GroundWater Quality Protection.* Lewis Publishers, Inc. Michigan, USA.

Clark, L., 1993 - *Monitoring and protecting groundwater quality*. Pollution Prevention, Vol. 3, Issue 4, p. 51-56.

Javandel, I. and Tsang, C. F., 1986 - *Capture-zone type curves: a tool for aquifer cleanup*. GroundWater, Vol. 24, nº 5.

Leitão, T.E., Zungailia, E. Jr. and Lobo-Ferreira, J.P., 1992 - *Metodologias para a Recuperação de Águas Subterrâneas e Solos Contaminados. Partes A e B.* Lisboa, Laboratório Nacional de Engenharia Civil and Direcção-Geral da Qualidade do Ambiente.

Leitão, T.E., Lobo-Ferreira, J.P. and Inácio, M.M., 1994 - *Metodologias para a Recuperação de Águas Subterrâneas e Solos Contaminados. Partes C, D, E e F.* Lisboa, Laboratório Nacional de Engenharia Civil and Direcção-Geral do Ambiente.

McDonald, M. G. and Harbaugh, A. W., 1988 - MODFLOW: *A Modular Three-dimensional Finite Difference Ground-Water Flow Model.* U.S. Geological Survey, Washington, D.C..

Some Experience in Rehabilitation of Urban Drainage Infrastructure and Sewage Treatment Systems in Germany

Lothar Fuchs
Institut für technisch-wissenschaftliche Hydrologie
Engelbosteler Damm 22
30167 Hannover
Germany

1. Introduction

After the reunification of Germany (1989) the difference in drainage systems and treatment plants between the former German Democratic Republic (DDR) and the Federal Republic of Germany (BRD) became obvious. The structural and technical conditions of the drainage systems as well as most of the sewage treatment plants in the former DDR were extremely poor. The existing drainage systems had hydraulic as well as structural problems. In some large cities such as Dresden parts of the sewer systems were blocked by sediments, so that the hydraulic capacity was in some parts reduced to the dry weather flow.

The basic figures about the drainage systems and treatment of wastewater are given in the following tables.

	total length of sewer systems (km)			
	total	combined	sanitary	rainwater
Germany	358,860	205,606	88,053	65,201
former BRD	320,161	178,808	80,758	60,595
former DDR	38,699	26,798	7,295	4.606

Tab. 1: Total length of sewer systems in Germany distributed between the former BRD and DDR (1991, data after "Statistisches Bundesamt")

NATO ASI Series, Partnership Sub-Series, 2. Environment – Vol. 45
Environmentally Devastated Areas in River Basins in Eastern Europe
Edited by Alfons Georges Buekens and Vasily Victorovich Dragalov

Inhabitants	total	connected to treatment plant		mechanical treatment only		biological treatment without nutrients removal		biological treatment with nutrients removal	
	1,000	1,000	%	1,000	%	1,000	%	1,000	%
Germany	80,275	68,298	85.1	5,822	7.4	26,876	33.5	35,580	44.4
former BRD	64,484	59,288	91.9	1,343	2.1	22,470	34.8	35,475	55.0
former DDR	15,789	9,010	57.1	4,490	28.4	4,406	27.9	114	0.7

Tab. 2: Inhabitants connected to treatment plants in Germany subdivided between the former BRD and DDR and classified according to the extent of treatment (1991, data after "Statistisches Bundesamt")

Separate collection of sewage and rainwater was introduced in most new developments schemes in the former Federal Republic, but was somewhat less wide-spread in East Germany.

The technical regulations for sewage drainage and treatment were suspended after reunification for 5 to 10 years in the former DDR, but after this period the German or EU regulations have to be fulfilled. Key figures of these regulations are for drainage systems a maximum periodicity of flooding of once every 20 years. For combined sewer overflow every state can decide whether to accept the national regulations or to define its own. In general, the regulations are fulfilled in case there is a certain sedimentation volume per ha of impervious drainage area or the loads/concentrations of the overflow is less than a certain limit. The key discharge limits for treatment plants are BOD 15 mg/l, COD 75 mg/l, P_{tot} 1 mg/l. Elimination of phosphorus and nitrogen is mandatory.

Because of the extremely poor situation a major effort in rehabilitation of sewer systems and treatment plants was implemented. In the former DDR the limited knowledge and experience about the possibilities and costs of advanced drainage and treatment together with the introduction of a new free market and some unscrupulous consultants led to a lot of misinvestments, especially in small communities.

In the following a case study is presented which serves as a general example and leads to some results, that are typical also for some other studies.

2. The Case Study Dresden

2.1 Development of Dresden's Drainage System

The basic concept for the Dresden drainage network was already conceived by the City Building Office (Stadtbauamt) in 1867 and is still valid today. The expedited "Beschleusungsplan" (lock management) foresaw in the construction of multiple main sewers located almost at right-angles to the river Elbe. Simultaneously it was planned to interdict all main sewers emptying into the river in parallel to the riverbank and to discharge all the waste water downstream of the city limits into the Elbe. The construction of the intercepting networks parallel to and on both sides of the Elbe, which direct the waste water to the waste treatment facility, began in 1898. The sewage treatment plant Dresden-Kaditz became operational on July, 15. 1910. While there have been no major changes to the drainage network, except for the inclusion of new catchment areas in the suburbs, (which were mainly implemented as separate sewer and rainwater systems) the waste treatment facility has undergone three major construction stages:

1. Stage 1908 - 1910: mechanical screening facility
2. Stage 1950 - 1954: primary settling facility
3. Stage 1986 - 1993: partial biological treatment stage

The waste water treatment facility at Dresden-Kaditz is the most important municipal treatment facility on the German side in the upper Elbe catchment area, including the cities of Dresden, Freital, Radebeul-Ost and some additional surburban communities. Figure 1 gives a rough idea of the sewer system of the city of Dresden.

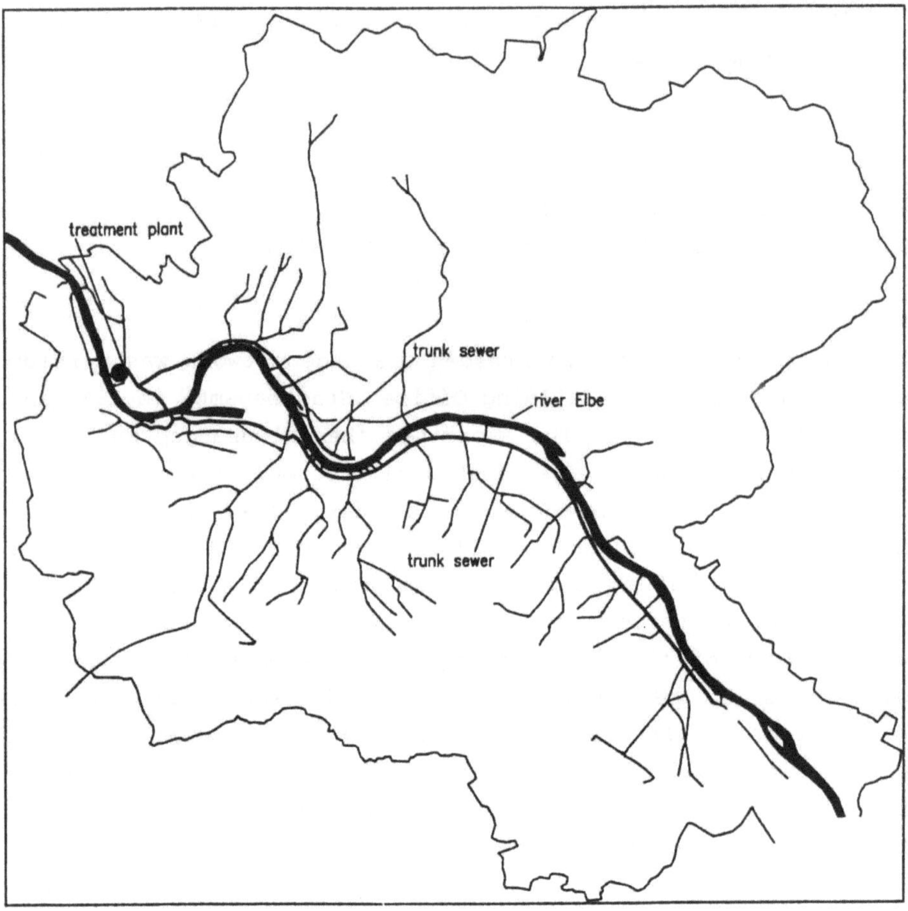

Fig. 1: The sewer system of the city of Dresden

2.2 Additional Expansion of the Treatment Facility Dresden

2.2.1 Development of the Time and Legal Constraints

After the Reunion the laws from the Federal Republic of Germany became valid in Saxonia (1990/91); hence all existing plans, had to be reconsidered. Essentially the following legal considerations had changed:

- instead of the sole treatment of the amount of sewage (Qs) it is now required to treat a mixed water flow of Qm = 2 Qs + Qf during precipitation,
- after nitrification a denitrification step is required,
- phosphorus must also be eliminated.

In order to implement the "First Action Program of the International Commission for the River Elbe for Reduction of Pollution in the Elbe and its Catchment Area" and the EC guidelines for the treatment of municipal sewage (EC AB1 L 135/40.), the preliminary planning for the expansion of the treatment facility (nitrogen and phosphorus elimination) began in 1992. During the planning phase, in the spring of 1993, the administrative guideline of the regional Ministry for Environment and Land-Development, "Incremental Expansion of Waste Water Treatment", took effect. This regulation prescribes a short-term reduction of total organic carbon (TOC) and an extended time-period for implementing measures to eliminate nutrients.

2.2.2 Conceptual Goals of the Pre-Planning

A flexible expansion concept was required in order to be able and react to the future evolution of waste water quantities and compositions, as well as to possible further restrictions of discharge limits or conditions. This concept had to ensure a gradual step-by-step reduction of the waste water burden for the Elbe by 1998, make use of the already existing facilities and take into account new areas of development. This applied for both pure sewage as well as for the treatment of mixed sewage/rainwater during periods of precipitation.

During the planning of the entire concept consideration was given to a major reduction of the burden on the Elbe, which would be achieved within a short time by incremental expansion of the facility. For this reason mature methods and proven technology were chosen while still allowing the future use of techniques which are still being developed.

During the conception of the facility, the following assumptions about the population to be serviced were used:

near future (by 2010)	1.0 million population equivalents
long term	1.5 million population equivalents

2.3 Expansion Goals for the Waste Water Treatment

The existing waste water treatment essentially includes the inlet facilities with mechanical sewage treatment and a high load partial biological waste water treatment.

This facility was originally conceived for the treatment of only the dry weather-flow. The earlier plans envisioned a two-stage biological treatment facility with nitrification when completed. Of these two stages, only the first, high load step together with a sludge sedimentation tank was completed. The original function of this stage implies that only a partial biological treatment of the dry weather-flow of waste water is possible.

The new requirements for the treatment led to a new concept. A cost-benefit analysis as well as the short-term implementation deadline of 1998 made an incremental construction of a new facility seem appropriate. Taking into consideration a future oriented land usage, it was recommended that the further expansion of the waste water treatment facility took place on a neighbouring piece of land, next to the present plant.

When completed the new waste water treatment plant will consist of the main components: inlet group, mechanical waste water treatment, one-stage biological facility and filtration.

However, until the completion of the third expansion stage, the existing treatment facility will remain an essential component of the waste water treatment concept. A comparison of appropriate methods and treatment combinations resulted in the concept depicted in Figure 2.

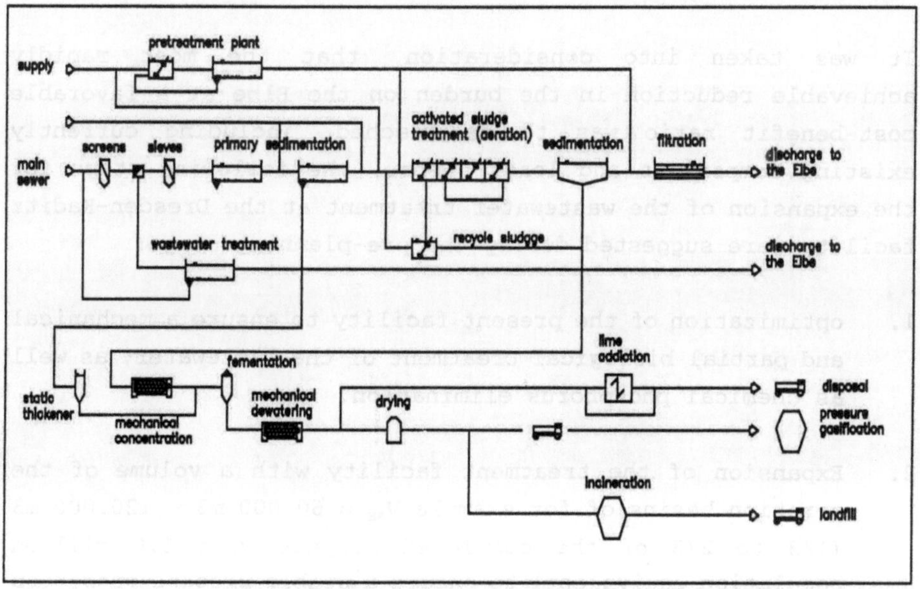

Fig. 2: Simplified schematic of the facility when completed (1.0 M inhabitants)

2.4 Concept for further Expansions

The final expansion goal of the treatment facility is shown in Figure 2. The investments for the entire project are huge and must be financed through waste water treatment fees. However the financial burdens upon the citizens, the industries and trades must remain bearable. Therefore, based upon the state of the art, the expansion concept was developed with the following principles:

- continued temporary usage of the existing facility components;
- incremental improvement of the treated-water quality and a reduction of the load on the Elbe;
- a flexible expansion capable of reacting to varying constraints;
- socially acceptable fees.

It was taken into consideration, that the most rapidly achievable reduction in the burden on the Elbe at a favorable cost-benefit ratio was to be reached, including currently existing components and loads. Hence, the following steps for the expansion of the wastewater treatment at the Dresden-Kaditz facility were suggested during the pre-planning stage:

1. optimization of the present facility to ensure a mechanical and partial biological treatment of the wastewater, as well as chemical phosphorus elimination.

2. Expansion of the treatment facility with a volume of the aeration basins of for example V_{AB} = 60.000 m3 - 120.000 m3 (1/3 to 2/3 of the completed project with 1.0 million population equivalent) to ensure a stable process operation with a mechanical/biological basic treatment and a partial nitrification/partial denitrification and phosphorus elimination. Usage of the existing biological stage for treatment of part of the waste water.

The existing mechanical treatment stages will continue to operate, so that the required mixed waste water quantity $Q_m = 2\ Q_s + Q_f$ can be treated to ca. 70 bis 80 %.

The average values of Table 3 should not be confused with the limit-values (2 hour mixed sample) required according to the Waste Water Ordnance (Rahmen-Abwasser-VwV). Generally these values are only satisfied when the average value is 50% to 100% below the limit values.

Parameter	Average Discharge Concentrations (mg/l)		
	measured	V_{AB}=60.000 m³	V_{AB}=120.000 m³
BOD_5	15	13	10
COD	69	65	55
NH_4-N	22	11	1
NO_3-N	-	11	7
P_{tot}	1.5	0.9	0.8

Tab. 3: Average discharge concentrations at the waste water treatment plant Dresden Febr.-May 1994 including its simulated projections for further expansion

The operation and optimization of the existing waste water treatment facility achieves a noticeable reduction in the burden on the Elbe. The COD-load is currently reduced by over 80% and the phosphorus has been reduced by approx. 70%. A minor reduction of nitrogen has occurred by its incorporation in the sludge. After the new biological treatment begins operation, the burden on the Elbe due to oxidigable compounds and phosphorus will be further reduced. However it is decisive, that both a nitrification of oxygen-depleting nitrogen compounds as well as a partial elimination of nitrogen through denitrification occurs. At current loads the treatment facility can meet the requirements of the Waste Water Ordnance and the EC-guidelines during dry-weather, assuming favourable conditions and a biological treatment volume of V_{AB}=120,000 m³.

The limited hydraulic capacity in the existing mechanical treatment stage will lead to frequent mixed waste water discharges into the Elbe with as a consequence an increased pollution load until the completion of the new facilty.

Thus for the long term mixed waste water treatment, a new inflow group and additional clarification ponds have been planned. An increase in the biological treatment capacity can be achieved by an expansion of the biological treatment facility or by use of biological filtration as well as possibly with the construction of a filtration.

The pre-planning recommendation for the concept is mainly based on the expansion of the treatment plant by a one-stage facility. The existing biological treatment basins can be used for the treatment of partial waste water quantities and/or the waste water of the Arzneimittelwerke Dresden GmbH (AWD).

Alternatively, technical experiments are to determine whether the expansion of the treatment facility with biological filtration offers advantages over the activated sludge technology.

2.3 Development of Dresden's Waste Water Treatment Concept

2.3.1 Concept of the Simulations

The concept for Dresden required research on the mixed waste water overflows. Since the reduction in the quantities and loads of combined sewer overflows are directly related to the capacity of the treatment plant, an integrated analysis of the current condition of the drainage network and the treatment plant as well as for various stages of completion of the new facilities was carried out using appropriate simulation models. The hydraulic response of the system was modelled with HYSTEM-EXTRAN.

HYSTEM-EXTRAN is an advanced hydraulic model, which consists of the surface runoff model HYSTEM and the hydrodynamic sewer model EXTRAN. With this model it is possible to simulate the hydraulic behaviour of complex sewer systems with special structures like weirs, detention basins, pumping stations, etc. under various conditions like free surface flow, surcharge and flooding. It is especially designed for the simulation of single events, however continous simulation of the rainfall-runoff process is possible.

Because of time and financial constrains a different continous model (KOSIM) was used to simulated the annual discharge loads and volumes. Based on the detailed simulation of the sewer systems with HYSTEM-EXTRAN a lumped system approach was setup. Because no measurements of flow and concentrations of the entire system were available, the lumped system for the simulation with KOSIM was calibrated with the data of the detailed simulation.

The resulting sewer system output - quantities and concentrations - from KOSIM served as the input for the continous waste water treatment plant model GESIM. GESIM was calibrated using the available measurements of flow and concentration at the inflow and at the outflow of the treatment plant.

Results of the simulations are available for the various projected stages of the treatment plant expansion. The results show the simulated discharge quantities and concentrations or loads of mixed waste water overflows as well as the corresponding information about the treatment facility discharges. Based upon these results, various plans were developed according to qualitative aspects, examined by means of simulations and evaluated on a basis of the fulfillment of various criteria.

2.3.2 Results of the Study

The study was essentially to the parameters BOD_5, COD, nutrients (N) and phosphorus (P_{tot}). During the calibration of these parameters a difference of approximately 5% between the simulated and measured values in treatment plant discharges was measured.

No guidelines for the discharge loads or concentrations are currently available in Saxony. However, the maximum values of the regulations for former West-Germany are exceeded by approximately 100%. The simulated discharges loads for the present state are listed in Table 4.

Parameter	Discharge/Combined sewer overflow loads (t/a)		
	treatment plant	sewer system	total
BOD_5	1,025	333	1,358
COD	4,926	1,319	6,245
N	1,342	42	1,384
P_{tot}	79	16	95

Tab. 4: Average annual discharge and combined sewer overflow loads in the present state (max. dryweather inflow: 136,000 m^3/d)

The studies of the various treatment plant expansion stages, which result in an increase in the inflow quantities to the treatment plant from 136,000 m3/d to 170,000 m3/d, show that for the first expansion stage the maximum values according to the Waste Water Ordinance can not be achieved. Additionally, the discharge quantities and loads during overflow events will increase due to the reduced hydraulic capacity of the treatment plant.

Based on the final concept of the treatment plant as envisioned in the pre-planning, various improvements and repairs in the sewer system were examined. The stages were conceived so that first a secured flow of the dry-weather flow to the treatment plant is ensured and that mildly contaminated stormwater and runoff no longer enters into the sewer system. The next stage incrementally reduces the values for the mixed waste water discharges first for the sewer system as a whole and then for each individual overflow. Essential measure for this are a decoupling of flowing waters, creation of ca. 20,000m3 additional storage volume, activation of storage volume in the sewer system by means of real-time control, adjustment of throttle capacities, decoupling of areas and closing combined sewer system overflows. Based upon the final conception, the discharge freights for the five stages of improvements can be found in Table 5 or Figure 3.

	Discharge/Combined sewer overflow loads (t COD/a, max dryweather inflow 170,000 m³/d)		
	treatment plant	sewer system	total
no expansion	4,748	2,080	6,828
1. Stage	5,423	1,453	6,876
2. Stage	5,208	1,284	6,692
3. Stage	1,736	1,284	3,020
4. Stage	1,139	1,284	2,423
5. Stage	1,759	567	2,326

Tab. 5: Discharge loads in the Elbe after various repair /improvement stages

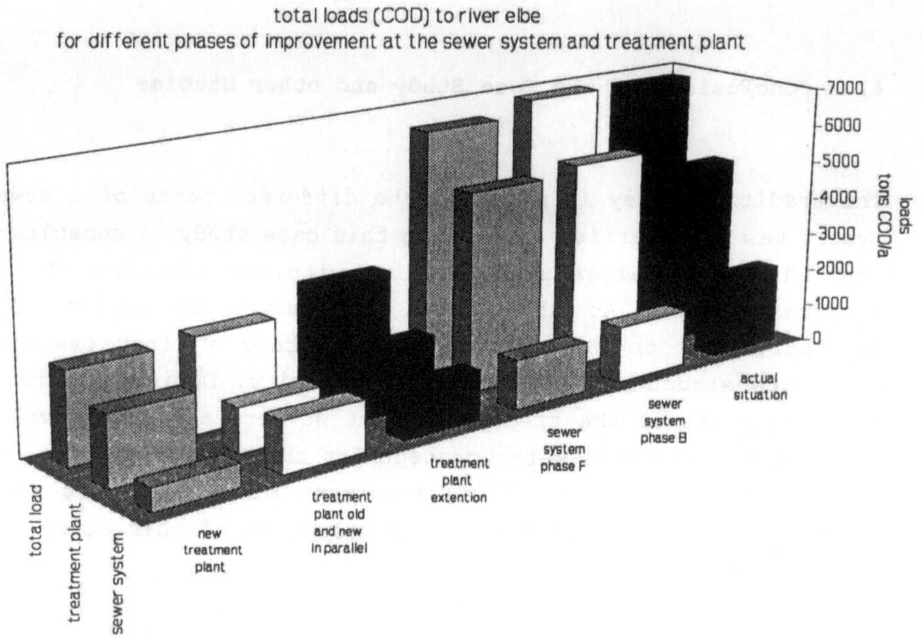

Fig. 3: Discharge loads (COD) in the Elbe for different phases of improvement

The measures resulting from looking at the total emissions show a clear reduction in the burden on the river Elbe from nearly 7000 t COD/a to ca. 3,500 t COD/a, which means a reduction of the discharge freight from the sewer system of ca. 50% despite an increase in the waste water quantity. Some direct influences result from actions in the sewer system and the waste treatment plant. Only after analysing the various stages of treatment plant and sewer system improvements it became possible, to prove that the sedimentation pond of the pre-planing was not optimally used. The usage of available storage volume within the sewer system by real-time control reduces the total emissions significantly.

During further work the results will be improved. The goal is to reduce the load on the Elbe without an increase in the necessary investment.

4. Conclusions of the Case Study and other Studies

The traditional way in planning the different parts of a sewer system was also carried out within this case study. A consultant started shortly after reunification with the planning of the treatment plant. At this phase we started to collect the necessary data on the sewer system to have a first overview about its structural and hydraulic capacity. During the final planning phase of the treatment plant we were asked to develop a comprehensive waste water concept for the city of Dresden and to evaluate the design of the treatment plant according to a general waste water concept. The conclusions of this study are now discussed:

Existing data

For the necessary simulations numerous data are needed. For sewer system simulation actual information about the physical system (diameters, elevations, connections, etc.) and the land cover (roofs, streets, pervious areas, etc.) are put into the simulation models. In the former BRD these data are normally available on maps or in data-bases with a sufficient high accurracy. In the former DDR the existing maps of the sewer system are in poor condition. Extensions of the systems are often not included. Normally all these data or parts of them have to be checked in the field. Also the information about the land cover can't be used out of the maps. Aereal photographs or satellite images give better and up to date information.

Information about rainfall is also very limited. If there is some data available, the measures were only untertaken during summer. Information about the areal distribution cannot be obtained.

Design regulations

The knowledge of the people at the communities about the existing design regulations is sometimes very limited. When a consultant comes up with solutions, these are not always the best ones for the given situation, but rather the best ones for maximising his fee. It happened that large treatment plants were planned for small communities with the result that the actual inflow was only 30% of the planned capacity.

Estimation of design parameter

Just from the beginning it is necessary to estimate a realistic increase (or sometimes a decrease) in the number of inhabitants and the industrial load equivalents. Further on in the former DDR there was no fee for drinking water or sewage treatment. Due to the upgrading of the infrastructure and industry and the downfall of some industries the water consumption was extremely reduced. For the city of Dresden the inflow to the treatment plant just before reunification was about 185,000 m^3/d. The actual inflow is only about 70 % of this value.

Integrated approach - Cost factors

The procedure in designing the treatment plant and the other necessary detention basins or sewers should be done according an integrated approach. The separation between treatment plant, sewer system and receiving waters leads to higher investment costs compared to an integrated design. But a totally integrated design is only possible, if this is foreseen in the corresponding regulations. The results of the case study show, that based on an integrated simulation of the treatment plant and the sewer system under qualitative and quantitative aspects a sizeable reduction of total investment costs for the treatment plant was possible. In the first stage we found that a sedimentation basin planned as a part of the treatment plant does not influence the discharge concentration significantly. The investment costs could be reduced by 35 Million DM. Assuming that the total load of the sewer system and the treatment plant should not be constant and an appropriate design of the plant according to a realistic increase of the inflow could reduce the total costs for the treatment plant and the sewer system in a range of 500 Million DM.

5. References

ATV (1993) Guidelines for the dimensioning and design of combined sewer overflows, ATV-A128, ATV St. Augustin.

Fuchs, L. et. al. (1994) HYSTEM-EXTRAN - Model description and documentation, ITWH Hannover.

Fuchs, L. et. al. (1994) Investigation of the total emissions of the city of Dresden (in German), part I and II, report for the city of Dresden, ITWH Hannover - Dresden.

Fuchs, L. et. al. (1995) Urban drainage and sewage treatment concept for the city of Dresden, report for the city of Dresden, ITWH Hannover - Dresden.

ITWH (1991) GESIM - Model description and documentation (in German), ITWH Hannover.

ITWH (1992) KOSIM - Model description and documentation (in German), ITWH Hannover.

Böhm, R.; G. Ermel, L. Fuchs (1994) Development of the river Elbe by the renovation of the sewage treatment plant Dresden, (in german), in: Die Elbe im Spannungsfeld zwischen Ökologie und Ökonomie, Teubner Verlag, Stuttgart.

Statistisches Bundesamt (1994) Development of sewage treatment and drainage in Germany, in Korrespondenz Abwasser, No. 1, 1994.

Ministry for Environment and Land Development of Saxony (1993): Incremental Expansion of Waste Water Treatment, administrative guideline.

Use of Remote Sensing and GIS for Environmental Modelling and Monitoring

John Elgy and Helen K Jones,
Department of Civil Engineering
Aston University,
Birmingham, B4 7ET,
UK

Abstract

Remote Sensing is proposed as a useful technique for monitoring environmental pollution, in particular landfill sites for landfill gas migration. Landfill gas has the potential to migrate from the site boundaries and cause damage either where it emerges into the atmosphere or in the root zones of vegetation It is also spatially and temporally very difficult to monitor and techniques developed for landfill gas can be applied to other sources of pollution such as ducts or pipelines. The presence of landfill gas in the root zone causes signs of stress to be visible within the vegetation. This stress takes the form of either chlorosis or dieback. Airborne thematic map data and low cost video imagery are used to classify the site according to vegetation stress. A comparison between the efficiency and costs of the two sources of imagery is made. It is recommended that remote sensing is used in conjunction with soil and geological maps in a geographical information system to identify areas of high landfill gas concentration and that this is combined with traditional field measurements to gain a better spatial and temporal understanding of gas migration. The future potential of the technique is disscussed

Introduction

The complementary technologies of geographical information systems and remote sensing are well suited to the modelling of environmental systems and monitoring the performance of those systems. Though remote sensing is entirely a tool for monitoring, GIS serves two functions: the first is almost a traditional role of recording spatial data for later recall in a structured fashion, the second is as a tool to model the system. In the inventory case the role is very closely related to that of a database management system where in the modelling case the GIS provides a method of modelling processes that vary over time and space. These technologies give us the way to record data about environmentally damaged land, relate that data spatially,

NATO ASI Series, Partnership Sub-Series, 2. Environment – Vol. 45
Environmentally Devastated Areas in River Basins in Eastern Europe
Edited by Alfons Georges Buekens and Vasily Victorovich Dragalov

model any rehabilitation programme proposed, and, using remote sensing to identify either potentially damaged land or to monitor the effect of changes to that land.

Remote Sensing as a Monitoring Methodology

In order to measure some parameter of environmental damage over quite large areas requires a very large number of measurements is required. For example, when monitoring the presence of methane and carbon dioxide in the soil arising from the decomposition of refuse in a landfill site, and if the area of influence of the site is 1 km square then monitoring a 5m resolution would require 40 000 observations. Even this resolution may be too large to detect some limited zones of gas migration.

Satellite remote sensing has long been used to observe areas of potential damage to large areas, for example forest fires, deforestation, drought monitoring, flooding, locus damage etc, (Dury ,1990). Satellite remote sensing is viable if the effect to be observed is over a large area, tens of kilometres, and the smallest feature to be observed is greater than about 30m. For smaller areas airborne remote sensing is required.

Groves (1989) used false colour infra red aerial photography to monitor the effect of air pollution on the pine forests of Bavaria. Jones and Elgy (1994) used an airborne thematic mapper to monitor the levels of gas emitted from a landfill site that had been covered and restored to agricultural use. The following relies heavily upon that paper.

The basic requirement for effective management of landfill gas migration is the satisfactory measurement of gas in the subsurface. Landfill gas is transient in character because of the way in which it is produced and migrates. Measured gas concentrations may be affected by a number of factors, including soil conditions, climatic conditions, especially atmospheric pressure variations and the rate at which the gas is actually being produced. Consequently the actual physical measurement of landfill gas is difficult, and the interpretation of sampling results must be undertaken with care. Sampling programmes may be further restricted by cost, time, and access.

Some of the problems can be overcome by monitoring from a different stand-point, that is, by examining the impact of the gas, rather than measuring its actual physical concentration. In other words we infer the concentrations from the effects the gas has upon vegetation and observe the stress in the vegetation by looking at the changes in reflectance of the near infra red from healthy growing vegetation. The approach outlined in Jones and Elgy (1994) is the analysis of vegetation health; changes in health being expressed as visible alteration in physiology or morphology of plants growing on gas affected soils. Furthermore, if gas in the root zone results in a change in the spectral response of the vegetation, remote sensing techniques may be used to infer the presence of gas.

Two sites were investigated to evaluate remote sensing techniques to detect vegetation health and infer landfill gas migration. The sites differ in severity of damage, types of vegetation, concentrations of landfill gas, and soil types. The results are used to develop a method for combining ground-based and remote sensing approaches to landfill gas monitoring.

Remote sensing does not detect gas migrating at depth below soil zone. However, it does provide valuable information concerning the presence of gas in the near-surface soils, no matter how distant from the point of generation. For a comprehensive description of the techniques and methods see Jones (1991).

Effects of Landfill Gas on Vegetation

Landfill gas in the root zone frequently results in a change in plant condition (Gilman *et al*, 1985; Leone *et al* , 1977; Arthur *et al* , 1981; Flower *et al* , 1981; Hewitt & McRae, 1985; Hill, 1985; Roberts, 1985; Stonell, 1985; Carpenter, 1986; Parsons & Smith, 1986; Spreull & Cullum, 1987; Tankard, 1987; and DoE, 1989). Despite the widely held conviction that landfill gas is detrimental to plant health, the complexity of the landfill gas - vegetation - soil relationship defies any attempt to identify the association with accuracy, or to predict the concentrations and combinations of gases that could be expected to result in vegetation injury. This lack of knowledge limits the application of remote sensing, but the information gathered still remains an important adjunct to field data.

Practical Integration of Remote Sensing and Ground Survey

Two landfill sites were studied to assess the use of remote sensing to monitor landfill gas migration. The experience gained from these studies highlights the problems of attempting to undertake a remote sensing survey to detect or monitor the migration of landfill gas. It must be remembered that the end point of the monitoring exercise is not the collection of data, but the resolution of a practical problem in a rational and cost-effective manner. This ideally requires that the data is transfered into a GIS for later storage and assessment of the risks. Any remote sensing survey must focus on assessing possible impacts and risks to local structures, and identify pathways of migration in order to assist effective siting of migration prevention systems and monitoring schemes.

Remote sensing is not an alternative to ground survey, and both techniques should be integrated to optimise their individual advantages, and minimise their disadvantages (Table 1). The following case studies show that the approach to each site is defined by a combination of factors, which include the availability of data, severity of vegetation damage, possible complicating factors, and ease of access.

Ground Survey	Remote Sensing Survey
Disadvantages	
i. Difficulties in obtaining accurate measurements due to fluctuating levels of gas and changing migration pathways. ii. No spatial perspective iii. Physical / political access to site.	i. Relies on the existence of a relationship between landfill gas and spectral response of vegetation ii. Cannot estimate gas concentrations iii. Will always require fieldwork to confirm presence of gas and establish concentrations
Advantages	
i. Quantitative measurement of gas (required to satisfy waste disposal authorities and legislative requirements)	ii. May identify previously unknown areas iii. May be able to infer information from spatial relations iv. Quick and unobtrusive method v. Spatial estimates of damage can be made vi. Marginal costs to survey additional sites would be low

Table 1 Advantages and disadvantages of remote sensing and ground based surveys

Ground based measurement of gas concentrations are required:

i. to establish the initial relationship between landfill gas and vegetation damage,
ii. to identify the relationship between vegetation damage and spectral response,
iii. to confirm landfill gas concentrations at locations suggested by image interpretation,
iv. to confirm the relationships were still viable if periodic monitoring was being undertaken, and to verify any additional areas apparently suffering from the effects of landfill gas.

The sensors and platforms available for the collection of remotely sensed data are many and varied. The choice of system for any application is limited by resolution of the sensor, the availability of suitable imagery, the techniques adopted and equipment available to analyse the imagery, the knowledge and skill of the interpreter, and the cost.

The spatial resolution of satellite data (the reflecance returned from the earth to the sensor is averaged over, typically, a 30m by 30m cell; to analyse this cell the data is converted into a computer image where each cell becomes a picture element or pixel) is too coarse to provide useful information for the study of small landfill sites. For these, a resolution of less than 5m is generally required. Although military satellites may give this resolution such imagery is not normally available, and when it is available the spectral range is often insufficient. An airborne platform is necessary. Although this may give the site operator more control over data acquisition, problems may occur due to geometric distortion caused by aircraft movement, and the relatively high cost of the data per unit area.

Numerous sensors may be mounted on airborne platforms, some of the most useful being the photographic camera, video camera, and multispectral line scanner (Table 2). The choice of system should account for the resources available to collect and process the data, the type of data required, and the frequency of data collection.

Sensor and Data Type	Recording media	Resolution	Spectral Range	Cost
Daedalus 1268 Multispectral	CCT	1 m upwards	Very wide	High
Remote Sensing Satellite	CCT or Print or CD	10 - 30 m	Very wide	Medium (for up to date data)
Monochrome Infrared camra	Video or photographic film	0.1 m and greater	Wide in visible Low-Med in the near infra red	Moderate to low, depending on technology

Table 2 Remote Sensing Data Sources

Remote sensing relies on electromagnetic radiation being detected at a sensor remote from the object being observed. The two classical versions of this are aerial photography and satellite remote sensing. Both cases are passive sensors, in that they rely upon the sun as the source of radiation, which is reflected (or re-radiated) back to the sensor from the area being observed. Active sensors, such as RADAR emit their own electromagnetic radaition and register its returning position and by measuring the time for the reflected wave to return can also record the range to the target object. Imaging and ground penetrating RADAR have considerable potential in monitoring covered landfill sites for depth to water table, and, in conjuction with boreholes, mapping and modelling the internal state of the landfill.

Photographic remote sensing builds an image by exposing a photosensitive emulsion to an area of the real world for a few micro seconds in the same way as a standard camara works. The aircraft or satellite flies on for a short distance and another image gathered. The spatial resolution of the image is calculated from the grain size of the photgraphic film and the focal length of the camara. It typically gives a resolution in the order of centimetres from aircraft and metres from space depending on the prevailing atmospheric conditions. Spectral resolution is the major disadvantage of photgraphic techniques. Colour films can only record the red, green and blue of the visible spectrum, while false colour composites record near infra red as red, red as green and green as blue and so allow for some vegatation studies to be carried out.

Scanning techniques record small discrete parts of the elctro magnetic spectrum, table 3 shows the discrete frequencies sensed by the Daedalus scanner used for this research. To obtain this much higher spectal resolution an image is built up by swinging the scanner from side to side whilst the sensor platform (aircraft or satellite) moves forward. At any one instant the value reflect at each of the frequencies is recorded onto magnetic tape and then the sensor moves onto the next part of the ground to record. When mounted in a satellite this type of sensor is capable of spatial resolutions of about 10m, though 30m is the usual pixel size. By flying the sensor lower, as in an aircraft, much higher resolutions are achievable, such as the 1m resolution used in this work.

The information provided by the greater spectral resolution allows different vegetation types to be detected and mapped and objective numerical values to be placed on the health of vegetation by using digital image processing techniques. Later studies by Ellis, Hedges and Elgy (1994) show that pre-dawn thermal imagery is very useful in detecting landfill gas in complex urban settings. ATM is arrived at by a line scanning technique and the aircraft slightly alters course and attitude between gathering one line and the next; georectifying these images is very difficult. It was found that after applying simple 'rubber banding' correction to match each image to the map then at some parts of the image, pixels from the visible flight were 5 pixels adrift from those in the corresponding thermal flight. This meant that both images could not be used in conjunction with one another. More sophisticated correction methods do exist, but they require ground control points at the edges of the image every 5 lines or so. This number of control points were not available so this technique could not be used.

The unique spatial information provided by remote sensing data may be utilised in a number of ways. It identifies variation in vegetation type and health to assist ground-based sampling, and to concentrate fieldwork in anomalous areas. However, there is risk that gas is present without affecting vegetation, and ground checks must be made. More specific surveys may be carried out in order to gather spatial information where gas discharges are likely, e.g. at the edges of sites or above geological / soil boundaries.

ATM Channel	Landsat TM Band	Spectral Response (μm)	Applications
1	-	0.42-0.45	Bathymetry; soil/vegetation differentiation
2	1	0.45-0.52	Coniferous/deciduous differentiation
3	2	0.52-0.60	Green reflectance by healthy vegetation
4	-	0.605-0.625	Secondary green reflectance band
5	3	0.63-0.69	Plant species differentiation by chlorophyll absorption
6	-	0.695-0.75	Geobotanical investigations
7	4	0.76-0.90	Biomass surveys; water body delineation
8	-	0.91-1.10	Secondary near infra-red band
9	5	1.55-1.75	Altered rock mapping; vegetation moisture measurement
10	7	2.08-2.35	Hydrothermal mapping
11	6	8.50-13.00	Plant heat stress detection; thermal mapping
12	6	8.50-13.00	Plant heat stress detection; thermal mapping - half amplification of band 11

Table 3 Electromagnetic Frequencies Recorded by Daedalus AADS 1268 Airborne Thematic Mapper

It is often necessary to undertake detailed measurement of gas concentrations prior to instigating a remote sensing survey. This establishes the relationship between landfill gas and the observed vegetation damage. If the relationship is weak, or complicated by other factors such as differences in soil texture, local contamination by heavy metals, etc., the benefits of a remote sensing surveys alone are limited. If the relationship is strong, however, remote sensing and field data can be integrated to provide:

An initial identification of further stressed areas, unknown from field work. This is of particular use in surveying large sites, or those with access problems (either political or physical). It may also identify problems occurring beyond site boundaries, often neglected by site operators.

i. An understanding of the way in which gas is migrating by observation of the spatial distribution of vegetation damage. For example, anomalously high concentrations of gas might be recorded along old site boundaries, at geological contacts, or above field drains.

The perspective gained from an airborne platform would provide understanding of these problems, and facilitate the siting of monitoring boreholes and gas colletcion wells.

ii. A baseline, against which to monitor change. That is, by comparison with known areas of high landfill gas concentrations to infer possible changes in the extent of migration, or identify new locations of vegetation damage.

iii. A rapid estimate of the extent of damage to agricultural areas for crop loss assessment, and subsequent compensation settlements.

A good relationship between landfill gas and vegetation health will occur when:

i. high concentrations of gas were affecting the vegetation, with severe vegetation damage resulting,

ii. the vegetation cover was homogeneous,

iii. the vegetation species was susceptible to damage by landfill gas,

iv. other environmental variables, such as soil moisture and texture, were relatively constant across the site.

Economics of Remote Sensing

The economics of information gathering can be divided into the elements of costs and benefits (Watkins, 1978), the benefit of the data in this case being represented by the possible reduction in field work that could be afforded by using remote sensing techniques. In practice, due to the difficulties in defining a single approach to the problem, the calculation of a 'benefit-cost' as a result of the reduction in sampling requirements is not possible. However, the relevant features of the different systems, and the ways in which they could affect the practical application of remote sensing, are briefly discussed.

Costs of remote sensing surveys include not only the costs of acquiring the data, but also the cost of preprocessing and processing. The least expensive data source may require the most expensive (and/or the least accurate) complete system, due to processing requirements. For example, although the cost of acquiring the video data is very low, a need for mosaicing of images to provide a single 'image' of the site, will increase processing costs, possibly substantially.

Jones (1991) gives estimates of the actual costs of acquiring the remote sensed data and processing it. The prices are very UK orientated but as a general principle heavy scanners in twin-engined aircraft are much more expensive to operate that light scanners in single-engined aircraft. The benifits of the line scanners over the less expensive use of near infra red sensors in photographic film or video cameras is very small and the problems of georectifying these images leaves their use questionable. However, the dawn thermal infrared images from the Daedalus scanner were much more useable than those from a thermal imaging video camera mounted in a helicopter. This was due partly to the better quality sensor used but mainly due to the very large field of view available from the Daedalus and the vertical mounting of the sensor.

Practical Application of Remote Sensing to Pollution monitoring

Two contrasting sites were selected to examine the application of remote sensing to the migration of landfill gas. Both sites had been quarried for sand and gravel, the pits were then infilled with domestic refuse with a view to final restoration for agriculture. No provision was made at the design stage for the production, migration or collection of landfill gas and both sites were to produce substantial quantities of gas in the following years.

At Site 1, vegetation on the restored landfill cap showed symptoms of stress, with extensive tracts of crop being generally unhealthy and chloritic. The poor growth was originally attributed to the effects of replacing top soil during site restoration, but subsequent monitoring for methane, carbon dioxide, and oxygen suggested that landfill gas could at least be partially responsible for the decline in vegetation health. Further detailed monitoring of gas, soil, and vegetation indicated that a number of factors were responsible, and multivariate statistical analysis of the data with discriminant analysis indicated that methane concentration, soil moisture, soil pH, and soil texture were significant contributors to vegetation stress (Jones, op cit). Multivariate discriminate analysis is a method of identifying to which of a number of nonmetric groups an object belongs based upon a series of measurements upon that object. The method usefully gives a measure as to the power of the discriminating function so we can state a reliability of the results obtained (see for example Hair et al, 1992). Overall, the relationship between landfill gas, plant health, and soil was not easy to establish at this site.

At Site 2, damage to vegetation had been observed adjacent to the landfill site. The damage included the death of a section of a hedgerow along the north-west site boundary, leafless branches on many of the trees in the adjacent woodland, and damage to the crops on the farm-land to the north west of the site. The situation deteriorated after the north west corner of the site was completed and restored with an impermeable clay cap. Although good crop yields proved successful restoration over the landfill, they also underlined the relative impermeability of the cap, a fact which resulted in increased lateral migration of landfill gas. The latter is also enhanced by the flattening action of surface compaction, resulting in a layered structure to the landfill

High concentrations of gas were found surfacing over relatively small, undisturbed areas of ground. Predictable variation in soil type allowed the relationship between vegetation dieback and landfill gas to be more readily established than at Site 1. These links were established using traditional GIS overlaying techniques (see Burrough, 1989)

Aerial photography, aerial video, and airborne multispectral data were used at both sites. The data was geometrically corrected (where necessary), and standard image processing and statistical techniques used to identify anomalous areas of vegetation, and to relate these to the presence of landfill gas in the shallow sub-surface.

Variability of the restored soils and settlement of the cap at Site 1, prevented the identification of a straightforward relationship between landfill gas and the spectral response of the crop. However, discriminant analysis revealed a relationship between gas concentration, soil condition, and spectral response. The remote sensing data was used to produce crop condition maps based on the spectral response of the crop and these maps were later used in conjunction with the results of the statistical analysis to identify probable soil and gas conditions.

In contrast, the relationship between landfill gas and vegetation damage at Site 2 was well established. The remote sensing data was successfully used to define the extent of crop damage and its relationship with high gas concentrations.

Some Other Applications of Remote Sensing to Pollution Monitoring

Infrared techniques have been used successfully in the past to detect the discharge of springs into water bodies such as streams or lakes (Brereton and Downing , 1975, and Brereton and Hall, 1983). Brereton and Downing (1975) listed two factors which determined whether a discharge can be identified;

i. the temperature difference between the groundwater and the surface water;
ii. the dilution ratio;

their work suggested a temperature difference of at least 2°C and a dilution ratio of 1:2 groundwater to surface water was required.

The temperature difference between the leaking pipeline and the normal ground surface temperature could also be used to identify either water leaking from pipelines or natural springs. Weil (1993), reports on detecting leaks in buried pipelines using an infra-red imaging camera - a plume of water flowing from a pressurised pipeline causes a reduction in ground temperature. Weil and Graf(1991), also reported the use of this technique, to detect the presence of hot water pipelines and in the detection of voids. Both these studies rely on either handheld or vehicle mounted cameras and work at very high spatial resolution. A more useful system would involve an aircraft and have a resolution of about 0.5m. The sensitivity of imaging thermographic cameras is around 0.1^{O}C, so these methods are only suitable for large leaks.

Conclusions

The traditional means of detecting near-surface migration of landfill gas is point sampling and direct measurement of gas concentrations. However, the problems of obtaining accurate gas concentrations are considerable, and the results of such surveys must be interpreted with care. The use of remote sensing to detect or monitor landfill gas migration relies on two dynamic systems that are difficult to quantify: the gas / plant, and the vegetation / sensor systems. The technique is successful wherever the underlying relationship between vegetation damage and high gas concentration is easy to establish. That is, in situations where there are high gas concentrations, homogeneous vegetative cover, and predictable variations in soil characteristics.

Use of the same techniques in complex urban environments where homogenious land cover is not normal have proved dissapointing, see Ellis *et al.* (1994). However, many restored lanfill sites in urban areas have been used for playing fields and other recreational land use. In these cases effect of landfill on the overlying vegetaion are more easily observed, particularly, as 'hot spots' within the landfill. When the gas migrates into the surrounding residential areas the effect of the gas migration are maked by the different land management practices applied to each garden and the land surrounding each residence.

Remote sensing techniques are a supplement to, rather than, a substitute for ground survey. The degree to which ground sampling requirements can be reduced varies according to specific site conditions. At sites where the gas is migrating vertically upwards through the cap, complications will be introduced due to variation in soil conditions as a result of the soil movement involved in restoration procedures. Such factors will frequently prevent identification of a decisive relationship between landfill gas and vegetation health, and hinder attempts to detect or monitor landfill gas using remote sensing. However, in these case thermal imagery may make it possible identify 'hot spots on the within the fill. It is impossible to define a single approach and each site must be evaluated on its own merits.

Remote sensing is most valuable where spatial relations, rather than absolute geometric accuracy, are important. Restricting the application of remote sensing to such situations would enhance the utility of the data, and reduce the costs of processing, as only simple geometric correction would be required.

In more general terms, the incorporation of remote sensing data, ground sampling information, general purpose maps, and thematic maps within a geographic information system would greatly enhance the usefulness of all forms of information to the landfill site manager. There are five important areas of further research:

i. The formation of spatial information systems or geographical information systems (the two terms are synonymoŭs here) for each landfill site, or group of sites incorporating: remote sensing data, ground sampling information, general purpose maps, thematic maps, and the history of the site. This data should be augmented at regular intervals so that not only does it form a register of contaminated land kept, but also updates the risks associated with that land. Spatial statistical analysis of all the data should be studied.
ii. Current GIS technology deals with 3 dimensional data very poorly and in reality can only be used to record what is happening at the surface. To model the processes occurring in a landfill site, the more complex 3D models and systems are required.
iii. Ground penetrating RADAR and other traditional geophysical study techniques can be used to monitor what is happening to the leachate formed within and surrounding the site.
iv. Rapid, low cost methods of georectifying ATM imagery are required so that images collected on different days can be overlain and change studies initiated.
v. Further low cost forms of airborne resote sensing, are required, preferably those that can be mounted in single engined aircraft (note that there are problems operating single engined aircraft over urban areas) with their lower operating cost so almost continous monitoring can be instigated.

A major advantage of remote sensing surveys is that large areas can be surveyed rapidly, and data acquired from otherwise inaccessible areas. However, the lack of an unambiguous relationship between landfill gas and vegetation injury prevents a direct estimate of landfill gas in the subsurface and a combination of ground-based and aerial approaches provides the optimum information regarding landfill gas migration.

Acknowledgements

This work was sponsored by the Science and Engineering Research Council and MJ Carter Associates. The authors wish to acknowledge the assistance of MJ Carter Associates, in particular Leslie Heasman, for supporting the research and Dr. T.R.E. Chidley for his considerable assistance.

References

Arthur, J. J., Leone, I. A., & Flower, F. B., 1981, Flooding and Landfill Gas Effects on Red and Sugar Maples, *Jour. Environ. Qual.*, **10** (4), 431 - 433.

Brereton N.R. and Downing R.A. (1975), Some Applications of Thermal Infrared Linescan in Water Resources Studies, *Water Services v79*, pp91-96.

Bretreton N.R. and Hall D.J. (1983), Groundwater Discharge Mapping at Altnabreac by Thermal Infrared Linescan Surveying, *IGS Report FLPU 83-7*, British Geological Survey, UK.

Burrough P.A. (1986), Principle of Geographical Information Systems for Land Resources Assessment, *Clarendon Press* Oxford.

Cairney, T., 1987, *Reclaiming Contaminated Land*, Blackie, 260 pp.

Carpenter, R. J., 1986, *Redevelopment of Land Contaminated by Methane Gas: The Problems and some Remedial Techniques*, Env. Safety Centre, Harwell Lab., Rpt No. R11900.

Department of Environment, 1989, *Waste Management Paper No. 27: The Control of Landfill Gas*, HMSO.

Dury S.A. (1990) A Guide to Remote Sensing, *Oxford Science Puplications*, Oxford, UK 1990

Ellis R.J. Hedges P.D. & Elgy J. (1994) Monitoring Landfill Sites Within an Urban Environment Using an Airborne Thematic Mapper, *Proc. 3 International Conference on Re-Use of Contaminated Land and Landfills* June 1994 London UK.

Flower, F.B., Gilman, E.F. & Leone, I.A., 1981, Landfill Gas: What it Does to Trees and how its Injurious Effects may be Prevented, *Jour. Aboriculture*, 7, 43 - 52.

Gilman, E.F.; Flower, F.B.; & Leone, I.D., 1985, Standardised Procedures for Planting Vegetation on Completed Sanitary Landfills, *Waste Man. Res.*, **3**, 65 - 80.

Groves M.A. (1989) The use of Remotely Sensed Data for monitoring Air pollution Related Damage to Forested Areas, *PhD Theseis*, Aston University, Birmingham, UK

Hair J.F., Anderson R.E., Tatham R.L. & Black W.C., 1992, Multivariate Data Analysis, Macmillan, New York.

Hewitt. A.K.J. & McRae, S.G., 1985, The Effects of Gases Emitted from Landfills on Soils and Crops, in *1st International TNO Conference on Contaminated Soil*, 11 -15 Nov. 1985, Utrecht, The Netherlands, 251 - 253.

Hill, C.P., 1985, Landfill Gas Migration from Operational Landfill Sites - Monitoring and Prevention, *Wastes Manage.*, **76** (4), 169 - 178.

Jones, H.K., (1991), The Investigation of Vegetation Change Using Remote Sensing to Detect and Monitor Migration of Landfill Gas, *PhD Thesis, Aston University*, 1991.

Jones, H.K., & Elgy J (1994), Remote Sensing to Assess Landfill Gas Migration, *Waste Management & Research (1994)* **12** pp 327-337

Leone, I.A.; Flower, F.B.; Arthur, J.J; & Gilman, E.F, 1977a, Damage to New Jersey Crops by Landfill Gases, *Plant Disease Reporter*, **61** (4), 295 - 299.

Leone, I.A.; Flower, F.B.; Arthur, J.J.; Gilman, E.F., 1977b, Damage to Woody Species by Anaerobic Landfill Gases, *Jour. Arboriculture*, **3** (12), 221 - 225.

Mancinelli, R.L., Shulls, W.A. & McKay, C.P., 1981, Methane Oxidising Bacteria used as an Index of Soil Methane Concentrations, *Applied Env. Microbiology*, **42** (1), 70 - 73.

Parsons, P.J. & Smith, A.J., 1986, The Environmental Impact of Landfill Gas, in *Proc. Energy from Landfill Gas Conf.*, 28 - 31 Oct.1986, Solihull, UK, 17 - 21.

Roberts, R.D., 1985, Biological Monitoring Techniques, *Proc. Symp. on Landfill Monitoring*, 23 May 1985, Harwell Lab., Oxon., 155 - 171.

Stonell, K., 1985, The Use of Wastes for Land Reclamation and Restoration, in *Agricultural, Industrial and Municipal Waste Management*, Publ. Mech. Eng. Publications Ltd., Bury St Edmunds, C85/85, 1 - 7.

Spreull, W.J. & Cullum, S., 1987, Landfill Gas Venting for Agricultural Restoration, *Waste Man. Res.*, **5** (1), 1 - 12.

Tankard, J.D., 1987, The Plight of Loscoe, Derbyshire, *Wastes Management*, 77 (3), 173 - 175.

Watkins, T., 1978, The Economics of Remote Sensing, *Photogramm. Eng. Remote Sensing*, **44** (9), 1167 - 1172.

Weil G.J. (1993), Non-contact, Remote Sensing of Buried Water Pipeline Leaks Using Infrared Thermography, *Proc. 20th Conf. on Water Management in the 90's, May 1-5 1993*, Seattle USA.

Weil G.J. and Graf R.J. (1991), Infrared Red Thermography based Pipeline Leak Detection, *Proc. 2nd National Speciality Conf. on Civil Engineering Applications of Remote Sensing and Geographic Information Systems.* Washington May 1991.

DETERIORATION OF THE WATER QUALITY IN THE RIVER SIMAV BY INDUSTRIAL ACTIVITIES

A. Türkman, F. Sengül, K. Kestioglu
Dokuz Eylül University
Department of Environmental Engineering
Bornova, Izmir
TURKEY

INTRODUCTION

In January 1992, Turkey has been accepted as a *newly developed* country. This development is associated with industrialization the origin of which goes back to 1930's. The contribution of industrialisation to development is partly offset by its polluting effects. Many rivers and lakes became heavily polluted as a result of rapid industrialization not followed by appropriate pollution control measures.

This paper investigates the change of the water quality of Simav River mainly due to industrial sources of pollution. The water quality has been determined in the upstream part of the river and also after receiving many types of industrial wastewaters. The quality of the river water with respect to time thas been followed up;

WATER QUALITY CRITERIA

Since Simav river is used for irrigation, fishery and industrial feed water purposes, its quality should meet specific requirements.

NATO ASI Series, Partnership Sub-Series, 2. Environment – Vol. 45
Environmentally Devastated Areas in River Basins in Eastern Europe
Edited by Alfons Georges Buekens and Vasily Victorovich Dragalov

During its flow of about 150 km, Simav River receives domestic and industrial wastewaters before it discharges into the Marmara Sea. The deterioration of water quality by industrial discharges has a special importance due to the presence of (phyto-)toxic substances.

In order to evaluate the quality of River water, the parameters measured in the study will be compared with Turkish water quality standards. The Turkish water quality requirements and classification is based on Water Pollution Control Regulation which has been published in the official gazette in 1988 and mentions four classes of water. Annex 1 gives the water quality classes for surface waters as given in this Regulation.

Since Simav river is mainly used for irrigation purposes, the water quality requirements are compared to the values in Annex 2, also published in Water Pollution Control Regulation.

WATER QUALITY OF SIMAV RIVER

The Simav River originates at Simav Village of Kütahya and joins Susurluk River after passing Kepsut village as indicated in Figure 1. Before confluence with the Susurluk river, the Simav river receives heavily polluted wastewaters of industrial origin (Table 2). Despite its heavy pollution load the River is still being used, mainly for irrigation purposes.

Water quality of Simav River at the upstream part is given in Table 1. Mean composition of river water (Hem, 1970) is also given for comparison.

Table 1 : Quality of Simav River at the Upstream Part of the River (DSI, 1984)

Parameter	Minimum	Average	Maximum	Average River Water
Flow (Q), m3/s	5.762	19.4144	48.500	
Temperature (°C) (T)	5.0	14.98	26.00	
pH	7.88	8.28	8.68	

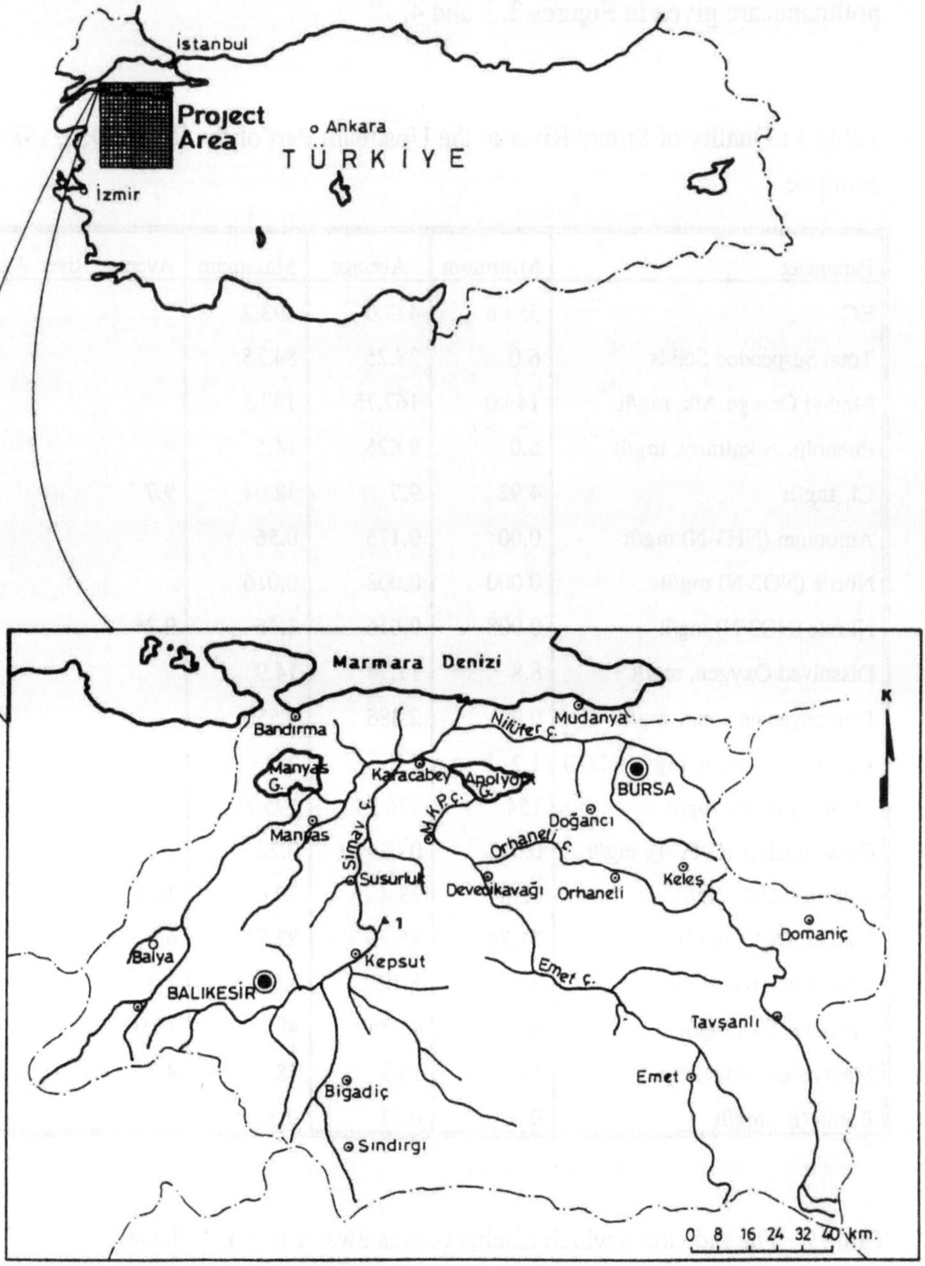

Figure 1. Simav River

Table 3 shows the river quality deterioration. The time evolution of certain pollutants are given in Figures 2, 3 and 4.

Table 1 : Quality of Simav River at the Upstream Part of the River (DSI, 1984) continue

Parameter	Minimum	Average	Maximum	Average River Water
EC	357.8	417.0	493.2	
Total Suspended Solids	6.0	38.25	84.25	
Methyl Orange Alk. mg/lt	149.0	167.75	197.5	
Phenolp. Alkalinity, mg/lt	6.0	9.625	14.5	
Cl, mg/lt	4.92	9.7	12.04	9.7
Amonium (NH3-N) mg/lt	0.00	0.175	0.56	
Nitrite (NO2-N) mg/lt	0.000	0.008	0.016	
Nitrate (NO3-N) mg/lt	0.008	0.616	1.46	0.24
Dissolved Oxygen, mg/lt	8.8	11.74	14.9	
Permanganate value, mg/l	0.9	2.386	4.556	
Biochem.oxy.Dem.mg/lt (BOD5)	1.2	1.6	2.0	
Total Hardness, mg/lt	154	176.75	195.75	
Orto-Phosfate (O-PO4), mg/lt	0.015	0.095	0.20	
Sulfate (SO4), mg/lt	12.8	23.4	30.6	11.4
Sodium (NA), mg/lt	11.73	15.75	22.73	6.3
Potassium (K), mg/lt	2.40	3.19	4.18	2.3
Calcium (Ca), mg/lt	36	43.35	48.2	15.0
Magnesium (Mg), mg/lt	11.0	15.9	18.93	4.1
Boron (B), mg/lt	0.00	0.11	0.5	

Table 2 : The industries which discharge wastewater to Simav River

Paper Industry	Sugar industry
Slaughterhouse	Tomato Paste Industry
Boron Industry	Animal Feed Industry
Dairies	Sunflower Seed Oil Industry

Table 3 : Water Quality of Simav River at Sampling Point 1

Parameter	15.9.1989	2.11.1989	11.12.1989	18.1.1990	13.3.1990
pH	7.9	6.5	8.0	8.0	8.3
Dissolved Oxygen, mg/l	-	9.0	12.1	11.8	-
Conductivity, µmhos/cm	620	950	290	410	390
SO3, mg/l	2.8	3.20	1.7	10.2	8.3
Turbidity	90	102	270	30	8.5
Boron, mg/l	3.7	4.3	18.9	3.6	3.1
Ca++, mg/l	47.0	52.1	32.1	51.2	50.3
Mg++, mg/l	22.0	20.3	17.1	21.9	19.3
Total Hardness	220	200	230	210	205
Cl-, mg/l	23.5	50.0	17.0	40.0	21.0
Al+3, mg/l	5.8	7.2	4.1	6.1	4.6
PO4-P, mg/l	1.20	1.80	0.4	0.4	0.5
SO4, mg/l	107.1	200.0	90.6	78.2	84.4
SS, mg/l	76.0	95.0	1390	64.0	24.0
COD, mg/l	-	-	120.0	92.0	64.0
BOD5, mg/l	-	-	100.0	50.0	24.0
Na+, mg/l	70.0	65.0	78.0	35.0	78.0
K+, mg/l	4.6	4.5	6.4	20.0	6.7
SAR	2.10	1.93	2.77	1.03	3.39
Parameter	**1.5.1990**	**17.5.1990**	**21.6.1990**	**31.7.1990**	**19.8.1990**
pH	8.2	7.6	7.7	7.6	7.5
Dissolved Oxygen, mg/l	-	9.4	8.6	11.6	-
Conductivity, µmhos/cm	380	380	400	390	370
SO3, mg/l	7.6	2.4	2.6	2.7	2.6
Turbidity	75	15	80	90	102
Boron, mg/l	2.8	2.6	3.0	2.9	3.1
Ca++, mg/l	55.8	55.3	57.3	50	28.3
Mg++, mg/l	26.0	24.8	26.1	28.1	26.4
Total Hardness	246	240	250	240	179
Cl-, mg/l	21.0	29.0	24.0	25.0	24.5
Al+3, mg/l	4.2	4.4	4.6	3.7	3.2
PO4-P, mg/l	0.35	0.33	0.32	0.29	0.12

Table 3 : Water Quality of Simav River at Sampling Point 1 (continue)

Parameter	1.5.1990	17.5.1990	21.6.1990	31.7.1990	19.8.1990
SO4, mg/l	82.0	74.1	78.2	82.0	94.7
SS, mg/l	16	40	32	44	80
COD, mg/l	40	60	304	76	96
BOD5, mg/l	28	35	170	26	35
Na+, mg/l	85	90	96	98	103
K+, mg/l	20.3	6.3	6.5	6.7	7.8
SAR	2.35	2.52	2.63	2.74	3.33
Parameter	**29.11.990**	**27.12.1990**	**8.2.1991**	**8.4.1991**	**20.9.1991**
pH	8.1	8.2	8.3	8.6	7.4
Dissolved Oxygen, mg/l	-	12	10.4	10.4	8.0
Conductivity, μmhos/cm	380	380	380	400	600
SO3, mg/l	1.8	0.5	2.3	1.3	1.9
Turbidity	90	110	180	32	35
Boron, mg/l	2.8	6.8	2.7	12.0	2.7
Cl-, mg/l	23	24	28	25	20.0
Al+3, mg/l	3.5	3.8	4.3	4.3	3.7
PO4-P, mg/l	0.09	0.10	0.18	0.29	0.40
SO4, mg/l	86.5	70.0	84.4	107.1	94.0
SS, mg/l	44	44	56	68	52
COD, mg/l	112	24	78	76	108
BOD5, mg/l	48	16	25	30	60
Na+, mg/l	120	131	151	165	44
K+, mg/l	8.4	9.5	9.5	10.6	4.0
SAR	3.58	3.74	4.13	5.54	1.31

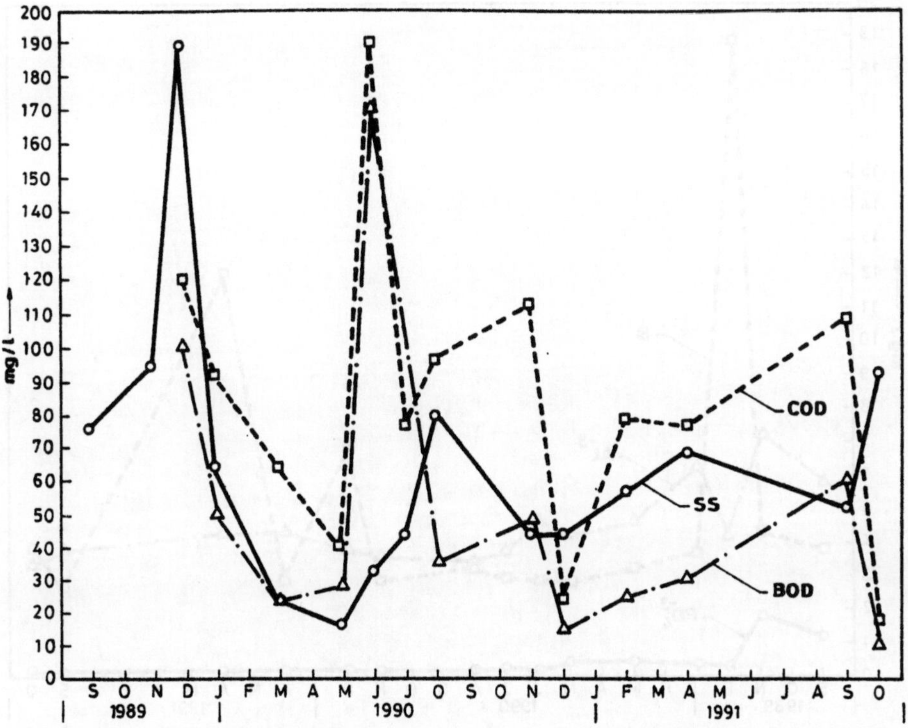

Figure 2. Variation of Simav River Quality With Respect to COD, SS, and BOD

DISCUSSION

Table 1 indicates the quality of Simav River at the upstream section, which is First Class according to the Water Pollution Control Regulation. In this part of the stream, the river water is used for irrigation and industrial water supply purposes. Only softening is applied as treatment for industrial use.

A sharp change is observed after receiving industrial discharges. This can also be confirmed by its aesthetic characteristics, such as bad odors, color, turbidity and floating substances.

The following observations have been obtained from the experimental values and figures 2, 3 and 4.

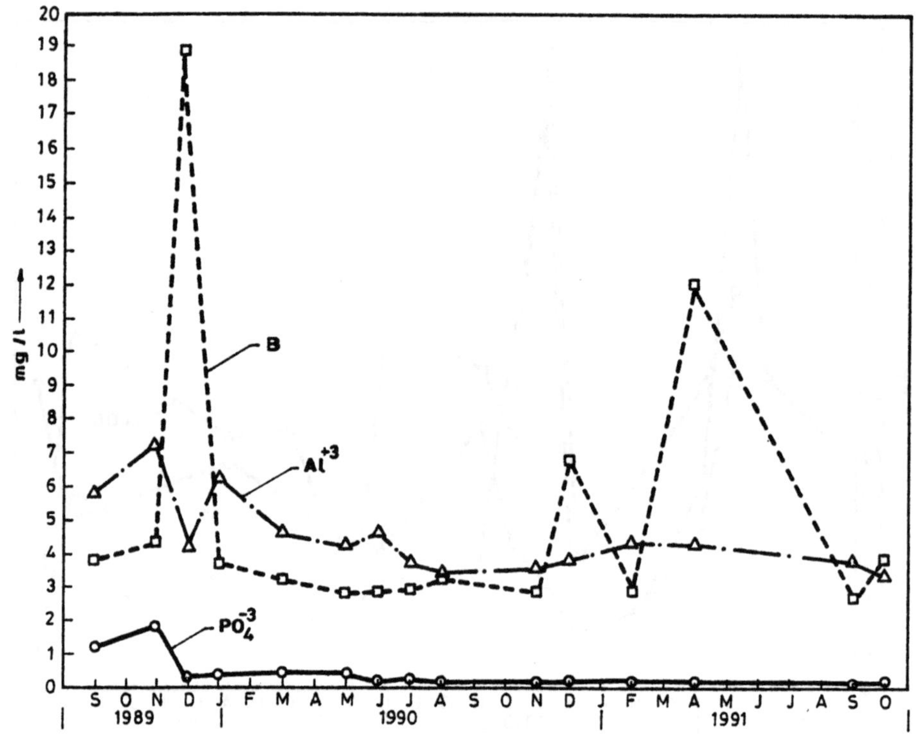

Figure 3. Variation of Simav River Quality With Respect to B, Al and PO4.

1) In general, the River water quality is highly variable for quality parameters like COD, Suspended Solids, BOD5 and Boron. Such variation of river water quality is a big disadvantage for downstream users.

The presence of high concentrations of pollutants in the effluents are symptomatic for the inefficiency of treatment in general. These problems have not yet been solved in Turkey and effluent discharge standards can rarely be met.

2) Boron is one of the important water quality parameters especially for irrigation since the element is phytotoxic. The area is a dense agricultural area, and whatever the water quality, it is used for irrigation. The crops produced in the area are; carrot, sugar beet, vegetables and melon. The irrigation of vegetables which are eaten raw also poses a microbiological problem.

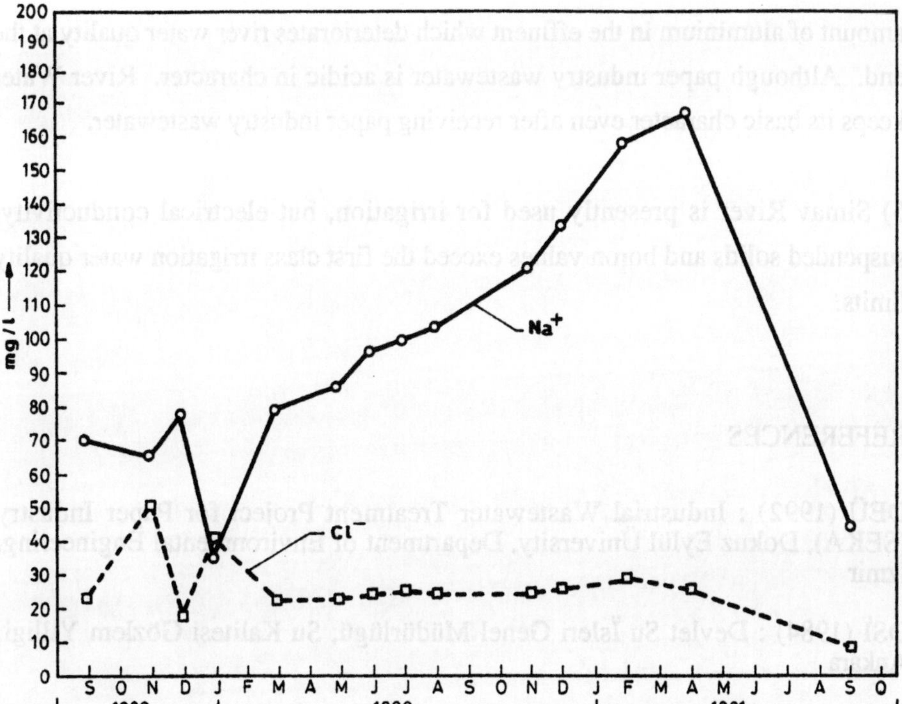

Figure 4. Variation of Simav River Quality With Respect to Na and Cl

At the source, the boron content of the river averages 0.11 mg/l. After industrial discharges this value increases to values as high as 19 mg/l, the average at sampling point 1 being 5.0 mg/l. Boron thus causes the water quality class to shift from 1 to 4.

3) Another important parameter is organic matter. The biochemical oxygen demand of Simav River before any pollution is varying between 1.2 and 2.0 mg/l (1.6 mg/l average), whereas it rises further to 50 mg/l. This value also shifts water quality from 1st class to 4th class according to Water Pollution Regulation.

4) Metals ions are another measure of River water quality. High concentrations of aluminium ion in the water originates from the industries which uses aluminium in the process. For example paper industry discharges considerable

amount of aluminium in the effluent which deteriorates river water quality at the end. Although paper industry wastewater is acidic in character. River Water keeps its basic character even after receiving paper industry wastewater.

5) Simav River is presently used for irrigation, but electrical conductivity, suspended solids and boron values exceed the first class irrigation water quality limits.

REFERENCES

DEÜ (1992) : Industrial Wastewater Treatment Project for Paper Industry (SEKA), Dokuz Eylül Üniversity, Department of Environmental Engineering, Izmir

DSİ (1984) : Devlet Su İsleri Genel Müdürlügü, Su Kalitesi Gözlem Yilligi, Ankara

Hem, J.D. (1970) : Study and Interpretation of the Chemical Characteristics of Natural Water, Geological Survey Water Supply Paper, United States Government Printing Office, Washington

Official Gazzette (1988) : Water Pollution Regulation, Official Gazzette 4 September 1988, Gazzette No. 19919, Ankara

ANNEX

Annex 1. Water Quality Classes According to Water Pollution Control Regulation.

Water Quality Parameters	Water Quality Classes I	II	III	IV
A) Physical and inorganic chemical parameters				
1. Temperature (°C)	25	25	30	>30
2. pH	6.8-6.5	6.5-8.5	6.0-9.0	other than 6.0-9.0
3. Dissolved oxygen (mgO2/l)	8	6	3	<3
4. Oxygen saturation (%)	90	70	40	<40
5. Chloride ion (mg Cl-/l)	25	200	400	>400
6. Sulfate ion (mg SO=4/l)	200	200	400	>400
7. Ammonium ion (mg NH4-N/l)	0.2	1	2o	>2
8. Nitrite-N (mg NO2-N/l)	0.002	0.01	0.05	>0.05
9. Nitrate-N (mg NO3-N/l)	5	10	20	>20
10. Total phosphorus (mg PO4-P/l)	0.02	0.16	0.65	>0.65
11. Total dissolved solids (mg/l)	500	1500	5000	>5000
12. Color (Pt-Co)	5	50	300	>300
13. Sodium (mg Na+/l)	125	125	250	>250
B) Organic parameters				
1. COD (mg/l)	25	50	70	>70
2. BOD (mg/l)	4	8	20	>20
3. Organic carbon (mg/l)	5	8	12	>12
4. Total Kjeldahl-N (mg/l)	0.5	1.5	5	>5
5. Emulsified oil and gres (mg/l)	0.02	0.3	0.5	>0.5
6. Methylene blue active substances(MBAS)(mg/l)	0.05	0.2	1	>1.5
7. Phenolic substances (vol) (mg/l)	0.002	0.01	0.1	>0.1
8. Mineral oil and derivatives (mg/l)	0.02	0.1	0.5	>0.5
9. Total pesticides (mg/l)	0.001	0.01	0.1	>0.1
C) Inorganic parameters				
1. Mercury (µg Hg/l)	0.1	0.5	2	>2
2. Cadmiyum (µg Cd/l)	3	5	10	>10
3. Lead (µg Pb/l)	10	20	50	>50
4. Arsenic (µg As/l)	20	50	100	>100
5. Copper (µg Cu/l)	20	50	200	>200
6. Chromium (total)(µg Cr/l)	20	50	200	>200
7. Chromium (µg/Cr)	n.d.	20	50	>50
8. Cobalt (µg Co/l)	10	20	200	>200
9. Nickel (µg Zn/l)	20	50	200	>200
10.Zinc (µg Zn/l)	200	500	2000	>2000
11. Cyanide (total)(µg CN/l)	10	50	100	>100
12. Fluoride (µg P-/l)	1000	1500	2000	>2000
13. Free chlorine (µg Cl2/l)	10	10	50	>50
14. Sulfide (µg S/l)	2	2	10	>10
15. Iron (µg Fe)	300	1000	5000	>5000
16. Manganese (µg Mn/l)	100	500	3000	>3000
17. Boron (µg B/l)	1000	1000	1000	>1000
18. Selenium (µg Se/l)	10	10	20	>20
19. Barium (µg Ba/l)	1000	2000	2000	>2000
20. Aluminium (mg Al/l)	0.3	0.3	1	>1
21. Radyoactivity (pCi/l)				
alpha-activity	1	10	10	>10
beta-activity	10	100	100	>100
D) Bacteriologic parameters				
1. Fecal coliform (EMS/100 ml)	10	200	2000	>2000
2. Total coliform (EMS/100 ml)	100	20000	100000	>100000

Annex 2. Irrigation Water Quality Requirements

	I.Class water (very well)	II.Class water (good)	III.Class water (can be used)	IV.Class water (should be used with precaution)	V.Class wat. (not appropriate)
EC25x106 (µmhos/cm)	0-250	250-750	750-2000	2000-3000	3000
Exchangeable sodium ratio (% Na)	20	20-40	40-60	60-80	80
Sodium absorption ratio (SAR)	10	10-18	18-26	26	
Sodium carbonate residue (RSC)meq/l	1.25	1.25-2.5	2.5		
mg/l	66	66-133	133		
Chloride(Cl-)meq/l	0-4	4-7	7-12	12-20	20
mg/l	0-142	142-249	249-426	426-710	710
Sulphate (SO4)meg/l	0-4	4-7	7-12	12-20	20
mg/l	0-192	192-336	336-575	576-960	960
Total dissolved solids (mg/l)	0-175	175-525	525-1400	1400-2100	2100
Boron (mg/l)	0-0.5	0.5-1.12	1:12-2.0	2.0	-
Irrigation water class	C1S1	C1S2.C2S2 C2S1	C1C3,C2C3,C3S3 C3S2.C3S1	C1S4,C2S4,C3S4 C3S4,C4S3,C4S2 C4S1	
NO3 or NH4	0-5	5-10	10-30	30-50	50
Fecal colif./100 ml	0-2	2-20	20-102	102-103	103
BOD5 (mg/l)	0-25	25-50	50-100	100-200	200
Susp. solids (mg/l)	20	30	45	60	100
pH	6.5-8.5	6.5-8.5	6.5-8.5	6-9	6 veya 9
Temperature	30	30	35	40	40

III

TECHNICAL ASPECTS OF REHABILITATION

TECHNICAL ASPECTS OF SOIL SANITATION

Alfons G. Buekens and Hilde De Lembre*
Department of Chemical Engineering and Industrial Chemistry, Free University of Brussels
B-1050 Brussels, Pleinlaan 2 Belgium
Research, Development and Consulting n.v. (RDC n.v.)*
B-9300 Aalst, Leo de Bethunelaan 80 Belgium

Several strategies can be used after identification of a contaminated site [1.2] :

- isolation of the site from the biosphere;
- in-situ immobilization of the contaminants;
- in-situ clean-up of soil and groundwater;
- excavation of the polluted soil followed either by its disposal in a well engineered site or by its treatment using appropriate methods.

This paper discusses the various treatment techniques available for contaminated soil, their operating principles, working methods, results obtained and limitations. Important parameters are the charcteristics of the site, the nature and amount of the pollutant(s) and the type of soil.

The soil can be classified as sandy, clayey, calcareous, bed rock, ... An important factor is the presence, position and flow of groundwater, the stratigraphy and heterogeneity and the presence of organic material (humus, peat, ...) since it strongly influences upon the partition and adsorption of pollutants.

The following **pollutants** are occuring frequently :

- light or heavy oil fractions, such as petrol, kerosene, gas-oil, fuel-oil, pitch. These pollutants are to be found in numerous places, such as oil refineries, petrol stations, the site of a spill, an accident, ...;
- specific hydrocarbons, such as benzene, toluene and xylenes (BTX), coal or wood tar fractions, or chemicals, such as creosote;
- halogenated solvents, such as trichlorethylene, perchlorethylene or 1,1,1,-trichloroethane (non-halogenated solvents, such as ethanol, acetone, or esters are readily biodegradable and hence less problematic);
- toxic and/or persistent chemicals, that may enter the food chain, e.g. PCB's, dioxins, DDT, lindane and many other pesticides;
- free and complex cyanides;
- heavy metals and their compounds.

The pollutants may be present under various forms :

- as interstitial gas, liquid or solid caught between soil particles;
- as a liquid, solid or gaseous phases in the pores of soil particles;
- adsorbed onto the surface of soil particles;
- absorbed, adsorbed or integrated in soil organics (humus, peat, ...) or even organisms;
- dissolved in the groundwater.

The basis for decontamination depends on the site, the local problem, risk analysis and technical possibilities. The following techniques may be used :

- physical or physico-chemical separation at the molecular scale (for homogeneous phases);
- mechanical or physico-chemical phase separation;
- thermal or chemical destruction;
- biodegradation or bioremediation;
- immobilization or solidification.

NATO ASI Series, Partnership Sub-Series, 2. Environment – Vol. 45
Environmentally Devastated Areas in River Basins in Eastern Europe
Edited by Alfons Georges Buekens and Vasily Victorovich Dragalov

Treatment may be applied in situ or, after excavation and transportation, ex situ. It may focus on the soil proper, but also on groundwater, which is normally the carrier of the pollution, or even on soil vapours. In principle all chemical unit operations can be used. The result obtained will rarely attain perfection. A major problem in soil treatment is hence how much imperfection can be tolerated : "how clean is clean ?".

1. Physico-chemical treatment methods

A frequently proposed method is soil washing, which proceeds in three steps :

- the soil is vigorously contacted with an extracting agent, normally water or an aqueous solution of suitable chemicals;
- the liquid and solid phases are separated;
- the liquid extracting agent is treated and recycled or disposed of.

The soil may be adequately cleaned or treated a second time.

When contaminated earth is contacted with water several contaminants may become dissolved, while others - those associated with fines - become more or less stably suspended in the washing liquor. Hence, the purpose of the operation is a phase transfer in which the contaminants almost entirely report to the liquid phase, leaving a clean soil and a contaminated liquor. The latter is separated from the soil by mechanical means (e.g. settling, hydrocyclones, filtration, ...). Afterwards the liquor is treated by appropriate means, such as :

- dissolved air flotation;
- solids classifiers;
- coalescers (for oil);

and the liquor is recirculated.

To increase the extraction rate and capacity various chemicals, such as acids, lyes or complexing agents may be used. The dispersion of fines is greatly aided by the presence of detergents and surfactants.

The method has mainly been applied to the washing of sandy soils, in which the contaminants are mainly associated with the clay or the silt fraction.

Since the latter is converted into a polluted sludge, requiring adequate disposal or treatment, the method is limited to soils with an organic and clay content of some 10 - 15 % [2]. Organic and inorganic pollutants can both be removed, but testing at a pilot scale is required, since the distribution of pollutants cannot be predicted on a basis of first principles.

In other cases the method has been applied in situ. The soil is treated in a vertical or in a horizontal direction by percolation of an aqueous phase, to which appropriate chemcials (detergents, acids, lyes, complexing agensts, ...) are added. The soil should be porous and relatively homogeneous, generally sand or gravel. The contaminants should not been tied or adhering too strongly. Examples are petrol or heavy metals, that can be leached out using appropriate chemicals.

In principle solvent extraction techniques can also be applied in soil treatment. The solvent facilitates the contaminant removal by an enhanced solubility and a lower viscosity. Organic solvents may be used with a sufficient volatility to allow for easy stripping. An alternative is supercritical extraction with carbon dioxide under elevated pressure.

2. Thermal treatment

Thermal treatment is based on the removal of contaminants by application of heat.

Raising the temperature enhances the volatility of the pollutants, which (when required) are desorbed and evaporate. The method is applied int. al. to solvents and light oil fractions (petrol).

At higher temperature (above 350 °C) thermal treatment leads to a partial or integral destruction of organic compounds; the latter may proceed under oxygen free (pyrolysis) oxygen lean (gasification) or fully oxidizing conditions and generally produces simpler compounds. Sometimes, however, tar is formed.

Since the organic vapours are liberated in the gas phase, possibly including degradation and oxidation products, the latter needs generally to be treated. Depending on the case this may proceed by :

- complete condensation of the vapours produced, sometimes with a recovery of lost or spilled solvents;
- a thermal or catalytic oxidation (the use of a catalyst normally is feasible only after complete removal of entrained dust);
- wet scrubbing of the vapours or fumes (in case the latter are well soluble);
- treatment in a biofilter, to remove the organics by absorption on biomass and biodegradation.

After thermal treatment the soil is sterile and its structure is often modified. It may become improper for use as a topsoil [3,4].

The temperature in some cases may be raised so high that melting and vitrification occur. Given their high cost and operating difficulties vitrification as well as plasma torch techniques are still at a development stage.

Wet air oxidation may be applied to a slurry of contaminated earth, which is heated in a heat exchanger up to such a temperature that the organic contaminants react with air under high pressure. The liberated heat is sufficient to raise the temperature further and destroy the contaminants.

Heat treatment also affects a variety of inorganic contaminants. Mercury can be volatilized, complex cyanides thermally destroyed. Thermogravimetric study, coupled with a suitable method of analysis of the evolving products, is eminently suitable as a method for study, evaluation and design.

The thermal treatment methods vary according to the method used for applying the heat.

Steam stripping allows to eliminate volatile organics, such as low boiling solvents, from the soil. The steam can be distributed by means of perforated tubes embedded in the material to be treated.

To eliminate high boiling compounds or destroy hazardous chemcials, such as complex cyanides or pesticides, much higher temperatures (400 - 600 °C) are required. In that case directly or indirectly heated rotary drums or fluidised bed units are used. Small units may be based on infra-red heating or even microwave heating. The products of degradation are generally postcombusted.

Some metals, such as mercury or thallium, may also be removed thermally. Their separation and recovery should be carefully studied to avoid air pollution. Mercury can be scrubbed using acidic liquors ($pH < 1$) or adsorbed onto activated carbon. Thermal treatment normally is not applied to halogenated solvents in order to avoid dioxin formation.

Thermal treatment in principle allows to eliminate all undesirable organics, either by evaporation, or by thermal destruction (pyrolysis or oxidation). The resulting vapours can sometimes be condensed (steam stripping) or destroyed by post-combustion. In cade of direct heating the gas stream to be treated is much more voluminous than in case of indirect heating; the latter, however, adds to plant complexity.

3. Microbial treatment

Microbial treatment is earmarked for biodegradable substances which can be converted by appropriate microorganisms into CO_2, H_2O, HCl, SO_4^{--}, ... and additional cell material. Important factors are the soil structure, the persistence of the contaminants and the possibility to create favourable surroundings for microbial growth (presence or addition of nutrients, of adapted bacterial strands).

The following conditions need to be optimised : temperature, moisture content, pH, redox potential, oxygen concentration, type of electron acceptor, presence of desired microorganisms and bioavailability [2], e.g. in order to transfer oil to the aqueous phase.

The method can be applied to hydrocarbons, including petrol, kerosene, fuel oil and the (rather refractory) polycyclic aromatics, and to certain metals, that can be assimilated by bacteria or converted into a soluble form (bacterial leaching). The latter method has been applied industrially to recover heavy metals from mining tailings. The bacteria oxidise pyrites and other sulphide minerals, generating dilute sulphuric acid that dissolves the desired metals.

Generally the treatment is aerobic, but for a complete conversion of chlorinated pollutants a combination of aerobic and anaerobic treatment is required [2].

The major limitations of the method are the long treatment periods required (generally years) and the resistance of certain contaminants or their metabolites. Biocides are obviously difficult to degrade. A high concentration of certain contaminants may lead to surroundings unacceptable for bacterial development.

Landfarming in an ex situ method, in which a layer of about 0.5 to 1 m of contaminated soil is spread onto an impervious bottom (e.g. a plastic liner) on which a drainage system is provided. The layer is ploughed or agitated periodically; nutrients and trace elements may be added and where desirable air can be distributed by means of a system of perforated pipes. Greenhouse techniques may be used to control or elevate the temperature. The leachate is generally recirculated. The treatment takes normally 1 to 3 years.

Bioreactors are used for treating slurried soil; the technology used can be compared to activated sludge treatment. The operating conditions (temperature, oxygen, nutrients, residence time, bacterial strains) can be controlled and higher rates achieved. On the other hand the plant is more expensive and its operation complicated than with landfarming.

Composting systems can be considered as a relatively dry form of bioreactor, in which the material is moved in the presence of air. They are designed for an intense aerobic treatment of the waste.

In situ **biorestauration** can be applied to permeable sandy soils, contaminated with mineral oil, low boiling and aromatic hydrocarbons and even certain chlorinated solvents or pollutants. Microbiological activity is stimulated by pumping water, possibly activated with nutrients, oxidants, or specific bacterial strains through the contaminated groundlayer.

Microbial treatment is a slow and normally incomplete process, operating on relatively easily degradable substances such as petrol and aromatic hydrocarbons. The bioavailability is often problematic due to negligible or limited miscibility with water, adsorption and mass transfer limitations.

To stimulate bioremediation it is possible to develop specific bacterial strains, to separate the soil in easily biodegradable and more refractory fractions, to stimulate mass transfer by appropriate physical techniques or to employ chemicals (e.g. oxidants) to convert poorly biodegradable pollutants into more easily convertible ones [2].

4. Electroreclamation

Electroreclamation is based on the application of electric current from electrodes through the contaminated soil, either in situ or ex situ. Ions as well as charged colloidal particles migrate in the electric field and are thus removed from the soil.

5. Air stripping

In this method volatile organics are purged form the soil by means of a carrier air stream. The latter should be collected and cleaned.

The air is aspired towards collection wells; it can be supplied (a) over the entire surface of the contaminated area to treat the unsaturated zone, (b) by dedicated air supply wells providairy - horizontal air flow towards the collection wells or (c) by supplying compressed air to the saturated zone.

The method can be applied to remove volatile solvents, int. al. petrol, aromatics, kerosene, trichloroethylene, perchlorethylene. The treatment is often continued for several years. Limitations are (a) too limited a volatility, (b) a strong binding to the soil or (c) a heterogenous soil structure impeding air flow through certain zones.

6. Problems remaining unsolved [2]

Even excavated soil treatment remains difficult for clay, loam and peaty soils, especially when polluted by both organic contaminants and heavy metals. In-situ treatment is generally more difficult and practically impossible for clayey or loamy soils, very heterogeneous soils and for non-soluble particulate pollutants. Experimentation is possible int. al. in temporary disposal sites.

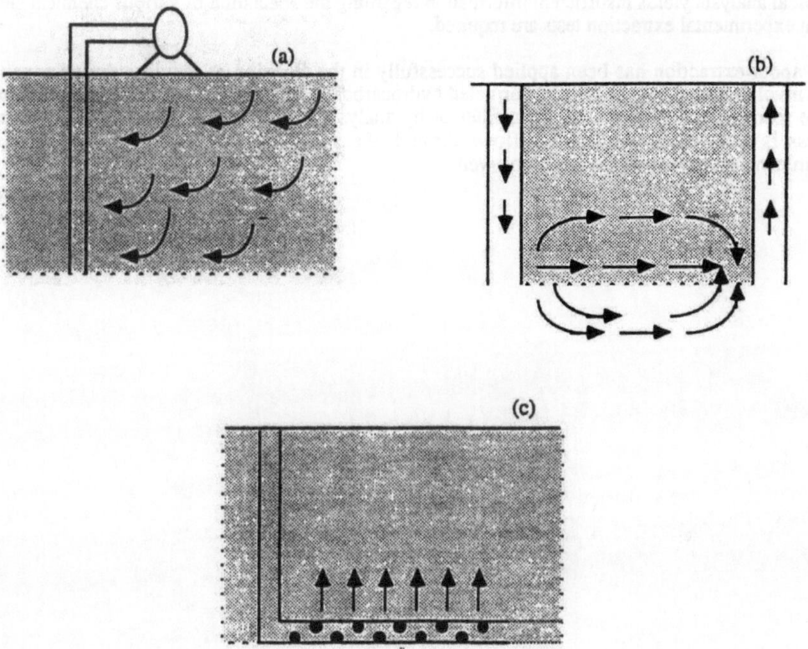

Figure 1 - Soil vapor extraction system : main devices and elements of the monintoring system

Rulkens et al. [2] suggest the following techniques :

- extraction of inorganic or organic pollutants using relatively mild extracting conditions;
- chemical degradation under relatively mild conditions;
- soil venting (vapour extraction);
- biological degradation by techniques similar to biorestoration, soil slurry reactor techniques or landfarming techniques;
- bioleaching techniques based on the use of *Thiobacilli* for removal of heavy metals.

In several heavily polluted regions clean-up techniques are required with following characteristics.

- relatively simple;
- treatment time of several years;
- extensive process with no negative environmental impacts;
- relatively cheap;
- not always necessary to achieve the standards for completely clean soil.

It is proposed [2] to use (a) green plants that accumulate pollutants, int. al. those which can be removed as plant sodes, (b) chemicals promoting biodegradation.

7. In situ techniques

Excavation and moving of earth is expensive. Hence increased attention goes to techniques that can be applied in situ. The main techniques are :

- soil washing;
- soil vapour extraction, and
- bioretoration;

but electroreclamation and even vitrification have been tested as in situ methods.

Chemical analysis yields insufficient information regarding the speciation of various chemical elements, so that experimental extraction tests are required.

Soil vapour extraction has been applied successfully in the cleaning of the unsaturated zone, e.g. in cases involving contamination by chlorinated hydrocarbons. The progress of remediation is followed by the recovery of contaminants by a filter or by analysis of soil vapour. A continuous extraction process is established as clean air flows through the subsoil and becomes loaded with volatile contaminants on its way to the extraction well

SOLUTIONS TO THE PROBLEMATICS OF DREDGED MATERIALS IN THE PORT OF ANTWERP

MAJOR WORKSHOP PRESENTATIONS AND DISCUSSIONS

S. D'haene
Soils n.v.
Scheldedijk 30
B-2070 Zwijndrecht
BELGIUM

DREDGING ACTIVITIES IN THE REGION OF ANTWERP

Antwerp is situated along the River Scheldt about 55 km away from the north.

The access channel to the port of Antwerp and Zeebrugge) in the North Sea is called "Scheur" and further upstream the "Western Scheldt". The Western Scheldt is situated on the territory of the Netherlands. The access channel from the border to the port of Antwerp is called the "Sea Scheldt" .

To guarantee the access to the different seaports about 40.000.000 m^3 has to dredged out of the maritime access channels (= maintenance dredging). Most of this volume is not contaminated and is dumped on defined dumping areas. Each year about 4.000.000 m^3 is removed permanently out of the system. The sandy dredged material is used for the construction of dikes, the creation of industrial areas, beach replenishment, etc ... The storage and/or treatment of the silty dredged material is of great concern.

The dredged materials from the **Scheur** are silty but not contaminated. They are dumped on dumping grounds in the North Sea.

NATO ASI Series, Partnership Sub-Series, 2. Environment – Vol. 45
Environmentally Devastated Areas in River Basins in Eastern Europe
Edited by Alfons Georges Buekens and Vasily Victorovich Dragalov

The dredged materials from the **Western Scheldt** are sandy and uncontaminated. They are dumped in shallow areas to narrow the channel.

The dredged materials from the **Sea Scheldt** are partly sandy and partly silty. In this part of the River Scheldt situates the contact point between salt and fresh water resulting in a flocculation process and thus a high turbidity. The silt forms a deposit at certain rises in the river or at places with low water velocities as access channels to the locks. Every year 3.000.000 m^3 is dredged in the Sea Scheldt. The material is partly dumped in the estuary, and partly removed tot prevent the silting up of the access channels. The removal of 300.000 tons D.M. (Dry Matter) dredged material is also a demand in the dumping permits of the dutch Government. As the silty dredged material is contaminated, the removal and sound disposal of this material is a permanent concern of the flemish Government.

In the docks of the port yearly 300.000 m^3 of contaminated silt is dredged. The silt is mainly stored on land.

Besides the maintenance dredging some capital dredging is executed. Since those sediments are clean they are used as construction materials.

STORAGE AND TREATMENT FACILITIES FOR DREDGED MATERIAL IN THE ANTWERP REGION

Contaminated dredged material can be harmful for the environment. In order to prevent these adverse effects it is necessary to take preventive actions towards the contaminated sediments. Basically two different actions are possible :

dredging of the contaminated material and disposal into a protected storage area (disposal option) dredging of the contaminated material and transformation of this material in a clean harmless product in an onshore installation (treatment option).

It should be stressed that in the long term there is but one solution to contaminated dredged sediments which is water purification. Through water sanitation plans the source of contamination should disappear in the future. This will create larger possibilities for the beneficial use of dredged material. In a short term contaminated dredged materials should be excavated since the contaminating elements will re-enter into suspension when the water quality of the overlying water layers will improve.

The handling of dredged material should be executed within a frame work of boundary conditions imposed by the geographical, technical hydraulic, geological, economical and social environments. The final choice of a disposal or treatment technique in a certain situation is frequently based on the BATNEEC-principle.

DISPOSAL OPTION

On land disposal

On land disposal is still a common practise although recently more restricted due to problems concerning environmental planning and permitting. A disposal site for contaminated sediments should be completely isolated from the environment. In order to prevent adverse effects attention should be paid to the location, construction, isolation and control of the dump site.

In the past some monodumpsites have been constructed using plastic membranes (polymers) as isolating material. As the life span of those artificial liners is not know research is done on active (natural) liners. The function of those liners is not only based on the waterimpermeability but mainly on their ability to adsorb leaching contaminants. In 1990 a research project "Geuzenhoek" started, aimed at evaluating the efficiency of glaucotlite, bentonite, dewatered sludge, peat and aluminosilicates as mineral liners. The results of this project will be presented during the workshop. For future on land disposals the dredged material will first be dewatered prior to disposal. In this way the disposal site will be used more efficiently and final capping and reintegration in the landscape will be accelerated.

It should be noted that on land storage has been chosen as a midterm solution in the Netherlands. With the "Sluftes" (90.000.000 m^3) "Papegaaibek" (1.500.000 m^3) and 2 other planned dumpsites (± 60.000.000 m^3) a solution is found till the year 2000. At that time treatment techniques should be available to handle at least the most contaminated dredged material.

Plans for the construction of an island depot offshore for the storage of dredged material have been cancelled due to public resistance.

Under water storage (capping)

Capping of contaminated dredged materials has been done in the docks of the left bank. During the past 2 years 1.100.000 ton D.M. has been stored under water. The technique consists of the creation of overdepths in the dock bottom, the refilling with contaminated sediments and the final capping of the area. The advantage of this technique is that the dredged material stays under the same environmental conditions which is favourable for the immobilisation of the contaminants. The final capping prevents the sludge from resuspension or contaminants from resolubilisation.

The Antwerp experience with capping will be communicated during the workshop. Under water storage is a widely used technique in the United States and has been used in Japan, the Netherlands and Denmark.

TREATMENT OPTION

The basic idea behind the treatment option is to recycle or to reuse the dredged material.

The characterisation of the material is a very important prerequisite before starting with the selection of the most appropriate technique for any treatment process. A detailed and accurate characterisation is also a basic factor for the success of any sediment cleaning project.

The characterisation should include :

- the geotechnical characteristics
- the mineralogical composition
- the contaminant characteristics
- the toxicity characteristics

The treatment techniques can be grouped in a few main classes :

- separation techniques
- dewatering techniques
- physico-chemical treatment techniques
- biological treatment techniques
- thermal treatment techniques

The treatment of dredged materials is mostly a combination of the above mentioned techniques. Such an approach consists of three steps :

- pre treatment phase : to subdivide the material into different parts with will defined geotechnical or chemical characteristics (separation techniques)
- treatment phase : to clean or to render harmless the toxical fraction (chemical, biological or thermal treatment techniques)
- post treatment phase : to make the treated fraction ready for recycling or disposal (dewatering techniques)

Up till now most processes do not include the second and main item because the appropriate techniques are not yet available or because they require excessive costs.

Separation techniques

The aim of this technique is the separation of the solid material into different fractions based on grain size. Since pollutants are preferably adsorbed onto the finer fraction separation results in a cleaner coarse fraction on a contaminated fine fraction.

Hydraulic separation fields:

These fields are based on the selective settlement process of material with a different grain size from a continuous water stream. The experience with separation fields in Antwerp will be communicated during the workshop. The separation field is cheap but relatively rough classification technique for dredged materials with a low contaminant level.

Hydrocyclonen:

Hydrocycloning is a separation technique, cut-off diameters between 10 and 25 µm. Some results with hydrocycloning will be presented during the workshop.

Dewatering techniques

The objective of the dewatering techniques is to enhance the natural dewatering process of fine grained material in order to produce a material with an acceptable cohesion which can be used for other purposes such as landscaping projects or safe disposal.

Dewatering fields:

The application of dewatering fields is a low cost method in which the natural dewatering process is enhanced by the optimisation of all boundary conditions and by the application of limited mechanical means in these fields. The use and effectiveness of dewatering fields will be demonstrated during the workshop.

Filter presses:

Belt filter presses (mostly used) are based on mechanical squeezing of the sludge between two belts. Dewatering is possible up to a dry matter content of 40 to 64 %. The technology has been applied on a industrial scale in many countries. Filter presses are mostly used in conjunction with thickeners as precond itioners. Some operational results will be presented.

Physico-chemical treatment techniques

The most important subgroups of this technology are :

- extractive methods
- immobilisation techniques

Both techniques can be considered as the two extremes of basically the same parameter which is the solubility of the contaminants. For the extraction technique it is the objective to increase the solubility of the contaminants to maximise the extracted quantity while in the immobilisation processes it is the objective to reduce the solubility as much as possible to minimise the potential leachability of the final material product. Uptil now immobilisation has not been used in Belgium for the treatment of dredged material. Research has been done on the extractive treatment of dredged material. A combined extraction and biological treatment technique has been developed. The process scheme will be presented together with some preliminary results.

Biological treatment techniques

Bioremediation attempts to enhance the natural destruction of organic pollutants by micro-organisms. Some results of a pilot project on the treatment of dredged material from the port of Zeebrugge will be presented during the workshop.

Thermal treatment techniques

A dewatering step is included in all operational thermal techniques.

The thermal techniques can be subdivided in three main groups :

- vaporisation : temperature range : 200- 800 °C
- incineration : temperature range : 800 - 1000 °C
- ceramic process : temperature : above 1000 °C

In each of these processes the post treatment of the exhaust gases is very important. Some testing has been done on the sintering of dredged material. The sintering process will destroy all organic pollutants and will bind the heavy metals in the matrix structure of the material. This will make the leaching of the heavy metals impossible. A second objective is the recycling of the dredged material into a product that can be used beneficially in the construction industry lilce artificial gravel.

FUTURE PERSPECTIVES FOR DREDGED MATERIAL, STORAGE AND TREATMENT

Since the dredging step is the cheapest activity compared to the treatment step, strict conditions concerning tolerances and selectivity should be imposed. Any saving in the volume to be treated will be much more important compared to the cost of accurate measurement or even the cost increase for dredging within very small tolerances. For this reason continuous developments will done on "ecological dredgers". The recently developed scoop-dredger will be presented.

The future dredged material storage and treatment can be summarised in the following scheme (addendum 1).

PRODUCT IDENTIFICATION CONTROL

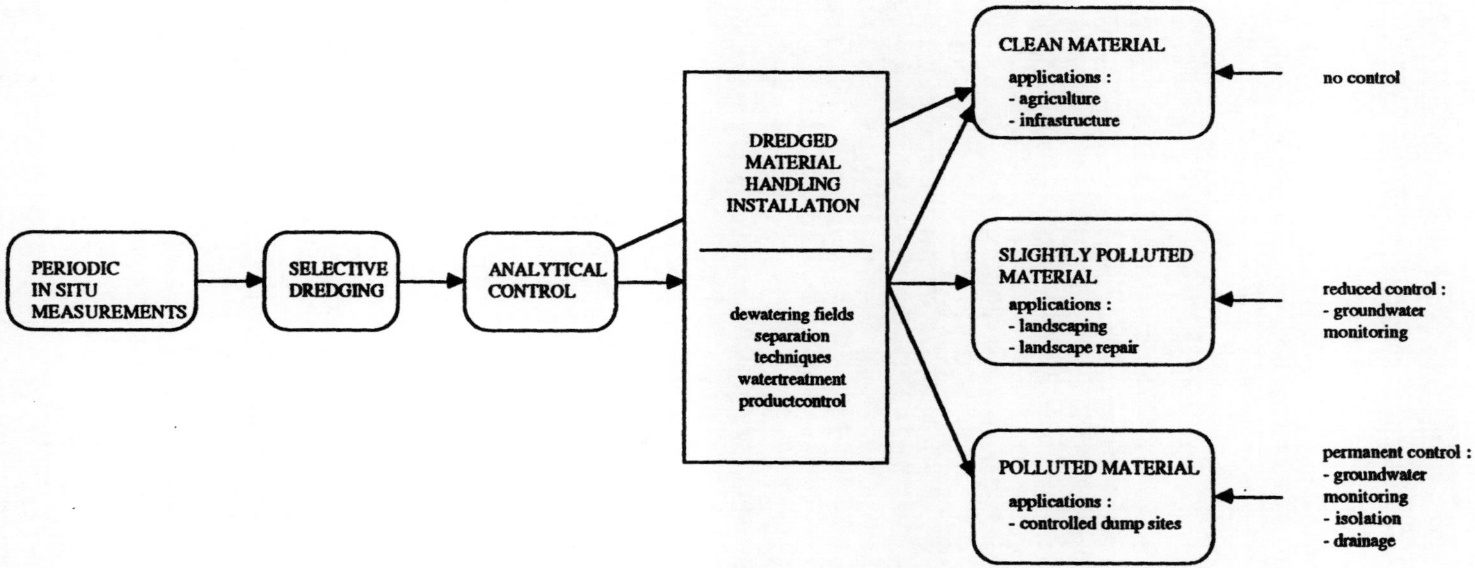

TREATMENT OF CONTAMINATED DREDGED SEDIMENTS - THE DUTCH APPROACH AND EXPERIENCE

Prof.dr.ir. W.H. Rulkens
Department of Environmental Technology
Wageningen Agricultural University
P.O. Box 8129
6700 EV Wageningen
The Netherlands

1. INTRODUCTION

Contaminated sediments cause serious environmental problems in The Netherlands. Annually, approximately 10 million cubic metres of polluted sediments are removed by maintenance dredging of harbours and waterways. In addition, remediation is required for more than 100 million cubic metres of polluted sediments, disposed mainly in the period 1960-1980. These sediments mostly contain various mixtures of heavy metals, polycyclic aromatic hydrocarbons (PAH), and organochlorine compounds. Remedial action to protect human beings and ecosystems from the negative influences of polluted sediments is necessary. Such a remedial action can consist of a controlled disposal of the sediment, an immobilisation of the pollutants in the sediment or a removal of the pollutants from the sediment.

In 1989 the Dutch government established the "Development Programme Treatment Processes for Polluted Aquatic Sediments in The Netherlands". The first phase of the programme was aimed at research into the development and demonstration of techniques and methods which would in particular offer practical possibilities for sediment clean-up. The research was carried out jointly by industry, research institutes and authorities.

NATO ASI Series, Partnership Sub-Series, 2. Environment – Vol. 45
Environmentally Devastated Areas in River Basins in Eastern Europe
Edited by Alfons Georges Buekens and Vasily Victorovich Dragalov

The aim of this paper is to present an overview of the state of the art of clean-up techniques for contaminated dredged sediments. Firstly the principles of clean-up techniques are discussed. Thereupon, the development stage and the practical experience obtained with these techniques are described. Finally, some general conclusions are presented. Most attention is paid to the results and experience obtained within the framework of the above mentioned programme on sediment treatment. However, results obtained within the "Dutch Integrated Soil Research Programme" are also used in the discussion of treatment techniques for sediments.

2. INTEGRAL REMEDIATION SYSTEM FOR DREDGED SEDIMENTS

An integral remediation process for dredged sediments consists of a number of logically arranged separate basic treatment steps. These treatment steps are the building stones of a complete processing plant. Each basic treatment step has its own characteristic possibility to remove, concentrate or convert one or more of the pollutants in sediments. The selection and the optimal arrangement of these basic treatment steps depends on the factors outlined in Chapter 3. Figure 1 gives an overview of the several basic treatment steps which can be applied. Four groups of processes can be distinguished:

* Classification processes, aimed at separation of the sediments into a heavy fraction (coarse fraction), mainly containing large particles, and a light fraction containing small mineral particles (silt) and organic matter. The pollutants are mostly bounded to the light fraction.
* Treatment processes for the light fraction. These include: a. mechanical dewatering and concentration into a small volume, followed by controlled disposal; b. thermal immobilisation by sintering or melting of sediment particles, possible in combination with other wastes; c. chemical clean-up (extraction, aimed at removal of contaminants such as heavy metals); d. biological clean-up aimed at biodegradation or biological mobilisation of pollutants.

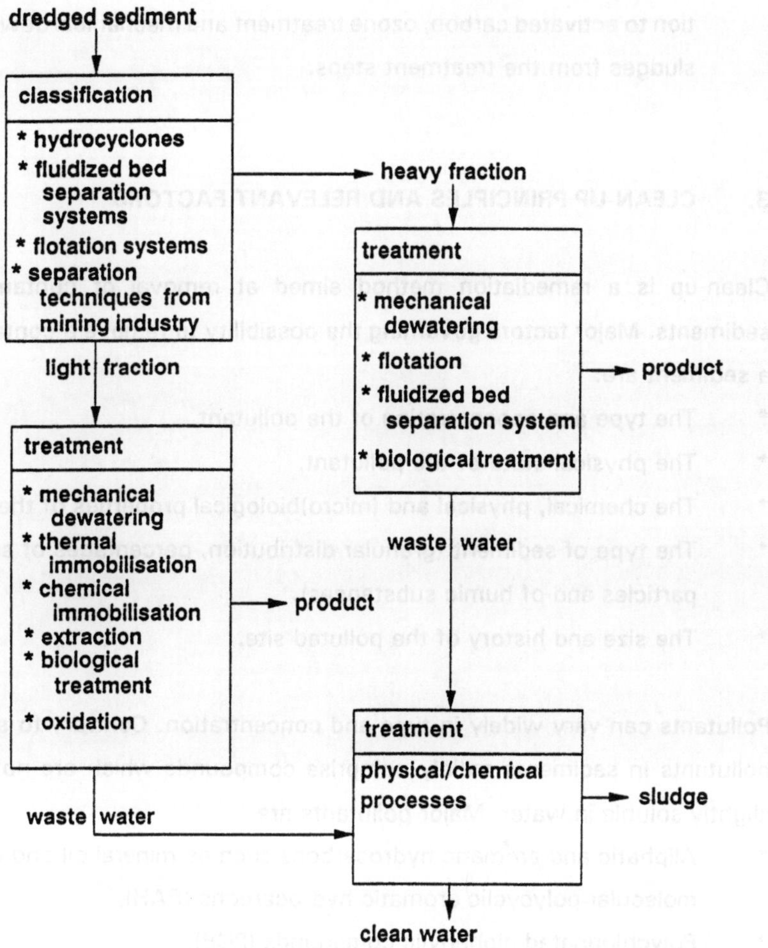

Figure 1 Integral remediation system for dredged sediments. Simplified process scheme.

* Treatment processes for the heavy fraction (sand fraction). These include: a. dewatering, followed by reuse or disposal; b. physical clean-up using a classification process (flotation); c. biological clean-up.
* Treatment of water flows. A treatment process of a dredged sediment always results in several waste water flows which in general are more or less polluted. To treat these waste water flows physical/chemical methods are used, such as coagulation/flocculation, sedimentation, flotation, adsorp-

tion to activated carbon, ozone treatment and mechanical dewatering of the sludges from the treatment steps.

3. CLEAN-UP PRINCIPLES AND RELEVANT FACTORS

Clean-up is a remediation method aimed at removal of contaminants from sediments. Major factors governing the possibility to remove a contaminant from a sediment are:

* The type and concentration of the pollutant,
* The physical state of the pollutant,
* The chemical, physical and (micro)biological properties of the pollutant,
* The type of sediment (granular distribution, percentages of sand and clay particles and of humic substances),
* The size and history of the polluted site.

Pollutants can vary widely in type and concentration. Contrary to soil pollution, pollutants in sediments mainly comprise compounds which are not or are only slightly soluble in water. Major pollutants are:

* Aliphatic and aromatic hydrocarbons such as mineral oil and low and high molecular polycyclic aromatic hydrocarbons (PAH),
* Polychlorinated biphenylic compounds (PCB),
* Heavy metals and heavy metal compounds,
* Pesticides.

Besides the type and concentration of the pollutants, their physical state is a very important factor. In Figure 2 the physical states which may occur are presented. These physical states are:

* Particulate pollutants of a size equal to, smaller or larger than the non-polluted sediment particles,
* Pollutants present as a liquid film around sediment particles,
* Pollutants adsorbed at the surface of sediment particles,

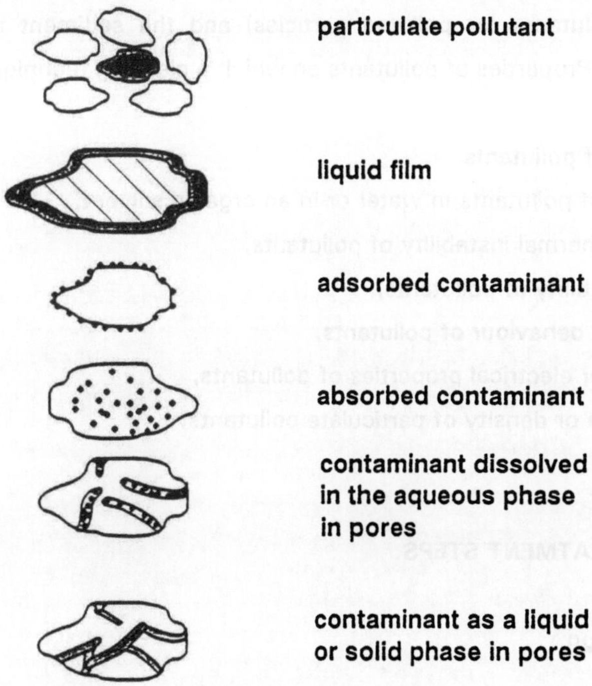

Figure 2 Physical state of pollutants

* Pollutants absorbed in (organic) sediment particles,
* Pollutants present as a solid or liquid phase in the pores of sediment particles,
* Pollutants dissolved in the aqueous phase in the pores of sediment particles.

As already mentioned, clean-up of contaminated sediments is aimed at the removal of pollutants from sediments. Basically, there are four different principles for removal: molecular separation, phase separation, chemical conversion and biodegradation. A clean-up technique is not always based on one single removal principle. Very often more than one principle is involved in a treatment process.

Each type of clean-up method makes use of the specific differences in properties between the pollutants (or polluted particles) and the sediment particles or sediment matrix. Properties of pollutants on which a cleaning technique could be based are:

* Volatility of pollutants,
* Solubility of pollutants in water or in an organic solvent,
* Chemical/thermal instability of pollutants,
* Biodegradability of pollutants,
* Adsorption behaviour of pollutants,
* Magnetic or electrical properties of pollutants,
* Size, shape or density of particulate pollutants.

4. BASIC TREATMENT STEPS

4.1 Classification

In most integral treatment systems, classification of the sediment plays a major role. Classification is a phase separation process aimed at the removal of a certain particle fraction from the sediment which is mostly present as an aqueous slurry. A classification process results in separation of the sediment into a clean or only slightly polluted fraction and a more or less strongly polluted fraction. Also separation into more than two fractions is possible. Very often the polluted fraction obtained after a classification process comprises the small particles or organic substances in the sediment. The main reason for this is that pollutants are preferentially adsorbed to either the small clay particles (large specific surface area and strong binding forces) or to the organic substances (strong binding forces). Therefore, the polluted fraction obtained after a classification process needs subsequent treatment. Such a treatment may be aimed at:

* Natural or mechanical dewatering for the purpose of final disposal at a controlled disposal site,

* Dewatering as a pretreatment step preceding a thermal process (in the case of organic pollutants) or an immobilisation process (in the case of heavy metals),
* Removal of the pollutants. Several methods are available:
  - Biological methods (biodegradation in the case of organic pollutants, bioleaching in the case of heavy metals),
  - Extraction of pollutants using water-based extracting agents or organic solvents (removal of organic pollutants or heavy metals),
  - Chemical destruction of pollutants using hydrogen peroxide, UV-light, ozone, or combinations of these treatment steps,
  - A second classification step different from the first one (for example the application of a flotation process after a hydrocyclone separation process).

Also these basic treatment steps mostly consist of several separate treatment steps. It will be clear that classification is only useful if the contaminants are mainly present in a specific fraction of the sediment particles. For example, when the pollutants are distributed more or less homogeneously over all sediment particles, classification is useless, unless it serves to prepare separate fractions for more specialised processing methods.
The major classification systems suitable to apply in clean-up processes of sediments are hydrocyclone systems, fluidized bed systems and flotation systems.

Hydrocyclones
A diagram of a hydrocyclone is presented in Figure 3. The slurry of sediment particles is fed under pressure to the tangential inlet of the hydrocyclone. Due to the centrifugal forces in the hydrocyclone the slurry is separated into an underflow containing the larger (heavy) particles, which leaves the hydrocyclone at the bottom, and an overflow, containing the smaller (light) particles, which leaves the hydrocyclone at the top.

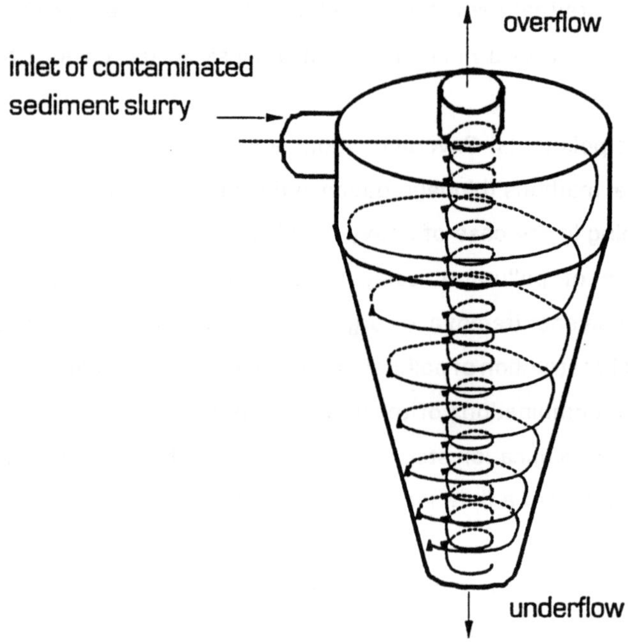

Figure 3 Hydrocyclone

The operation of a hydrocyclone can be characterised by three factors:

* The particles cut diameter d_{50}. This is the diameter of those particles of which 50% is present in the underflow and 50% in the overflow,
* The sharpness of the separation process governed by the percentage of the particles with diameter $< d_{50}$ leaving the hydrocyclone with the overflow, and the percentage of the particles with diameter $> d_{50}$ leaving the hydrocyclone with the underflow,
* The capacity of the hydrocyclone.

Hydrocyclones are rather simple separation systems. The particles cut diameter can easily be changed by changing the pressure drop over the hydrocyclone, the dimensions of the hydrocyclone, the dimensions of the inlet and outlet channels, and the shape of the hydrocyclone.

Hydrocyclones can also operate in series. In general, it can be stated that hydrocyclones make selective particles separation possible in diameter range above 10 to 20 μm.

Fluidized bed separation systems

A diagram of a fluidized bed separation system is presented in Figure 4.

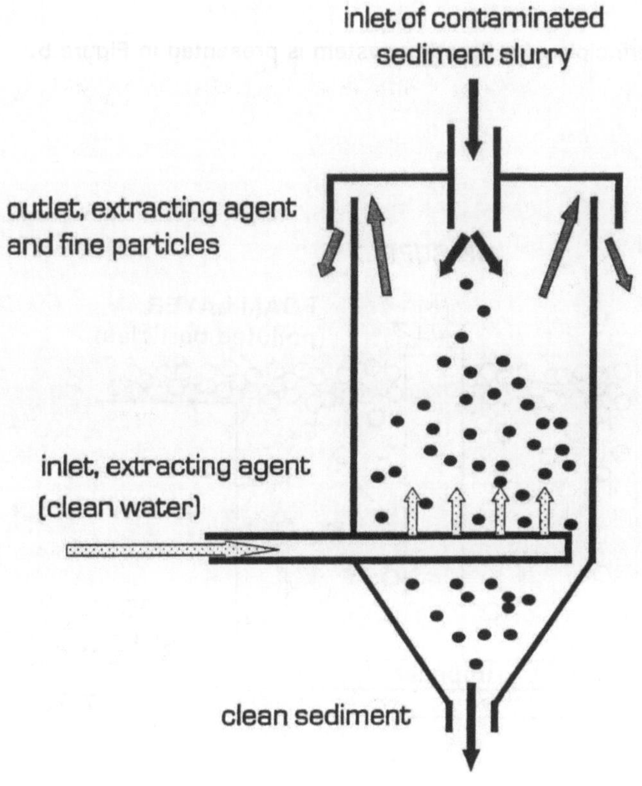

Figure 4 Fluidized bed separation system

This separation system is based on differences in gravimetric settling velocity between polluted particles and clean particles. A fluidized bed separator consists of a vertical cylindrical column with a conical bottom. The slurry of clean and polluted particles is fed at the top of the column in a direction opposite to a clean water flow. Adjustment of the superficial velocity of the liquid in the column makes it possible to remove the required coarse fraction of the sediment at the bottom of the column and the fine fraction of the sediment at the top. The fluidized bed separation process is also very simple. However, owing to the relatively small gravity forces on the small sediment particles, selective separation is only possible for particles with a diameter larger than 50 to 75 μm.

Flotation systems

A diagram of the principle of a flotation system is presented in Figure 5.

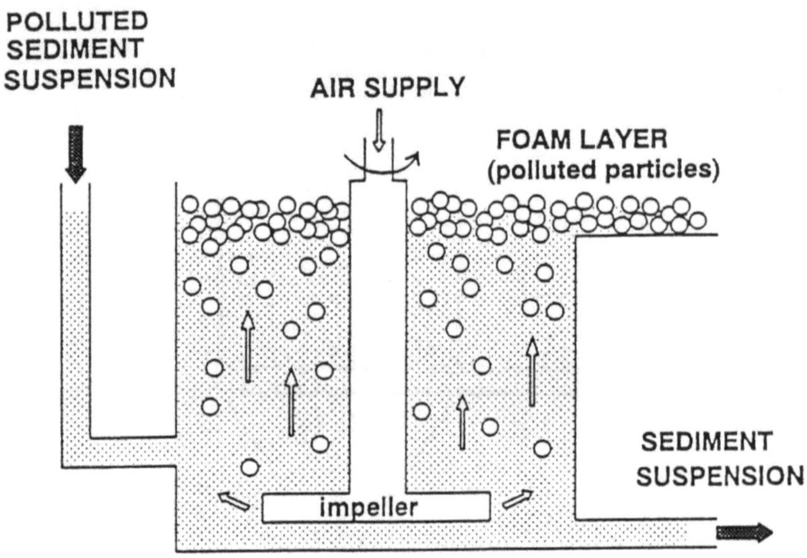

Figure 5 Flotation

The aqueous slurry to be separated is aerated with small air bubbles (or bubbles consisting of another gas, e.g. hydrogen and oxygen generated by electrolysis). The air bubbles attach to the sediment particles. Compared to an original sediment particle, a sediment particle combined with an air bubble has a much larger size and a much lower density. The consequence is that these sediment-air conglomerates float to the surface. At the surface, these particles form a foam layer which can be removed mechanically.

In order to achieve a selective removal of the polluted particles, it is important that the introduced air bubbles become only attached to the polluted particles. Attachment of air bubbles to a surface is only possible if this surface has hydrophobic properties. In many types of polluted sediments, the pollutants are attached to the surface of sediment particles or consist of particulate pollutants. Air bubbles will become selectively attached to these particles if the pollutants have a hydrophobic character. If this is not the case, there are possibilities to give these particles a hydrophobic character, namely by adding so-called collectors. Collectors are surfactants having at least one polar head and one hydrophobic tail in their molecular structure. The hydrophobic tail of the molecular structure is responsible for the hydrophobic character and makes attachment of air bubbles to sediment particles possible. Apart from collectors, other chemicals are also necessary such as modifiers or depressants to improve the collector attachment and the quality of the foam layer.

Compared to hydrocyclones and fluidized bed separation techniques, the application of flotation techniques requires more experience, more chemical knowledge, and empirical testing.

Using selective separation of sediment particles by flotation is possible for particles with diameters larger than 10 to 20 μm. However, in the range of small particle sizes sufficient selectivity can only be achieved if the non-polluted particles are substantially larger or smaller than the polluted particles.

Other classification techniques

Apart from the classification techniques mentioned above, there are some classification techniques which are not yet applied to sediments, but which may be of interest in the future. These are mainly separation techniques applied in the mining industry. While these techniques are not elaborated upon here, the following list is given: Humphrey-spiral separation techniques, jig techniques, shaking tables techniques, tilting frame separation techniques, magnetic separation techniques, including high gradient magnetic separation techniques.

Based on the experience obtained with these methods in the mining industry (ore benefication), it can be expected that in general a selective separation of a polluted fraction is possible for sediments containing relatively large particles (diameter larger dan 50 to 75 μm) or consisting of a relatively clean fraction of large particles and a polluted fraction of very small particles.

Experience with classification techniques

Large scale experience in classification of various types of sediments has mainly been obtained with hydrocyclones. The following results can be mentioned.

* Hydrocyclone separation plays a central role in most clean-up processes. In a first step it is aimed at a separation of the sediment into a clean or only slightly polluted fraction and a more or less strongly polluted fraction. Pollutants can be present both in the coarse and in the fine fraction.
* Hydrocyclone separation of a strongly polluted sediment only seldom results in a clean sediment fraction. In most cases, a separation into a less polluted and a more strongly polluted fraction is achieved. Subsequent cleaning of the less polluted fraction using a physical, chemical or biological treatment step is necessary.
* Hydrocyclone separation is only useful if a sufficiently large fraction of coarse material, for example sand or gravel particles, is present.
* Large-scale application of a treatment installation based on hydrocyclone separation requires an intensive pre-investigation. A simple sieving analysis is not enough. To save time and money, such a pre-investigation should preferably be executed with a specially developed flexible installation on a laboratory scale and not with the full-scale installation. Based on the experiences with the application of hydrocyclones the so-called fingerprint characterisation method has been developed.
* This characterisation method includes the following treatment steps:
    - Application of a series of subsequent hydrocyclone separations using hydrocyclones with decreasing size and, therefore, with decreasing particles cut diameter. The first step consists of applying the hydrocyclone to the original slurry. The successive hydrocyclone separations are always applied to the fine particles containing the overflow of the previous hydrocyclone separations.

- Application of a polishing step aimed at investigating the possibilities to polish the coarse fractions. Three types of techniques on a laboratory scale can be used: laboratory gravity separator (a small type of fluidized bed separator), laboratory multi-gravity separator (based on centrifugal forces), laboratory flotation unit.
- Sieving techniques.

4.2 Extraction

Extraction is a clean-up step based on molecular separation of pollutants from sediment particles and concentration of the removed pollutants into a small volume. Basically, the extraction process consists of three major treatment steps:

* Intensive mixing of contaminated sediment and extracting agent,
* Separation of extracting agent and clean sediment particles,
* Treatment and possibly recovery of the extracting agent.

An essential factor in the extraction process is the extracting agent. Two essentially different groups of extracting agents can be considered; water- based extracting agents and organic solvents. At this moment the extraction processes using water based extracting agents are the most important ones.

In the application of water-based extracting agents, two mechanisms regarding the transfer of pollutants from sediment particles to extracting agent can be distinguished. The first mechanism is aimed at dissolving pollutants in the extracting agent. This process may be enhanced by the addition of chemicals. For removal of heavy metals from sediments the following extracting agents can be used: inorganic acids (HNO_3, HCl and H_2SO_4), organic acids (lactic acid, acetic acid, oxalic acid, citric acid) and complexing agents (NTA, EDTA). The second mechanism is aimed at the dispersion of polluted particles into the extracting agent. Chemicals that can be used to promote this dispersion process are: sodium hydroxide, surfactants and detergents. It is clear that in the case of a sediment slurry, extraction using an aqueous extracting agent can be achieved by adding the required chemicals to the slurry.

To separate the extracting agent containing the dissolved and suspended pollutants from the clean sediment particles, several types of classification techniques can be used. These techniques are based on differences in particle size, shape and density, as well as in settling velocity and magnetic and surface properties. Separation techniques and systems which, amongst others, may be used are gravity settling techniques, centrifuges, hydrocyclones, fluidized bed separation systems, flotation techniques and Humphrey spiral separation techniques.

Several physical and chemical wastewater treatment systems are available for cleaning the separated extracting agent. In general, treatment of the extracting agent results in a relatively large amount of sludge.

Based on experience with the application of extraction for soil clean-up it can be expected that extraction of sediments using water-based extracting agents is only possible for sediments consisting of coarse particles (particle diameters > 50 to 75 μm). The extraction process is also suitable for treating sediments containing small particles such as clay particles if the pollutants completely dissolve in the extracting phase.

Experience with extraction

The applicability of extraction for the removal of heavy metals has been investigated on a laboratory scale for three types of sediments, polluted with copper, cadmium, zinc and lead. Two of these samples consisted of the overflow of a hydrocyclone separation process. In this investigation, the two above-mentioned types of extraction procedures were used:

* Extraction with inorganic acids (mainly HCl).

  To obtain sufficiently high extraction efficiencies (in general more than 70 to 80%) a low pH (<1) is necessary. Aeration of samples prior to the extraction step has a positive effect on the extraction efficiency. This effect is probably due the oxidation of sulphides to sulphates and the oxidation of organic matter to soluble compounds resulting in a release of the bound heavy metals. It is observed that, due to the high buffer capacity of the sediment (probably caused by the large amounts of lime compounds), a large amount of HCl is needed.

* Extraction with complexing agents.
  With EDTA (5% solution) similar extraction efficiencies are obtained as with HCl. Also, with this extracting agent, a positive effect of aeration is observed. However, a much longer treatment time is necessary with EDTA than with HCl (about 1 to 3 months). This longer treatment time makes it necessary to operate the system in a so-called heap-leaching mode. Prior to disposal into the heap, the small sediment particles have to be flocculated to larger particles. A specific advantage of the EDTA extraction process is that EDTA can be recovered from the extracting agent by electrolysis resulting in a possibility to reuse the EDTA. Due to the relatively mild pH-conditions, the sediment matrix is less disturbed than in the case of leaching with inorganic acids.

4.3 Biodegradation

In biodegradation processes microorganisms are used for the conversion of contaminants into non-polluting compounds such as water and carbon dioxide. Natural biodegradation of pollutants occurs in most soil environments. However, in general, the conditions for biodegradation are not favourable for achieving an effective clean-up. Bioremediation techniques are aimed at improving the conditions for microbiological degradation. Conditions which have to be optimized may be the temperature, the moisture content of the soil, pH, redox potential, concentration of oxygen, type of electron acceptor, presence of desired microorganisms, and the bioavailability of the pollutant to the microorganisms. Biodegradation can occur under aerobic or anaerobic conditions. In general, conditions are aerobic. However, in biodegradation of chlorinated hydrocarbons a combination of aerobic and anaerobic treatment steps is required for complete conversion of the pollutants.
With respect to biological treatment of sediments contaminated with organic pollutants, there are, at present, three major types of biological techniques available:

* Landfarming,
* Bioreactors,
* Aeration basins.

Landfarming

Landfarming (Figure 6) is a proven technique that is frequently applied in soil remediation. It can also be applied to aquatic sediments, especially to sediments with a relatively large fraction coarse material or to the coarse fraction obtained after classification of a sediment into a fine and a coarse fraction.

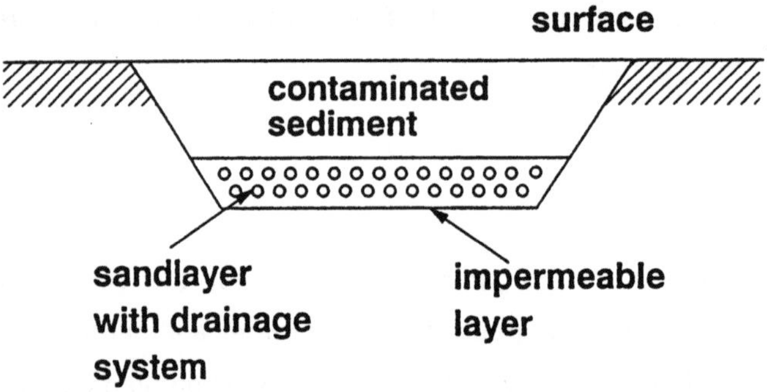

Figure 6 Landfarming of sediment

In landfarming the polluted sediment is spread out in a layer of 0.5 to 1 m in thickness on top of a specially constructed impermeable sublayer provided with a drainage system. The aerobic microbiological degradation of pollutants in this layer is stimulated by regularly cultivating the layer to enhance oxygen transfer, and adding water, nutrients and trace elements if necessary. Polluted drainage water is collected and can be recirculated over the layer. A further acceleration and improvement of the process can be obtained by the use of greenhouse techniques, forced aeration and increasing the temperature of the layer. Basically, landfarming is a relatively simple method and makes treatment times of 1 to 3 years acceptable. Landfarming has to be considered as an extensive, simple treatment method with a relatively little energy demand.

Bioreactors

Bioreactor techniques (Figure 7) involve the treatment of a slurry of contaminated sediment in a compact reactor system to which oxygen and nutrients are supplied.

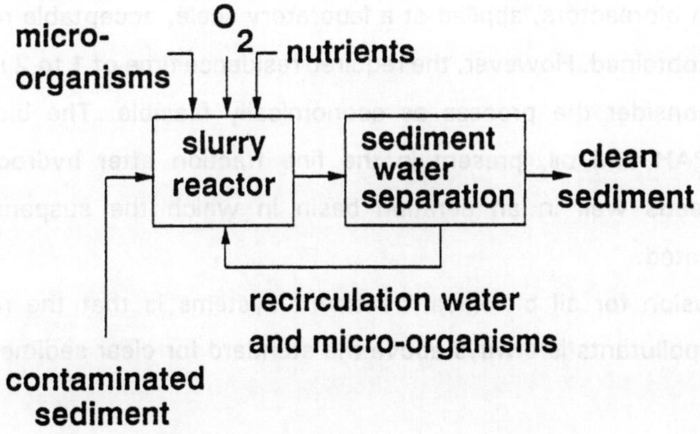

Figure 7 Bioreactor for sediment treatment

The slurry is intensively mixed to promote better contact between microorganisms, pollutants, oxygen and nutrients. Due to this intensive contact and the better possibilities for controlling the process, substantially higher biodegradation rates can be achieved than in landfarming. However, the process is more complicated than landfarming and requires a rather complex and expensive treatment installation.

Aeration basin

An aeration basin is a large scale facility for treating the fraction of fine sediment particles which is obtained, for example, by means of a hydrocyclone process. The suspension of sediment particles in an aeration basin is intermittently mixed and aerated. To that aim a pontoon provided with mixers and aerators is moved slowly over the length of the basin. Treatment time is relatively short but also the energy demand is low. The residence time of the sediment particles in the aeration lagoon varies between two and eight months.

Experience with biodegradation

Landfarming, bioreactors and aeration basins have been investigated intensively for treating sediment containing PAH and oil under aerobic conditions. In landfarming removal efficiencies of 60 to 80% were obtained. The final concentration of PAH (16 PAH or EPA) varied between 15 and 50 mg/kg dry matter. Also with bioreactors, applied at a laboratory scale, acceptable removal efficiencies were obtained. However, the required residence time of 1 to 2 months is too long to consider the process as economically feasible. The biological degradation of PAH and oil, present in the fine fraction after hydrocyclone separation, proceeds well in an aeration basin in which the suspension is intermittently aerated.

A general conclusion for all biological treatment systems is that the residual concentration of pollutants is always above the standard for clear sediment.

4.4 Bioleaching (bacterial leaching)

A special type of extraction of heavy metals is bioleaching. This process is based on the property of certain types of acidophilic bacteria to utilise reduced sulphur compounds as electron donors, which provide energy for their growth. Several of the most important species belong to the genus *Thiobacillus*. These bacteria live autotrophically under aerobic conditions. The optimal pH for growth is 1-4, the optimal temperature for growth varies between 15°C and 55°C. Due to the oxidation of sulphur compounds to sulphate, the pH decreases and heavy metals are mobilised. After mobilisation, the sediment particles and the solution containing heavy metals and sulphate are separated. The heavy metals can be removed from this solution by conventional physical/chemical wastewater treatment processes.

Experience with bioleaching

Research was carried out at laboratory scale into the removal of heavy metals from sediments. The investigated bioleaching process consists of two treatment steps. The first step comprises a treatment with lactic acid aimed at mobilising the

heavy metals as metal complexes. This process proceeds at a pH between 3.5 and 7. The necessary lactic acid can be produced microbiologically in a separate reactor and can then be added to the sediment slurry or it can be produced directly in the sediment slurry reactor. A typical property of the microbiological production of lactic acid is the relatively fast reaction rate. Besides the fact that lactic acid dissolves part of the heavy metals, its other function consists of reducing the buffer capacity. The second step in the bioleaching process is aerobic biological treatment of the sediment slurry with *Thiobacilli*. Due to the oxidation of sulphides or other reduced sulphur compounds, the pH decreases to about 1 and heavy metals dissolve as sulphates. Contrary to the microbiological production rate of lactic acid, the microbiological production rate of sulphuric acid is relatively low, resulting in a longer residence time in the reactor.
With the biological leaching process, removal efficiencies which are roughly comparable with those found in the extraction experiments with HCl and EDTA are observed. Only for Cd, somewhat higher removal efficiencies are obtained. In general, bioleaching is a much more complex process than a physical/chemical extraction process.

5. COSTS

It is possible to give an indication of the costs for each basic treatment step. These costs largely depend on the type of treatment and on the capacity of the treatment process. A rough indication of the costs of some important basic treatment steps is as follows:

* Hydrocyclone separation: NLG 8 to NLG 40 per tonne dry matter (1.8 NLG ≈ 1 US $)
* Biological treatment (degradation):
  - Landfarming: NLG 40 to NLG 100 per m^3
  - Aeration basin: NLG 12 to NLG 35 per m^3
  - Bioreactors NLG: 115 to NLG 640 per m^3

* Extraction:
  - With hydrochloric acid: NLG 10 to NLG 350 per tonne dry matter
  - With EDTA: NLG 20 to NLG 150 per tonne dry matter
* Flotation: ≈ NLG 80 per tonne dry matter.

When basic treatment steps are combined in an integral treatment system, the total costs are not a simple sum of the costs of the individual treatment steps. The costs must be converted into the quantity of the material which is treated in a particular stage.
A reasonable estimation of the costs of an integral treatment process consisting of separation and cleaning steps varies between NLG 50 to NLG 200 per m^3.

6. SOME GENERAL CONCLUSIONS

Research on labscale, pilot plant scale and practical scale, for a large part carried out within the framework of the "Development Programme Treatment Processes for Polluted Aquatic Sediments in The Netherlands" has shown that various processing techniques are in principle suitable for decontaminating and processing of sediments. However, at present only a few of these are operational.

Some techniques are at present not yet operational, but still are very promising. These techniques will be investigated and scaled up in the second phase of the above mentioned programme and evaluated in field scale, pilot plant projects, to demonstrate the performance of this type of sediment remediation. In the final assessment, financial as well as technical and environmental aspects will play a role.

7. REFERENCES

Anonymous. Development Programme Treatment Processes for Polluted Aquatic Sediments in The Netherlands, Phase 1 (1989-1990). Ministry of Transport, Public Works and Water Management. Directorate General for Public Works and Water Management, Institute for Inland Water Management and Waste Water Treatment. Lelystad, The Netherlands.

Bruggeman W.A. and G.N.M. Stokman. Remediation of contaminated sediments in the Netherlands. RIZA, Institute for Inland Water management and Waste Water Treatment, Lelystad, The Netherlands. In: The 2nd Japan/Dutch Workshop on Integrated Water Quality Management, Japan, October 15-25, 1993.

Rulkens, W.H. and Bruggeman, W.A. Thermal physical/mechanical, microbiological and in-situ clean-up techniques (in Dutch). Bodem, nr. 4, 1993.

Rulkens, W.H., J.T.C. Grotenhuis, E.R. Soczó: Remediation of contaminated soil: state of the art and desirable future developments. In: F. Arendt, G.J. Annokkée, R. Bouman and W.J. van den Brink (eds.), Contaminated Soil '93, Kluwer 1993.

Rulkens, W.H. Classification and extraction techniques for contaminated sediments. Department of Environmental Technology, Wageningen Agricultural University, The Netherlands. In: The 2nd Japan/Dutch Workshop on Integrated Water Quality Management, Japan, October 15-25, 1993.

Veen, H.J. van, and J. Joziasse. Bioremediation of contaminated sediments in the Netherlands. TNO Institute of Environmental and Energy Technology, Apeldoorn, The Netherlands. In: The 2nd Japan/Dutch Workshop on Integrated Water Quality Management, Japan, October 15-25, 1993.

Thermal Treatment methods for Contaminated Dregded Sludges

Ronald TIZE
SEGHERSengineering NV
Molenweg 107 Bus 2
B-2830 WILLEBROEK
BELGIUM

SUMMARY

1. Introduction : Belgian situation
2. Properties of dredged sludges
3. Place of thermal processes in the global treatment
4. Sludge drying : process description
5. Sludge decontamination in a fluidised bed reactor
6. Costs for both treatment methods
7. Bibliography

1. Introduction - Belgian Situation

Maintainance dredgings is that produced in routine dredging in navigations whilst clearance in that produced from non navigable waterways.

Pollution of surface water in both natural drainage and canals can result in (severe) pollution of waterway bottoms. Removal of sludge is therefore becoming more and more an environmental problem.

The main problem is maintainance dredgings. Landfill with essentially non-polluted material is not considered as a waste disposal problem.

NATO ASI Series, Partnership Sub-Series, 2. Environment – Vol. 45
Environmentally Devastated Areas in River Basins in Eastern Europe
Edited by Alfons Georges Buekens and Vasily Victorovich Dragalov

In 1988 these amounts of maintanance dredgings were dumped within Flemish Belgium.

Maritime channels and harbours	2.8 million m^3
Canals and waterways	1.1 million m^3

This 3.9 million m^3 is not all polluted in the sense of the various classifications.

Yearly some 35 to 40 million m^3 (density 1.6) of maintainance sludge is dredged from the approaches to the harbours in Flemish Belgium. The greater part is dumped at sea.

By mid 1989 there were 28 sites permitted to receive maintainance dredging within the framework of the co-ordinated waste-plan. These sites have a minimum capacity of 12.1 million m^3 and cover in total 450 hectare. The majority are above ground holding basins. Underground dumping, mainly in worked-out sand/gravel pits, does play a significant role.

There are no equivalent data for clearance dredging, niether amount nor degree of pollution, for the some 12000 km of Flemish waterways, of which some 1000 km are class 1.

It is estimated that about 185000 m^3 of clearance dredging come annually from class 1 waters, of which about 42000 m^3 are dumped.
By the end of 1989 there were only 11 permitted sites (covering 47 hectare) for clearance dredgings.

In a very few cases are new techniques for dredging handlings applied, and then mostly as experimental pilot projects.
It is important to note that dredging companies are more and more prepared to take an environmental viewpoint into account.

The problem has not only the aspect of space required out also that of those involved : Authorities, Dredging companies; Harbour authorities and the like; agreeing to a coherent pollution classification framework as well as to exploitation conditions for dredging dumps.

2. Properties of Dredged Sludges

Dredged sludge generally comprimes fine matter with a dry solids content of 20-50 %, depending on the dredging method. Depending on circumstances, the mineral fraction < 16 µm will vary between 10-70 %. The fraction > 210 µm is usually not higher than a few percent. The commonest contaminations are compounds of the heavy metals (Zn, Cd, Cr, Cu, Pb, Hg) , as well as organic compounds [oil, polycyclic aromatic hydrocarbons (PAH), organic chlorine compounds]. The metals are probably present as sulphides and carbonates, or complexly bound in the organic matter present, or bound to the minerals.

From analyses of sieve fractions it appears that the metal compounds, but with organic contaminations also, are especially found in higher concentrations in the fines (< 20 µm). Organic contaminations are often present in increased concentration in a coarse fraction (> 210 µm).

Contaminated dredged sediment is in many instances being formed in water systems with floating components (organic matter and sludge particles). When contaminations are added to the system they will, with regards to the organic microcontaminants, the poorly soluble, lyophilic compounds such as PCB, pesticides, PAH and heavier aliphatic compounds, showing a preference to be compounded with the solid phase. Reasonably soluble organic compounds such as benzene, toluene and chloro-methanes will remain partly in the water phase, and these compounds will, together with the floating particles, settle to a considerably lower degree to the bottom.

With regard to further processing, the following properties of dredged sediments are important :

1. Dredged sediment contains a great deal of water (on average approx. 75 %).
2. The bond of contamination to sludge is often strong.

3. Dredged sludge contains a relatively high amount of fine mineral matter (fraction < 16 µm is between 10 and 80 %)
4. Fines have a higher contamination content than the coarse fraction.

When looking at thermal treatment methods, the organic content in the sediments and the heating value of this organic matter becomes also very important.
This organic content is low compared to sludge generated in wastewater treatment plants (only 10 to 35 % instead of 60 to 80 %). The heating value of the organic matter is expected to be almost comparable and will be in the range of 5.000 to 5.500 kcal/kg.

3. Place of Thermal Process in the Global Treatment

3.1. Global treatment process

In essence many unit operations are available for the treatment of contaminated residues such as dredged sediments. However, evidently not all unit operations are applicable or feasible in any particular situation. For example, some operations are especially designed to treat very large quantities at low costs. Other operations, in contrast, are preferably designed for small-scale applications at relatively high costs.
Processes for the treatment of large quantities of residues will generally be built up of several unit operations combined in such a way that optimum results are realized at a minimum of operating costs. In reference to environmental technology, an "optimum result" means that the environmental merit of applying a process to a residue is as large as possible, within the accepted conditions of reasonable costs. In a generalized scheme, the scenario for the treatment of residues may be visualized as indicated in Fig. 1, where generally two categories of techniques can be distinguished :

a) Large-scale concentration techniques. These techniques are characterized by large-scale application, low costs per unit of residue to be treated and a low sensitivity to variations in circumstances. It is advantageous that these techniques may be constructed in mobile or transportable plants.

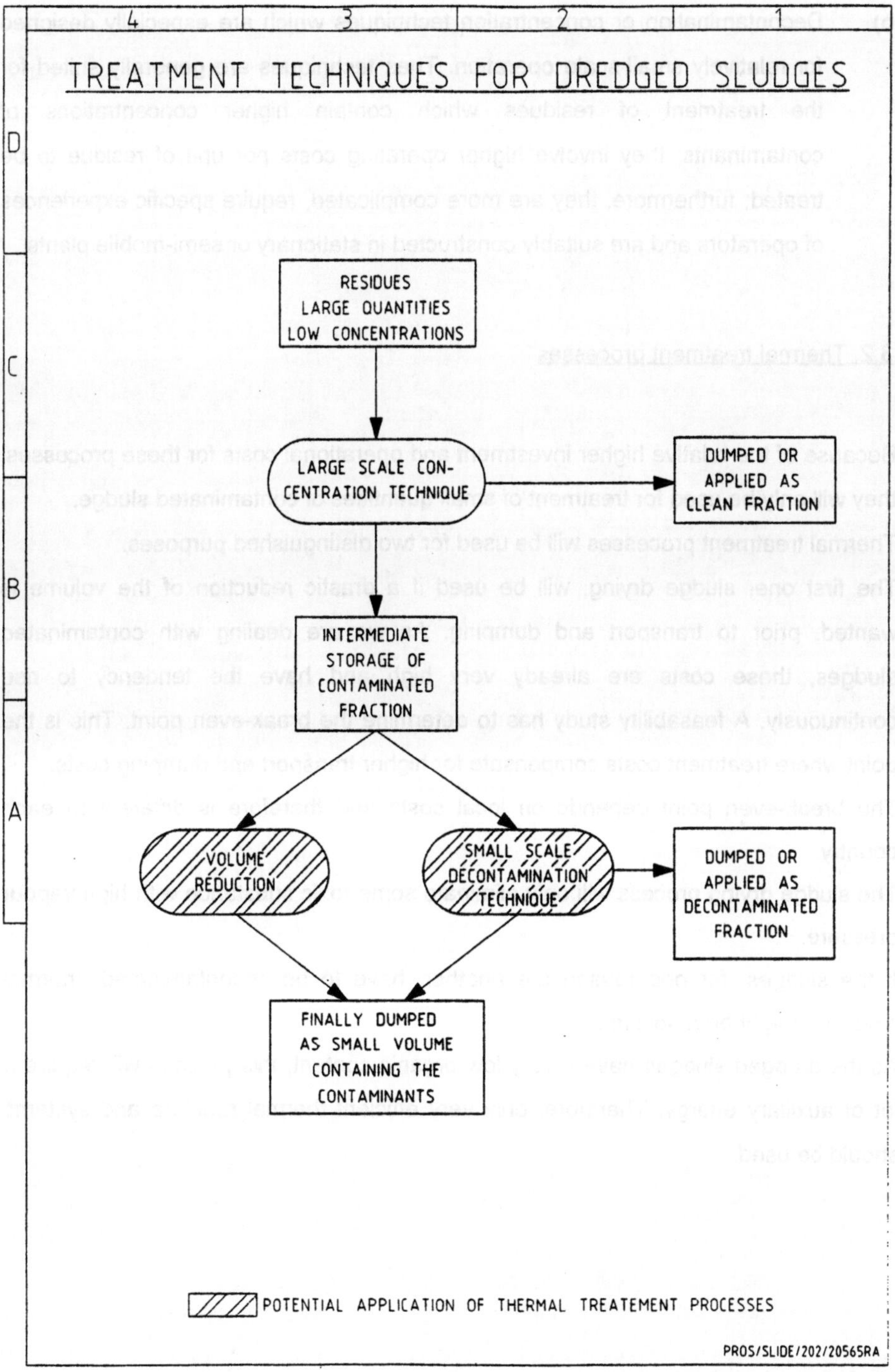
4
3
2
1
D
C
B
A
TREATMENT TECHNIQUES FOR DREDGED SLUDGES
RESIDUES LARGE QUANTITIES LOW CONCENTRATIONS
LARGE SCALE CONCENTRATION TECHNIQUE
DUMPED OR APPLIED AS CLEAN FRACTION
INTERMEDIATE STORAGE OF CONTAMINATED FRACTION
VOLUME REDUCTION
SMALL SCALE DECONTAMINATION TECHNIQUE
DUMPED OR APPLIED AS DECONTAMINATED FRACTION
FINALLY DUMPED AS SMALL VOLUME CONTAINING THE CONTAMINANTS
POTENTIAL APPLICATION OF THERMAL TREATEMENT PROCESSES
PROS/SLIDE/202/2056SRA

b) Decontamination or concentration techniques which are especially designed for relatively small-scale operation. Thes techniques are generally suited for the treatment of residues which contain higher concentrations of contaminants; they involve higher operating costs per unit of residue to be treated; furthermore, they are more complicated, require specific experiences of operators and are suitably constructed in stationary or semi-mobile plants.

3.2. Thermal treatment processes

Because of the relative higher investment and operational costs for these processes, they will only be used for treatment of small quantities of contaminated sludge.

Thermal treatment processes will be used for two distinguished purposes.

The first one, sludge drying, will be used if a drastic reduction of the volume is wanted, prior to transport and dumping. As we are dealing with contaminated sludges, these costs are already very high and have the tendency to rise continuously. A feasability study has to determine the break-even point. This is the point where treatment costs compensate for higher transport and dumping costs.

The break-even point depends on local costs and therefore is different to each country.

The sludge drying process will only eliminate some toxic substance with high vapour pressure.

If the sludges, for one reason are another, have to be decontaminated, thermal oxidation might be a solution.

As the dredged sludges have a very low organic content, this process will require a lot of auxiliary energy. Therefore, only very efficient thermal reactors and systems should be used.

4. Sludge Drying - Process Description

4.1. Introduction

When focusing on the sludge particle that has to be dewatered, four phases for the water content can be distinguished :

- free water
- capillary water
- colloidal water
- intracellular water.

The free water can easily be separated from sludge by gravity. Capillary and colloidal water can partially be removed by mechanical forces. This is mostly preceded by chemical conditioning. The intracellular water is most difficult to remove and can only be eliminated by breaking the cell structure through thermal treatment.

The purpose of thermal sludge drying technology is to minimize the sludge volume as far as possible, without destroying the organic part of the sludge. Sludge drying can be desired to reduce the total sludge volume and to minimize transport costs or to obtain an-easy- to-handle product which can be landfilled or co-disposed.

To dry sludge with the help of thermal energy, different systems are in use, namely direct dryers and indirect dryers.

In direct drying systems the drying medium is in direct contact with the sludge. The flue gases supply their heat directly to the sludge.
Generally, the temperature of the gases or the air are about 400 °C.

In indirect drying systems the drying medium is separated from the sludge by a wall. The heat of the transporting medium (steam or thermal oil) is transferred through the wall. The temperature level is lower (about 200 °C for the medium).

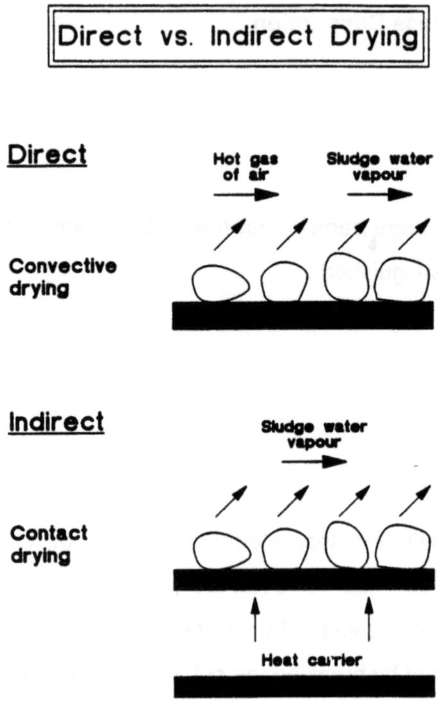

In general indirect drying is preferred :

In the indirect system, only a small amount of sludge vapours is produced. This small quantity has to be treated further on. It can be sent back to a postcombustion chamber or to the freeboard of a furnace in which the high temperature, turbulence and long enough residence time will crack down the smelling aromates. This will provide an odourfree operation. In the case of direct drying, the complete mixture of flue gases and sludge vapours has to be sent back to the postcombustion chamber. This volume is quite important and disturbs the flow pattern in the chamber. Moreover it will cool down the furnace.

Also in the case the vapours have to be condensed (drying on itself), the amount to be treated is limited. In case of indirect drying the heat carrying medium can be heated in a downstream boiler or heat-exchanger. In this way the furnace temperature is not influenced. Only the small amount of sludge vapours will have to be heated up again to 850 °C.

In practice, three different types of indirect sludge drying are on the market available: disk dryers, film dryers and multi stage dryers. Each type has is own particularities.

4.2. The drying proces

See attached Typical flow diagram N° 19201.

a) Sludge intake

The mechanically dewatered sludge can either come from an adjacent waste water treatment plant, from external waste water treatment plants, or both.

External sludge which is brought in by means of dump trucks, can best be received in a completely enclosed silo. The silo can be positioned inside the drying building in order to avoid odour problems.
The sludge is extracted from the reception silo by means of two twin extraction screws. The screws are of the shaftless type, which guarantees constant extraction.

The sludge can be fed either to a wet sludge storage silo (1.1) or to the hopper of the sludge pump (1.2).

Sludge from an adjacent WWTP is collected in a shaftless transport screw as it comes out of the centrifuges (or any other mechanical dewatering equipment) and is brought to the storage silo or to the hopper of the sludge pump.

The hopper of the pump is provided with a paddle mechanism to assure continuous feed to the pump.
The motor of the paddle mechanism is equiped with power detection. Whenever sludge feed is interrupted the power detection will automatically shut down the pump to prevent it running empty.
Pump (1.2) is a multiple stage monopump able to pump the sludge up the sludge mixer, which is positioned on top of the dryer.

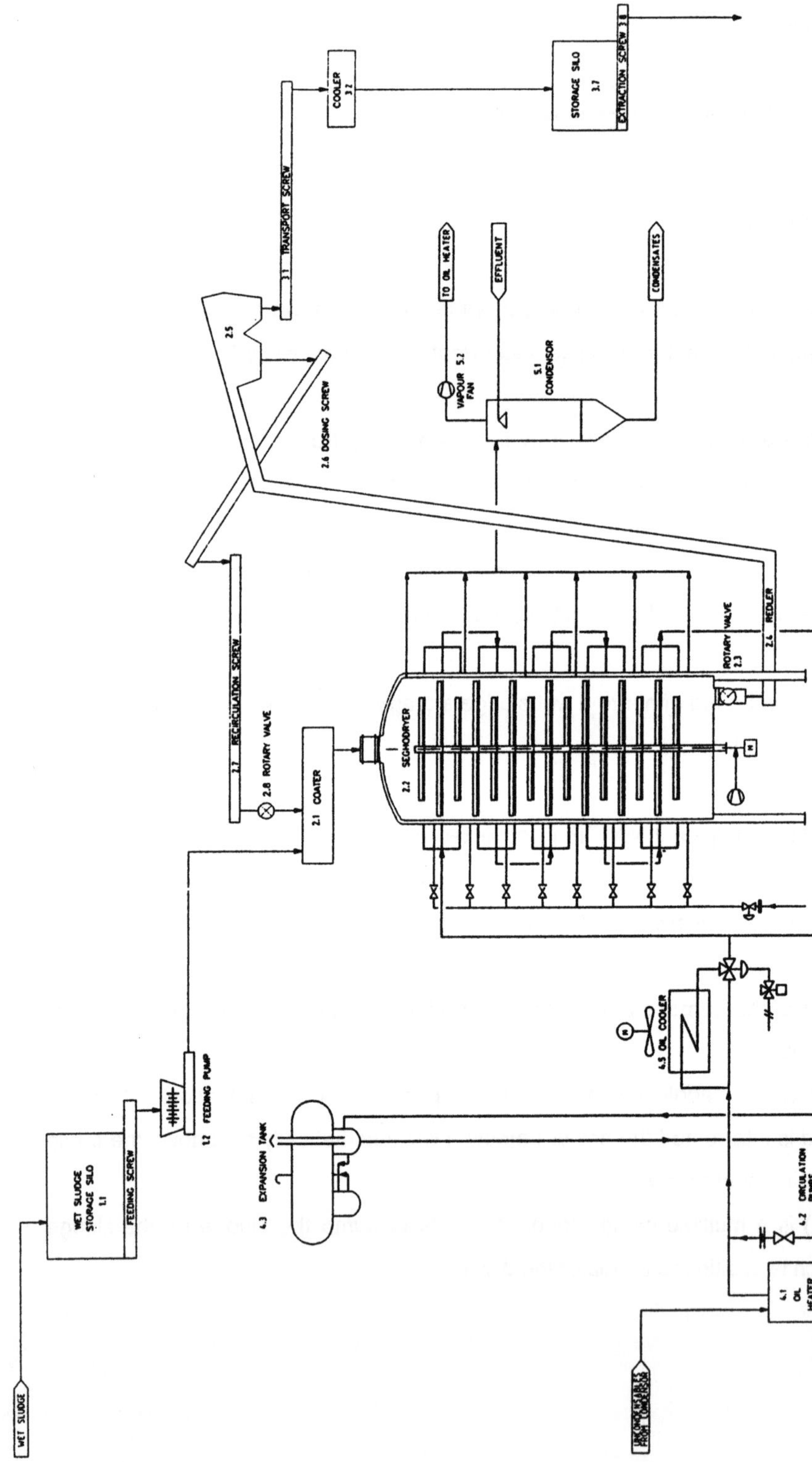
WET SLUDGE
WET SLUDGE STORAGE SILO 1.1
FEEDING SCREW
1.2 FEEDING PUMP
2.1 COATER
2.8 ROTARY VALVE
2.7 RECIRCULATION SCREW
2.2 SEGHODRYER
2.6 DOSING SCREW
2.5
3.1 TRANSPORT SCREW
COOLER 3.2
STORAGE SILO 3.7
EXTRACTION SCREW 3.8
TO OIL HEATER
EFFLUENT
CONDENSATES
VAPOUR 5.2 FAN
5.1 CONDENSOR
ROTARY VALVE 2.3
2.4 REDLER
AIR
4.5 OIL COOLER
4.6 STORAGE TANK
4.3 EXPANSION TANK
4.2 CIRCULATION PUMPS
4.1 OIL HEATER
UNCONDENSABLES FROM CONDENSOR

b) Sludge drying

The SEGHODRYER is a horizontal tray dryer on which the sludge is dried and pelletized at the same time.

In Sludge Conditioner (2.1) wet sludge is mixed with recirculated dried sludge. The sludge conditioner feeds directly into the dryer.

The SEGHODRYER (2.2) is a vertical multi-stage tray dryer. It consists of horizontal trays which are heated by thermal oil in a closed loop or by steam.
By means of a raking mechanism, connected to a central rotating shaft, the sludge is moved over the upper plate and pushed to the edge of the upper tray from where it falls onto the second plate.
The continuously rotating scraping arms move the sludge back from the outside to the inside of the tray, from where it falls onto the next lower tray.
This way, the sludge trickles down from tray to tray until the bottom tray is reached.

The sludge is dried by direct contact with the trays. Because of the ploughing action of the raking arms, the sludge particles are continuously mixed and turned over, which results in a high drying efficiency.

The dried sludge (more than 90 % DS) is collected at the bottom of the dryer and transported by means of a redler (2.4) to a separation hopper (2.5).

From this separation hopper part of the sludge is recirculated to the sludge conditioner by means of a conveying screw (2.6 & 2.7).
Necessary seal valves (2.3 & 2.8) are provided to prevent air from entering the SEGHODRYER.

c) Dried sludge handling

The dried sludge that is not reciculated (> 90 % DS) are transported by means of a screw (3.1) to a pellet cooler (3.2). The cooler consists of a water cooled screw (the shell as well as the bladed shaft are cooled). In the cooler the sludge is cooled to 40

°C. This prevents auto-ignition of the sludge in the storage silo. Cooling also avoids absorption of humidity in the air, which can lead to clogging.

The product is transported to the storage silo (3.7).
Dried sludge can, via screw 3.8, be brought back into the system when starting up the plant.

d) Thermal oil circuit with control loop

The thermal oil is heated in a thermal oil heater (4.1) to a temperature of 240°C. Circulation pumps (4.2) pump the thermal oil to the sludge dryer. The oil flows through the hollow trays from the top of the dryer to the bottom. At the exit of the last plates the oil has been cooled to about 210°C. The oil is then pumped back to the oil boiler via the expansion / deaerator unit (4.3).

A vessel for emptying the complete oil circuit is provided for use during maintenance (4.4).
The main control loop is a control of the oil flow to the dryer as a function of the oil temperature at the exit of the dryer.

An emergency control loop is provided to activate the oil cooler (4.5) in case of either too high a vapour temperature too high an oil temperature.

The heater itself is provided with an integrated controller in order to achieve a constant oil outlet temperature of 240°C.

e) Vapour treatment

The complete dryer is kept under vacuum to prevent odorous gas from escaping to the atmosphere. Vapour fan 5.2 pulls the vapours through a contact condensor (5.1). The air, with uncondensables, collects at the top of the condensor and is sent to the thermal oil boiler for destruction. The condensates can be fed back to the WWTP. The only exhaust flow are the gases from the thermal oil boiler. This flow is very small and virtually harmless.

If requested the energy in the vapours can be recovered by using a sludge / vapour contact condensor.

In the sludge / vapour contact condensor the vapours are brought in direct contact with the incoming sludge water before being centrifuged.
During contact the vapours are cooled to a temparture of about 45°C. At the same time the sludge water is heated which results in a higher separation efficiency in the centrifuges.

4.3. The SEGHODRYER

The SEGHODRYER is an indirect contact sludge dryer which largely reduces the initial sludge volume and produces biologically sterilized sludge. The pathogen-free sludge, containing less than 10 % moisture, can be landfilled without further treatment as a free flowing product.

The SEGHODRYER is a vertical, multi-stage tray dryer, requiring minimal floor space. The modular desgin allows the evaporating capacity to vary from 1000 to 7000 liter of water per hour and per unit, depending on the number of trays and on the cross section of the dryer.

The indirect drying process ensures that there is not any contact between the sludge and the heating medium. In the case of the SEGHODRYER, the trays are heated with steam or thermal oil in a closed loop. A central shift with raking arms ensures even distribution of the sludge over the trays and displaces the sludge from one tray to the other. The raking arms of THE SEGHODRYER are equipped with specifically designed ploughs turning the sludge on each tray in such a way that a free flowing powder of sludge particles is obtained.
The dry sludge particles are collected at the bottom and partly recirculated to a mixer mounted on top of the dryer, where they are coated with incoming wet sludge.
This mixing of dry sludge with a cutain quantity of wet sludge improves the physical appearance of the dryer feed. Easier transport through the dryer and more contacts between the sludge particles and the plates, enhances the water release. The

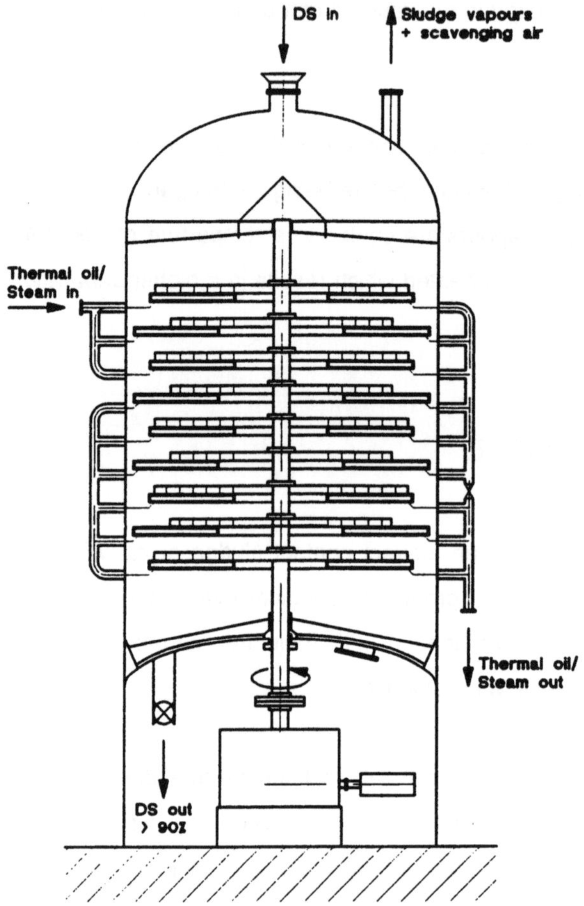

SEGHODRYER

dynamical renewal of the contact zone between the sludge particles and the heating surface is controlled by the well designed ploughing action, resulting in a very high heat transmission coefficient.

The gentle and low-friction raking motion results furthermore in a low torque anc hence in a low specific power consumption, in comparison to conventional sludge drying systems. Instead of increasing the power consumption during drying, the flue transition thus rather decreases the power needed for the transport of the drying sludge over the trays inside the SEGHODRYER.

Finally, continuous operation since 1985 has proven the SEGHODRYER to be highly reliable and virtually maintenance-free.

4.4. Heat- and massbalance

MASS AND HEAT BALANCE SLUDGE DRYING

INPUT DATA

BASIC PARAMETERS

DS flow to system	kg/h	1000.00
DS content system in	%	40.00
DS content system out	%	95.00

BASIC TEMPERATURE DATA

Temp sludge in	°C	10.00
Temp vapours	°C	120.00
Temp dried sludge	°C	100.00
Temp sludge out silo	°C	30.00
Temp cooled sludge	°C	30.00
Temp scavenging air	°C	10.00

PROCESS PARAMETERS

DS content dryer in	%	78.00
DS content dryer out	%	95.00
Air % vol in total vapour	%	30.00
Temp loss in redler	°C	3.00
Temp loss in screw	°C	0.00

THERMODYNAMIC DATA

Spec heat DS	kcal/kg°C	0.35
Spec heat H2O	kcal/kg°C	1.00
Enthalpy air in	kcal/Nm3	3.10
Enthalpy air at vap temp	kcal/Nm3	37.50
Enthalpy H20 vapour	kcal/kg	650.50

DRYER CHARACTERISTICS

Outside surface dryer	m2	320.00
K value of insulation	Kcal/m2.h.°c	0.33

CALCULATED DATA

Heat and mass flows of the drying circuit

NR	DESCRIPTION	TEMP (°C)	DS (%)	FLOW (kg/h)	FLOW (lbs/h)	ENTHALPY (kcal/kg)	ERMAL FLUX (kcal/h)
1	Dewatered sludge	10	40.00	2 500	5 511	7.400	18 500
2	Dried sludge to user	30	95.00	1 053	2 321	11.475	12 079
3	Dried sludge to mixer 1	30	95.00	1 053	2 321	11.475	12 079
4	Dew. sludge to mixer 1	10	40.00			7.400	
5	Vapour (air incl.)	120	-	2 445	5 390	396.939	970 441
5.1	Water (air excl.)	120	-	1 447	3 191	650.500	941 513
5.2	Scavenging air out	120	-	997	2 199	29.002	28 928
6	Dew. sludge to mixer 2	10	40.00	2 500	5 511	7.400	18 500
7	Dried sludge to mixer 2	97	95.00	5 588	12 320	37.103	207 338
8	Feed to dryer	57	78.00	8 088	17 831	27.922	225 838
9	Dried sludge from dryer	100	95.00	6 641	14 640	38.250	254 013
10	Dried sludge to cooler	97	95.00	1 053	2 321	37.103	39 055
11	Cooled sludge to silo	30	95.00	1 053	2 321	11.475	12 079
12	Scavenging air in	10	-	997	2 199	2.398	2 391

13	ENERGY DEMAND DRYER	kcal/h	980 865
		BTU/h	3 892 323
	ENERGY EXCHANGE FOR EVAP H2O.		
		Kcal/kg	678
		BTU/lb	1220

CALCULATED DATA

Heat balance around dryer system

	FLUX IN Kcal/h	FLUX OUT Kcal/h	DEMAND Kcal/h
Sludge	18 500		
Sweep air	2 391		
Sludge vapour		970 441	
Pellets product		12 079	
Loss dryer and mixer		11 616	
Loss conveying equipment		7 620	
Total	20 891	1 001 757	980 865

Energy exchange in dryer

Kcal/kg water evaporated	678
BTU/lb water evaporated	1 220

5. Sludge Decontamination in a Fluidised Bed Reactor

5.1. Different Process Schemes

In all considered schemes a fluidised bed type incinerator is chosen because of its specific advantages. Its high combustion efficiency, resulting in low excess air coefficients, its controllable elutriation process and its optimal process fluegas quality make it the best reactor type to incinerate sludge.

Since mechanical dewatering requires less energy than thermal drying, the most economic solution is to dewater as far as possible mechanically.

The dry solid content of the sludge needed to be autothermal depends on the organic content in the dry solids, the heating value of the sludge and on the characteristics of the system used, f.e. the higher the preheating temperature of the combustion air, the lower the dry solid content needed and the lower the fuel consumption if authothermic conditions can be reached.

Dredged sludges have a very low organic content.

In that case the final dry solid content needed (more than 35 % DS) cannot be reached by mechanical dewatering only. Thermal drying becomes a must. It can be combined with any mechanical dewatering equipment. This decision on the type will be made based on technical, environmental and economical considerations.

It is possible to use an apparatus in line which dries all the sludge up to the needed dry solid content (second scheme) or to use a dryer type which dries part of the sludge stream to 90 % DS, and to reach the same DS content by mixing.

This last solution is more flexible than the first one since the dry solid content of the sludge fed in the furnace can be adjusted easily and fast.

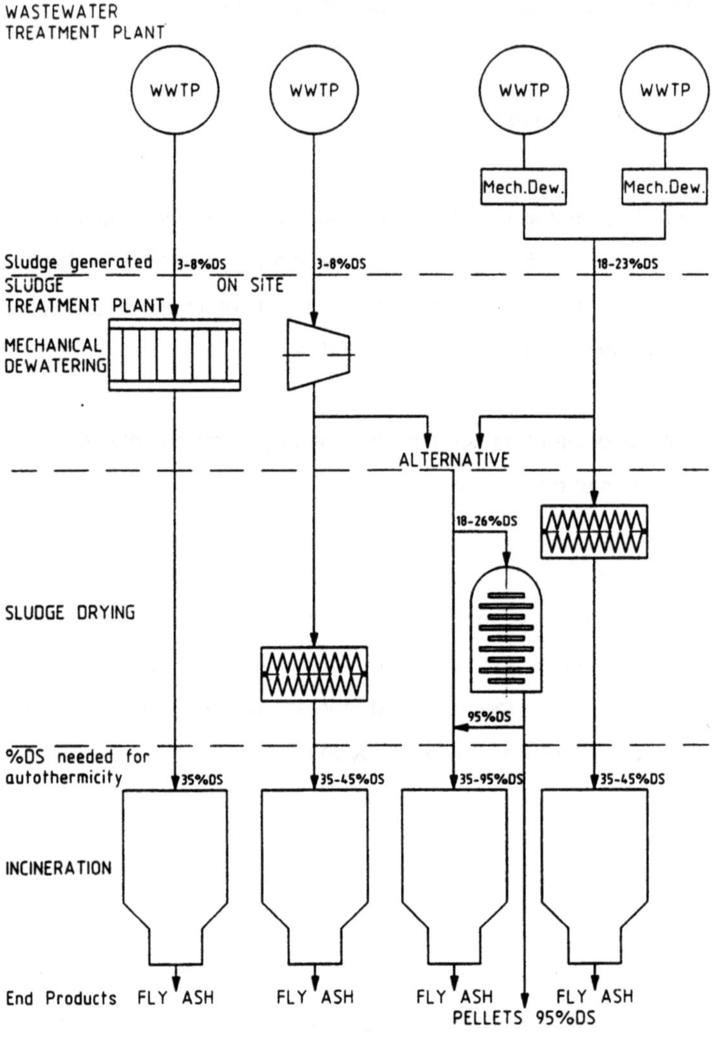

5.2. Fluidised Bed Reactor

The sludge, suitably dried, is introduced into the fluid bed reactor through a special rotating dispersor and spread evenly over the bed. The combustion and fluidization air is fed along one of two compressors and controlled adequately on the basis of the oxygen content of the flue gas. The decontaminated sludge is discharged from the bed by means of an overflow. This sludge will have to be cooled before further handling.

The fluidisation air is preheated in the gas-gas heat exchanger to a temperature of 65O°C. It passes the start-up burner and enters the furnace along its wind-box and its high-technology-air-distributor. In this way the combustion air brings in turbulent motion and created thus a fluidized layer.

In this fluidized layer the sludge fed by the dispensor is immediately granulated and intensively burnt. At the furnace temperature of 85O°C the combustion is achieved mainly in this layer and completed in the post-combustion zone in the freeboard above the fluidized layer. In this freeboard the sludge vapours are completely burnt too, as is also the case for the uncondensables penetrating the condensor handling the vapours of the dryer (if any). These uncondensables are introduced in the furnace just above the fluidized layer in order to have a few seconds residence time in the free board. (for complete burn-out and destruction of malodorous or hazardeous components).

The ash of the burnt material is elutriated from the fluidized bed furnace and is entrained by the flue gases. These ash particles are then trapped in the electrostatic precipitator.

The advantages of this fluid bed incinerator are :

- excellent energy distribution over a fluidized layer resulting in uniform temperature in the fluid bed.
- sufficiently long residence time of the sludge in a hot area with turbulent contact between sludge and combustion air in such a way that the combustion is complete.
- the ash particles are only elutriated after sufficient combustion.

The main operating parameter for an optimal combustion is the temperature of the fluidized layer. This temperature is controlled in different phases. A first control occurs by adapting the temperature of the combustion air with the help of a by-pass on the air preheater making use of a temperature control circuit. In this way, the bed temperature is kept within the temperature range of 84O-860°C. In the case that the sludge feed results in a temperature outside this range, the second phase is put in

action. This means that the bed temperature gets controlled automatically by adapting the quantity of wet sludge fed into the inlet mixer. If the sludge quality becomes that bad that the bed temperature drops under 82O°C, the third phase of temperature control starts by feeding fuel into the furnace along the mixer.
In case of a temperature above 860 °C water is fed into the mixer.

To cope with the requirement of 6 % oxygen in the flue gases, the oxygen content of the flue gas is continuously measured. The signal of the oxygen instrument activates the rotation speed of the ventilator of the combustion air.
A secundary oxygen indicating measurement controls the gases at outlet of the stack.

5.3. The combustion air preheater

The air preheater is specially designed to recover the heat of the flue gases of a fluidized bed furnace to preheat the combustion and fluidization air to the high temperature of 650° C. Heat is transferred by a combination of radiation and convection, and the major design requirements are :

- Resistance to the high metal temperatures.
- Resistance to corrosion.
- Avoidance of fouling by the dust and by chemical compounds in the flue gas.
- Easy assembly with the furnace freeboard.

5.4. Flue gas cleaning system

The flue gas cleaning system should be designed in such a way that contaminants removed from the sludge are captived.

The flue gas leaving the boiler enters the electrostatic precipitator at a controlled temperature of about 22O°C. The dust trapped is collected at the bottom of the filter and transported by a screw into the collecting screw. The latter screw thus sends the

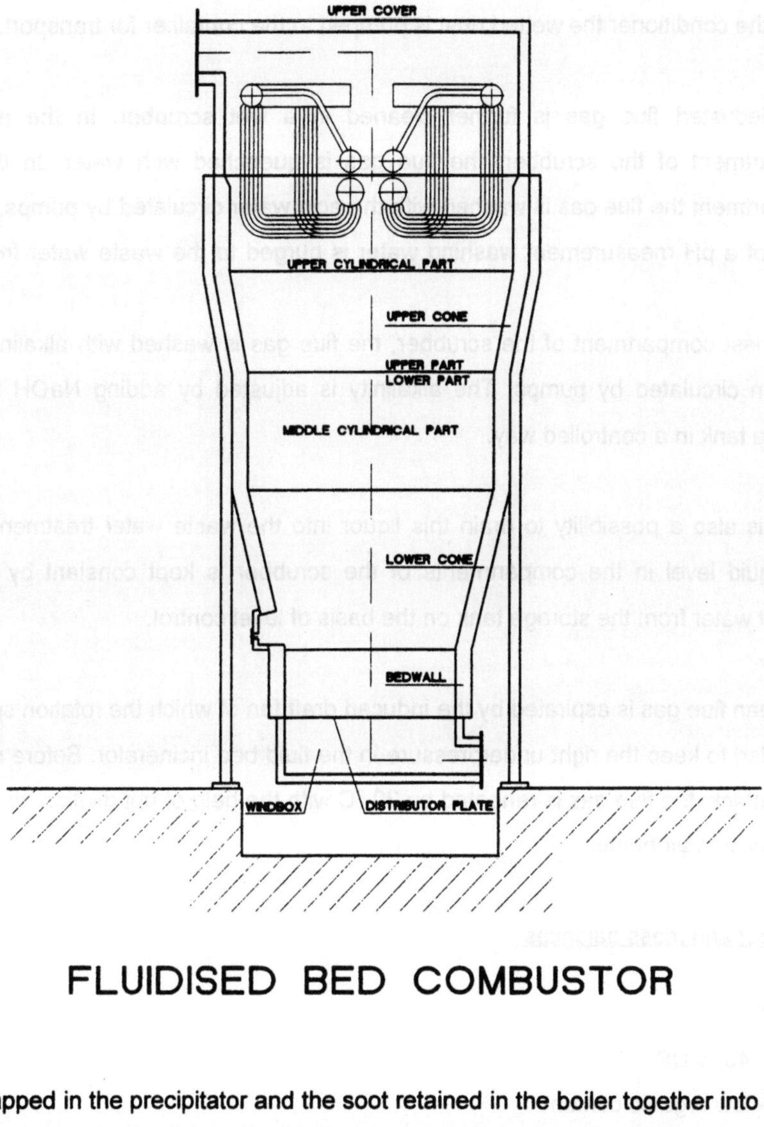

FLUIDISED BED COMBUSTOR

dust trapped in the precipitator and the soot retained in the boiler together into the fly ash bin. This bin is evacuated according to the readings of a weight cell. The dust is pneumatically transported to the storage silo or immediately to a special filter equipped container with the help of the ventilator. The screw extracts the dust from the storage silo and feeds the conditioner. In the conditioner the dust can be wetted using effluent from the effluent storage tank.

The dust extraction from the silo is manually controlled whilst automatic control is foreseen for the wetting in the conditioner.

From the conditioner the wetted dust is pumped to the container for transport.

The dedusted flue gas is further cleaned in a wet scrubber. In the entrance compartment of the scrubber, the flue gas is quenched with water. In the next compartment the flue gas is washed with the acid water circulated by pumps. On the basis of a pH measurement washing water is purged to the waste water treatment plant.
In the last compartment of the scrubber, the flue gas is washed with alkaline water solution circulated by pumps. The alkalinity is adjusted by adding NaOH from its storage tank in a controlled way.

There is also a possibility to drain this liquor into the waste water treatment plant. The liquid level in the compartments of the scrubber is kept constant by adding effluent water from the storage tank on the basis of level control.

The clean flue gas is aspirated by the induced draft fan of which the rotation speed is controlled to keep the right underpressure in the fluid bed incinerator. Before release by the stack, the flue gas is reheated by 20 °C with the help of hot dilution air heated in the dilution air heater.

5.5. Heat and mass balances

5.5.1. Case 1

- 40 % DS
- low organic content
- preheat to 650 °C
- fuel adduction

5.5.2. Case 2

- 40 % DS
- higher organic content
- preheat to 650 °C
- sludge predrying
- no auxiliary fuel

CASE I

PROJECT : **DREDGED SLUDGE INCINERATION (no drying -auxiliary fuel)**

CHARACTERISTICS	SLUDGE	GAS	FUEL	COMBUSTION PARAMETERS	
DS in system (%DS)	40.0				
DS in furnace (%DS)	40.0	100.00	100.00	Ambient air temperature in °C	20
ODS/DS in %	10.0	100.00	100.00	Sludge temperature in °C	20
Composition ODS in %				Combustion air temp. after fan in °C	20
C	53.300	69.75	85.30	Preheated comb. air temperature	650
H	8.100	23.25	11.20	Flue gas temp. out furnace in °C	850
O	31.330			Minimum flue gas temperature in °C	220
N	5.000	7.00		Lime content in % of dry solids	0.00
S	1.300		3.50	Quantity of vapour to furnace in %	0.00
Cl	1.000			Fraction of dryer heat by boiler	1.00
Rem	0.000			Steam circuit condensation heat	470.00
Total	100.030	100.000	100.000	Steam circuit temperature difference	50.00
DS flow in KG/HR	1000.000	0.00	182.87		
Scavenging air in KG/HR	0.000			0	MCAL/H <- FURNACE HEAT BALANCE SALDO
Moisture in air in KG/KG	0.013			779	MCAL/H <- EXHAUST HEAT BALANCE SALDO
Excess air coefficient	1.742	1.742	1.742		

FLUEGAS TABLE

	C-CO2	H-H20	O-O2	N-N2	S-S02	Cl-Cl2	REM	MINERALS	WATER	TOTAL
FUEL MIXT.COMPOSITION	0.0780	0.0107	0.0117	0.0019	0.0029	0.0004	0.0000	0.3355	0.5591	1.0000
O2 coefficie	2.6667	8.0000	-1.0000	0.0000	1.0000	0.0000	0.0000			
KG O2 STOCH	0.2080	0.0852	-0.0117	0.0000	0.0029	0.0000	0.0000			0.2844
KG AIR STOCH	0.8966	0.3674	-0.0503	0.0000	0.0124	0.0000	0.0000			1.2260
NM3 AIR STOCH	0.6943	0.2845	-0.0390	0.0000	0.0096	0.0000	0.0000			0.9494
NM3 REAL	1.2092	0.4954	-0.0679	0.0000	0.0167	0.0000	0.0000			1.6534
FLUE GAS										
coeff weight	3.6667	9.0000	0.0000	1.0000	2.0000	1.0000	1.0000	1.0000	1.0000	
KG elements	0.2860	0.0959	0.0000	0.0019	0.0057	0.0004	0.0000	0.3355	0.5591	1.2845
excess air			0.2109	1.6398					0.0278	1.8785
scavenging air			0.0000	0.0000					0.0000	0.0000
TOTAL(weight)	0.2860	0.0959	0.2109	1.6417	0.0057	0.0004	0.0000	0.3355	0.5869	3.1630
% WEIGHT	0.0904	0.0303	0.0667	0.5190	0.0018	0.0001	0.0000	0.1061	0.1855	1.0000
specific weight	1.9643	0.8036	1.4286	1.2500	2.8571	3.1654	1.5000	2500	0.8036	
TOTAL(volume)	0.1456	0.1193	0.1476	1.3133	0.0020	0.0001	0.0000	0.0001	0.7303	2.4585
% VOLUME WET	0.0592	0.0485	0.0601	0.5342	0.0008	0.0000	0.0000	0.0001	0.2971	1.0000
% VOLUME DRY	0.0905		0.0918	0.8163	0.0012	0.0001	0.0000	0.0001		1.0000
FLUE GAS TEMP 850.00	°C or	1123.16 °K								
ENTHALPY (KCAL/NM3 GAS)	441.53	827.43	295.66	279.58	425.00	266.09	280.50	5.31E+05	827.43	508.62
ENTHALPY (KCAL/KG SLUDGE)	64.29	98.73	43.65	367.18	0.85	0.03	0.00	71.29	604.28	1250.30
ENTHALPY (KCAL/KG GAS)	224.78	1029.69	206.96	223.66	148.75	84.06	187.00	212.50	1029.69	395.30

MASS BALANCE SYSTEM	VOL IN	WGHT IN	VOL OUT	WGHT OUT
SLUDGE + AUX FUEL		2683		
COMBUSTION AIR	4436	5728		
HUMIDITY IN COMB.AIR	93	74		
SCAVENGING AIR	0	0		
HUMIDITY IN SC.AIR	0	0		
FLUE GAS			6596	8486
CONDENSATES			0	0
TOTAL	4528	8486	6596	8486

CASE II

PROJECT : DREDGED SLUDGE INCINERATION (drying -35 % ODS/DS)

CHARACTERISTICS	SLUDGE	GAS	FUEL
DS in system (%DS)	40.0		
DS in furnace (%DS)	54.0	100.00	100.00
ODS/DS in %	35.0	100.00	100.00
Composition ODS in %			
C	53.300	69.75	85.30
H	8.100	23.25	11.20
O	31.330		
N	5.000	7.00	
S	1.300		3.50
Cl	1.000		
Rem	0.000		
Total	100.030	100.000	100.000
DS flow in KG/HR	1000.000	0.00	0.00
Scavenging air in KG/HR	129.449		
Moisture in air in KG/KG	0.013		
Excess air coefficient	1.780	1.780	1.780

COMBUSTION PARAMETERS	
Ambient air temperature in °C	20
Sludge temperature in °C	20
Combustion air temp. after fan in °C	20
Preheated comb. air temperature	650
Flue gas temp. out furnace in °C	850
Minimum flue gas temperature in °C	220
Lime content in % of dry solids	0.00
Quantity of vapour to furnace in %	100.00
Fraction of dryer heat by boiler	1.00
Steam circuit condensation heat	470.00
Steam circuit temperature difference	50.00

0 MCAL/H <- FURNACE HEAT BALANCE SALDO

298 MCAL/H <- EXHAUST HEAT BALANCE SALDO

FLUEGAS TABLE

	C-CO2	H-H20	O-O2	N-N2	S-S02	Cl-Cl2	REM	MINERALS	WATER	TOTAL
FUEL MIXT.COMPOSITION	0.1007	0.0153	0.0592	0.0094	0.0025	0.0019	0.0000	0.3508	0.4603	1.0001
O2 coefficie	2.6667	8.0000	-1.0000	0.0000	1.0000	0.0000	0.0000			
KG O2 STOCH	0.2685	0.1224	-0.0592	0.0000	0.0025	0.0000	0.0000			0.3342
KG AIR STOCH	1.1573	0.5276	-0.2551	0.0000	0.0106	0.0000	0.0000			1.4404
NM3 AIR STOCH	0.8962	0.4086	-0.1975	0.0000	0.0082	0.0000	0.0000			1.1154
NM3 REAL	1.5951	0.7272	-0.3516	0.0000	0.0146	0.0000	0.0000			1.9853
FLUE GAS										
coeff weight	3.6667	9.0000	0.0000	1.0000	2.0000	1.0000	1.0000	1.0000	1.0000	
KG elements	0.3692	0.1377	0.0000	0.0094	0.0049	0.0019	0.0000	0.3508	0.8096	1.6836
excess air			0.2606	1.9691					0.0333	2.2630
scavenging air			0.0162	0.0537					0.0009	0.0708
TOTAL(weight)	0.3692	0.1377	0.2768	2.0322	0.0049	0.0019	0.0000	0.3508	0.8438	4.0174
% WEIGHT	0.0919	0.0343	0.0689	0.5058	0.0012	0.0005	0.0000	0.0873	0.2100	1.0000
specific weight	1.9643	0.8036	1.4286	1.2500	2.8571	3.1654	1.5000	2500	0.8036	
TOTAL(volume)	0.1880	0.1714	0.1938	1.6257	0.0017	0.0006	0.0000	0.0001	1.0501	3.2314
% VOLUME WET	0.0582	0.0530	0.0600	0.5031	0.0005	0.0002	0.0000	0.0000	0.3250	1.0000
% VOLUME DRY	0.0935		0.0964	0.8088	0.0009	0.0003	0.0000	0.0001		1.0000
FLUE GAS TEMP 850.00	°C or	1123.16 °K								
ENTHALPY (KCAL/NM3 GAS)	441.53	827.42	295.66	279.58	425.00	266.09	280.50	5.31E+05	827.42	520.15
ENTHALPY (KCAL/KG SLUDGE)	82.99	141.80	57.29	454.51	0.73	0.16	0.00	74.55	868.89	1680.93
ENTHALPY (KCAL/KG GAS)	224.78	1029.68	206.96	223.66	148.75	84.06	187.00	212.50	1029.68	418.41

MASS BALANCE SYSTEM	VOL IN	WGHT IN	VOL OUT	WGHT OUT
SLUDGE + AUX FUEL		2500		
COMBUSTION AIR	3678	4750		
HUMIDITY IN COMB.AIR	77	62		
SCAVENGING AIR	100	129		
HUMIDITY IN SC.AIR	2	2		
FLUE GAS			5987	7443
CONDENSATES			0	0
TOTAL	3857	7443	5987	7443

6. Costs for both treatment methods

6.1. Capacity of the plant

All cost calculations are based on a plant capacity of 1 TDS/hr.

Characteristics of the sludge to be treated

DS content after mechanical dewatering by centrifuges : 40 % DS

ODS content : 10 % ODS/DS and 35 % ODS/DS

lower heating value organics : 5.500 kcal/kg ODS

DS content after drying : 95 % DS

6.2. Costs for drying plant

1447 kg H_2O/h evaporated

dryer type : 5207

Estimated cost :

1.	Dryer	32.900.000 BEF
2.	Sludge feeding system	6.600.000 BEF
3.	Vapour condensation	1.400.000 BEF
4.	Dried sludge handling	7.300.000 BEF
5.	Thermal oil generator	4.100.000 BEF
6.	Electricity and instrumentation	4.000.000 BEF
	Total Price	56.300.000 BEF

6.3. Costs for sludge incineration plant

6.3.1. Case 1 : auxiliary fuel firing

1000 kg DS/h

40 % DS

10 % ODS/DS

40 % DS to furnace

650 °C air preheating temperature

183 kg fuel/h auxiliary fuel

Condensed vapours

Beddiameter : 3.1 m

Freeboard diameter : 3.7 m

Estimated cost :

1.	Sludge feed to furnace	5.290.000 BEF
2.	Sludge incineration	47.800.000 BEF
3.	Flue gas cleaning and evacuation	35.200.000 BEF
4.	Electricity and instrumentation	10.500.000 BEF
	Total Price	98.790.000 BEF

6.3.2. Case 2 : predrying of sludge

1000 kg DS/h

40 % DS

35 % ODS/DS

54 % DS to furnace

650 °C air preheating temperature

Vapours to furnace

Beddiameter : 2.95 m

Freeboard diameter : 3.50 m

Estimated cost :

1.	Sludge feed to furnace	5.290.000 BEF
2.	Sludge drying	8.300.000 BEF
3.	Sludge incineration	46.800.000 BEF
4.	Flue gas cleaning and evacuation	33.900.000 BEF
5.	Electricity and instrumentation	11.500.000 BEF
	Total Price	105.790.000 BEF

7. Bibliography

- OVAM : Afvalstoffen plan 1991-1995

- W.J.Th. Van Gemert, J. Quakermaat and HJ. Van Veen.
  Methods for the Treatment of Contaminated Dredged Sediments

An Integrated Approach in Urban Drainage Infrastructure Design and Performance Assessment

Roumen Arsov
Higher Institute of Architecture and Civil Engineering
Faculty of Hydrotechnics
1 Chr. Smirnensky blvd,1421 Sofia, Bulgaria

I. Introduction

In global aspect about 50 to 90 percents of the sewer systems in operation are combined with overflows, which reduce the design flow load and the sewer network cost, but cause serious pollution to the receiving waters during rains (Fig. 1).

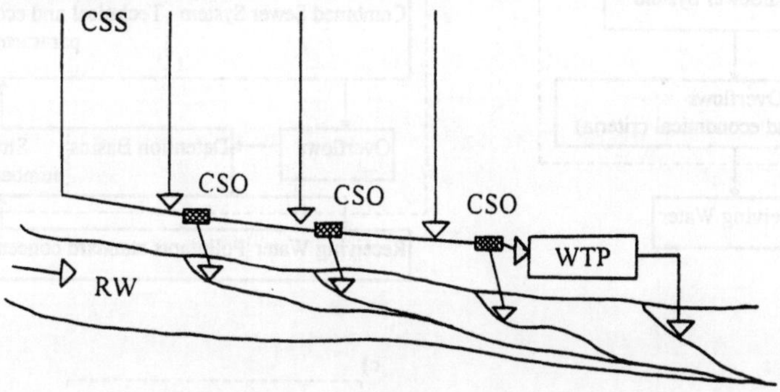

Fig 1

Scheme of the base elements of the Urban Drainage Infrastructure (UDI); CSS - combined sewer system; CSO - combined sewer overflow; WTP - waste water treatment plant; RW - receiving water.

Some rough estimations show that in this way about 3% of the global annual sewer dry weather volume falls into the rivers in the urbanised territories, which is very serious pollution load. In a result of the overflows operation associated with impulsive delivering of significant amount of contaminated mixture of sewage and runoff water, local spot depositions of heavy organic and mineral particles occur on the bottom of the receiving waters around these points. The spot depositions in their turn are sources of further river's pollution when spread down the streams by wet weather flushes.

The capacity of the natural water streams and basins in the urbanised territories to receive the wastewaters overflowing from the sewer during the rains, without violating the sanitary standards is limited. It depends on their morphology, flowrates or volumes, natural impurities, as well as on the intensity, duration, frequency of the rains and runoff water contamination.

Obviously the capacity of combined sewer overflows (CSO) have to be determined not only by technical and economical reasons, regarding the sewer network (as it is the traditional practice), but have to consider the real buffering capacity of the receiving waters (RW) in respect of their ability to accept additional pollution load without water quality standards violation. In this respect many administrations at the level of municipalities, river basins and states have established recently special programs for urban areas ecology improvement considering the CSO - RW interactions, as well as the waste water treatment plants performance and impact on the river water quality.

It is well known, that the waste water treatment plants (WTP), associated with combined sewer systems (CSS) have their own specific contribution to the receiving water quality regulation which adequate assessments demands relevant modelling, considering the sophisticated interactions between these objects.

Due to definite technological reasons, the waste water treatment plants are capable to receive only a limited part of the wet weather flow transported by the sewer networks. A significant part of this mixture,

NATO ASI Series, Partnership Sub-Series, 2. Environment – Vol. 45
Environmentally Devastated Areas in River Basins in Eastern Europe
Edited by Alfons Georges Buekens and Vasily Victorovich Dragalov

contaminated with anthropogenic wastes is directed into the receiving waters without purification just before the inlets of the treatment plants. In addition, the efficiency of the WTP's decrease during the rain, because they are designed to purify only the dry weather inflow and normally are only checked up for hydraulic conductivity exceeding during the rain no more than two or three times the dry weather flowrate, as it is by rule of thumb.

It is clear that the wet weather wastewater treatment plant capacity depends at grate extent on the mixed sewage and runoff water quality as well, which changes during the rain and demands adequate assessment.

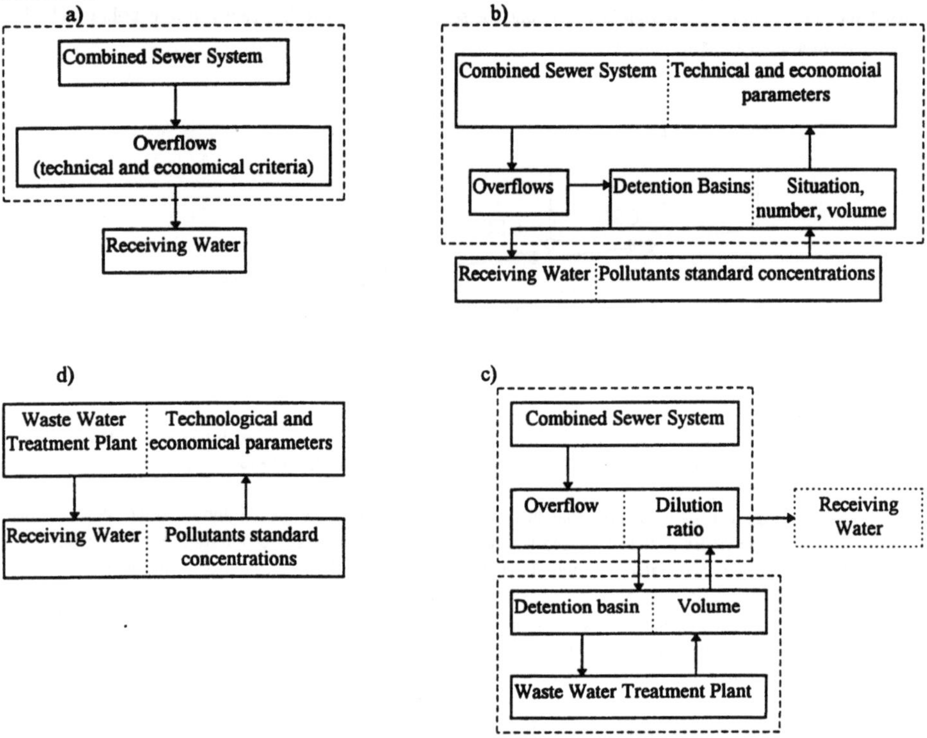

Fig.2
Scheme of the classic approach considering the bilateral interactions between the base elements of the of the urban drainage infrastructure.

Only bilateral relations between the three basic elements of the urban drainage infrastructure - CSS, WTP and RW are considered in the current practice up to now in their design, maintenance and performance assessment. These relations in the classical approach are normally "one way" directed or consider their reciprocal impact only of the two of the relevant elements (Fig.2).

The classical approach normally excludes the reciprocal impact of the combined sewer system (CSS) and receiving waters (RW), since the overflows' number, location and capacity determination considers only the technical and economical parameters of the sewer system, without adequate relation to the receiving water quality (Fig. 2a). Special attention is paid recently on the retention basins' role in regulation not only the sewer network capacity but the receiving water quality as well which demands considering of the CSS and RW parameters reciprocal impact (Fig. 2b).

In the classical approach the CSS' parameters impact on the parameters of the WTP is limited by the runoff dry weather sewage dilution ratio n_0 at the overflow just before the treatment plant inlet, which is normally fixed to 1 or 2, or vary in relatively narrow limits if there are detention basins at WTP (Fig.2c).

Because of the role of the wastewater treatment plants, their parameters are always related to the receiving waters' parameters, i.e. a reciprocal impact between these two types of objects exist in the classical approach (Fig. 2d).

Obviously, the design, maintenance, performance assessment and rehabilitation of the three basic elements of the urban drainage infrastructure (UDI) - RW, CSS and WTP must not be considered separately and independently any more. The contemporary general level of knowledge and technology achieved is a reliable base for an integrated approach, development and implementation in solving the problems in this field.

This paper is an attempt for contribution to such integrated approach development, trough revealing the RW, CSS and WTP reciprocal impacts, based on their contribution to the pollutants mass balance and transfer.

II. Pollution Mass Balance in the Receiving Waters

The receiving water quality standards keeping is based on the following general mass balance equation, taking into account the contribution of the three main elements of the urban drainage infrastructure - RW, CSS and WTP:

$$Q_{RW}.C_{RW} + Q_{CSO}.C_{CSO} + Q_{WTP}.C_{WTP} = \left(Q_{RW} + Q_{CSO} + Q_{WTP}\right)C_S \quad (1)$$

where Q_{RW}, Q_{CSO} and Q_{WTP} are the flowrates of the receiving water combined sewer overflows and waste water treatment plants, respectively; C_{CSO}, C_{RW} and C_{WTP} are the relevant pollutants' mass concentrations; C_S is the standard value (upper limit) of the relevant pollutants' concentration in the RW.

Normally the parameters Q_{RW}, C_{RW} and C_S are known. It is important to be noticed that the pollutants' concentrations C_{RW} are considerably affected by the organic sediments accumulated on the bottom of the receiving water around the combined sewer overflows, which could be resuspended by the flashes in the river bed. The adequate considering of a such events demands modelling of hydraulic and pollutants mass transfer processes in the receiving water, which is rather sophisticated and is beyond the scope of this paper. Taking into account the runoff/dry weather sewage dilution ratio n_o at the overflow, situated just before the WTP inlet, the Q_{WTP} during the rain is determined by the following formula:

$$Q_{WTP} = \left(n_o + 1\right)Q_d \quad (2)$$

where Q_d is a dry weather sewage flowrate.

The following balance equation is valid for the combined sewer system during a rain:

$$Q_{CSO} + Q_{WTP} = Q_r \quad (3)$$

where Q_r is the runoff flowrate.

Considering the equations (2) and (3), the mass balance equation (1) can be rewritten as follows:

$$Q_{r,l} = \frac{Q_{RW}(C_S - C_{RW}) + \left(n_0 + 1\right)Q_d\left(C_{CSO} - C_{WTP}\right)}{C_{CSO} - C_S} \quad (4)$$

where Q_r is replaced by the limiting runoff flowrate $Q_{r,l}$, such as the standard pollutant concentration C_S not to be violenced.

Every rain, producing a runoff flowrate Q_r higher than the limiting value $Q_{r,l}$ will cause an inadmissible pollution load to the receiving water. If fixed values for $Q_{R,W}$, $C_{R,W}$ and Q_d are adopted during a rain, the limiting runoff flowrate $Q_{r,l}$ will depend on the values of the pollutant concentrations C_{WTP} and C_{CSO}, as well as the dilution ratio n_0, affected by the same parameters.

As it was discussed in the chapter I, the waste water treatment plant effluent pollution concentration C_{WTP} is affected during the rain by the relevant influent waste water pollutant concentration (which is actually equal to the C_{CSO}) and by the technological parameters determining the WTP capacity in respect to the BOD, phosphorus, nitrogen and suspended solids removal.

III. Wastewater Treatment Plant Impact

The general symbol C_{WTP} could be attached either to the WTP effluent BOD concentration - C^L_{WTP}, or to the WTP effluent suspended solids (SS) concentration- C^{SS}_{WTP}, which are chosen here among the other possible pollutants (P, N, etc.) because of their main significance in the case and for simplicity of describing. The value of the C^{SS}_{WTP} could originate only by the inefficient performance of the secondary clarifiers during the rain and therefore depends on the WTP wet weather flowrate - Q_{WTP} and on the dilution ratio n_0, respectively.

As we show earlier [2], the WTP wet weather effluent BOD concentration - C^L_{WTP} is determined as follows:

$$C^L_{WTP} = 0{,}5\left[C^L_{CSO} - K_S - \frac{T_d.\mu_0.x_0}{(n_0+1)Y}\right] + 0{,}5\left\{\left[K_s + \frac{T_d.\mu_0.X_0}{(n_0+1)Y} - C^L_{CSO}\right]^2 + 4C^L_{CSO}.K_s\right\}^{0{,}5} \quad (5)$$

where T_d is the time of aeration in the bioreactor during the dry weather period, n_0 is the maximum value of the relative rate of BOD removal, h^{-1}, K_S is a saturation constant (equal to BOD concentration in the bioreactor at 0,5 μ_0), g/m^3, X_0 is the organic suspended biomass (activated sludge) concentration in the bioreactor, g/m^3; Y is the yeld coefficient (regarding the amount of organics transferred into the biomass cells), gBOD/g biomass.

The aerated bioreactor technological parameters T_d, μ_0, x_0, K_S and Y are known and therefore are fixed constants in equation (5). The analyses of the later show, that the value of the limiting runoff flowrate $Q_{r,l}$ depends on two variable parameters- C^L_{CSO} and n_0. The wet weather effluent BOD concentration C^L_{CSO} could not be affected by operational actions at the treatment plant. The dilution ratio n_0, if chosen (as it is normally) by the classical rule of thumb in the narrow interval, cited before, could hardly be accepted as an adequate process regulator. The value of n_0 however could be regulated in more rational base, regarding WTP, CSS and RW reciprocal impact.

It is clear by the equation (5) that the grater the dilution ratio is, the grater is the value of the wet weather effluent BOD concentration, which reflects negatively on the receiving water quality. But small dilution ratio means that greater part of the sewage during the rain will be directed into the receiving water without treatment. Obviously, the rational value of n_0 have to be related to the minimum value of the pollutants' mass releasing in the RW during a rain. As we show in [2], this approach leads to the following formula for optimal value of the dilution ratio n_0:

$$n_o = \frac{T_d.\mu_0.x_0}{Y\left(C^L_{CSO} - K_S\right)} - 1 \quad (6)$$

Assuming the usual mean values and ranges for $T_d = 5h$, $\mu_0 = 0{,}03h^{-1}$, $X_0 = 2000$ g/m^3, $Y = 0{,}6$, $K_S = 50$ g/m^3 and $C^L_{CSO} = 50 \div 550$ g/m^3, the relevant values of n_0 vary in the range from $+\infty$ to 0 respectively (Fig. 3). At $C^L_{CSO} > 550$ g/m^3 and $X_0 = 2000$ g/m^3, the dilution ratio n_0 becomes negative, which means that the aerated bioreactors are overloaded and can not manage such overpolluted waste water. The maximum value of the wet weather influent BOD concentration -$C^L_{CSO,max}$ could be obtained by formula (6), replacing $n_0 = 0$ in it:

$$C^L_{CSO,\max} = \frac{T_d.\mu_0.x_0}{Y} + K_S \quad (7)$$

The bioreactors performance and capacity in respect to a definite inlet pollution concentration C^L_{CSO} are considerably affected by the value of the biomass concentration X_0, which is known as an efficient and powerful remedy for the system's regulation. When the inlet pollutant concentration C^L_{CSO} is low and approaches the K_S value, the dilution ratio n_0 becomes unusual high and can approach to the infinity. Of course the high values of n_0 are inadmissible because of a limited capacity of the secondary clarifiers, which violencing could cause the biomass losses and additional receiving water pollution by organic suspended solids. As it is well known, the secondary clarifiers' performance can be regulated on the base of the Solids Flux Theory [4] and State Point Concept [3], which could be applied for the dilution rate upper limit determination as well.

During the rain, when the WTP inflow increases an additional and uncontrolled biomass transfer occur from the bioreactors to the secondary clarifiers. If the WTP operator does not undertake definite measures and the water flowrate exceeds a definite value (depending on the secondary clarifiers performance conditions and state) the biomass concentration in the bioreactor - X_0 decreases and the effluent BOD concentration C^L_{WTP} increases accordingly. In this situation an operator's intervention is necessary in other the biomass concentration X_0 to be kept at constant level. The operator's intervention concerns the activated sludge recirculation ratio, which usually have to be increased at proper level. These situations determine in a definite way [4, 5] the limit values of the biomass solids flux which could be applied to the secondary clarifiers and therefore reflects to the WTP wet weather inflow upper limit or relevant dilution ratio n_0 upper limit, respectively.

As it is shown in [2], using analytical solutions for some well - known graphical operations on the limiting solids flux function [5], the following formulas for the dilution ratio upper limit - $n_{0,l}$ are obtained for the cases with and without operator's intervention, respectively:

$$n_{o,l} = \alpha F / Q_d . X_0^{\beta} - 1; \tag{8}$$

$$n_{0,l} = \frac{X_0}{X_u - X_0} - \frac{1}{X_u - X_0}\left[\frac{\alpha(1-\beta)(X_0 - X_u)F}{Q_d X_0}\right]^{\frac{1}{\beta}} + \frac{\alpha}{X_0 Q_d}\left[\frac{\alpha(1-\beta)(X_0 - X_u)F}{Q_d X_0}\right]^{\frac{1-\beta}{\beta}}, \tag{9}$$

where F is the clarifiers water surface area; α and β are empirical constants, determining the solids flux function, with values in the ranges 19 - 30 and 1,5 - 2 respectively, according to Pitman [7]; X_u is the thickened biomass concentration at the bottom of the clarifiers. The second and the third additives in formula (9) have not significant influence on the $n_{0,l}$ value, which allows this formula to be simplified for rush calculations (with error about 6%) as follows:

$$n_{0,l} = \frac{X_0}{X_u - X_0} \tag{10}$$

The graphical presentations of the equations (6), (8) and (9) are shown in Fig.3 as an illustration of the optimal dilution ratio n_0 and it's upper limit - $n_{0,l}$ variations at different operational strategies concerning the bioreactors and secondary clarifiers.

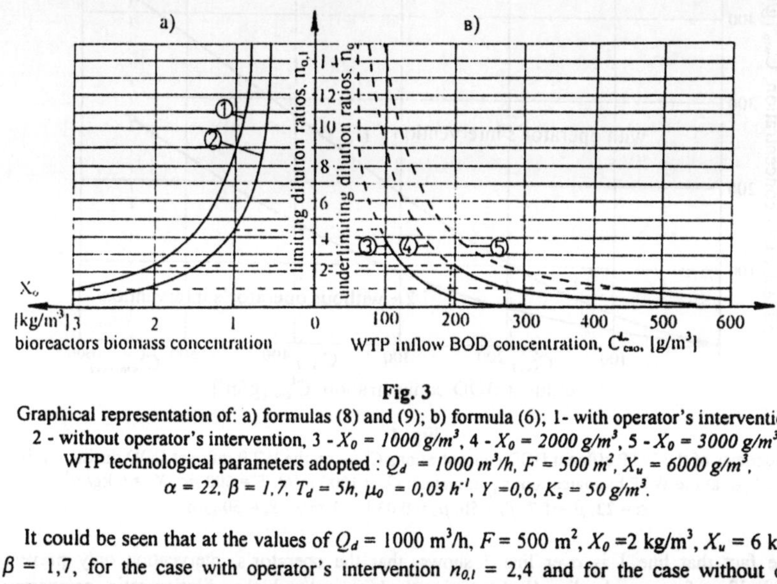

Fig. 3

Graphical representation of: a) formulas (8) and (9); b) formula (6); 1- with operator's intervention, 2 - without operator's intervention, 3 - X_0 = *1000 g/m³*, 4 - X_0 = *2000 g/m³*, 5 - X_0 = *3000 g/m³*. WTP technological parameters adopted : Q_d = *1000 m³/h*, F = *500 m²*, X_u = *6000 g/m³*, $\alpha = 22$, $\beta = 1{,}7$, $T_d = 5h$, $\mu_0 = 0{,}03\ h^{-1}$, $Y = 0{,}6$, $K_s = 50\ g/m^3$.

It could be seen that at the values of $Q_d = 1000\ m^3/h$, $F = 500\ m^2$, $X_0 = 2\ kg/m^3$, $X_u = 6\ kg/m^3$, $\alpha = 22$, $\beta = 1{,}7$, for the case with operator's intervention $n_{0,l}$ = 2,4 and for the case without operator's intervention $n_{0,l}$ = 1,03. The corresponding limiting values of the wet weather WTP influent BOD concentrations - C^L_{CSO} are 197 kg/m³ and 296 kg/m³ respectively and vary with the different values of the biomass concentration X_0 adopted. When the wet weather inflow BOD concentration C^L_{CSO} exceeds these limiting values, the dilution ratio limit values $n_{0,l}$ depends on the C^L_{CSO} according to the formula (6). When the wet weather influent BOD concentration C^L_{CSO} is lower than the relevant limiting values, $n_{0,l}$ does not depend on it and is determined by formulas (7) or (8), respectively. Therefore the limiting concentration $C^L_{CSO,l}$ corresponding to the crucial point in respect of the way of $n_{0,l}$ determination could be obtained by formula (6), rearranged as follows:

$$C^L_{CSO,l} = \frac{T_d . \mu_0 . X_0}{Y(n_{0,l} + 1)} + K_s \tag{11}$$

It is clear, that the limiting concentration $C^L_{CSO,l}$ depends entirely on the WTP technological parameters as well as on it's mode of operation. The higher is the value of X_0, the higher are the relevant upper limit values of C^L_{CSO}, which the bioreactors could manage. At the same time the higher is the value of X_0, the lower are the relevant dilution ratios upper limits $n_{0,l}$, which means that the biological processes of BOD removal deteriorate faster and easier during a rain at higher initial biomass concentration X_0.

Therefore, taking into account formula (6) and (11), the formula (5) could be rearranged as follows:

$$C^L_{WTP} = 0{,}5\left(C^L_{CSO,l} - C^L_{CSO,l}\right) + 0{,}5\left[\left(C^L_{CSO} - C^L_{SCO,l}\right)^2 + 4C^L_{CSO}.K_S\right]^{0,5} \tag{12}$$

The analyses of the formulas (11) and (12) show, that at definite WTP's biological units design and operating parameters, the wet weather effluent BOD concentration - C^L_{WTP} depends entirely on the inflow BOD concentration C^L_{CSO} as it could be expected. This dependence is illustrated graphically in Fig.4, where curve 1 reflects the case "with operator's intervention", and curve 2 reflects the case "without operator's intervention".

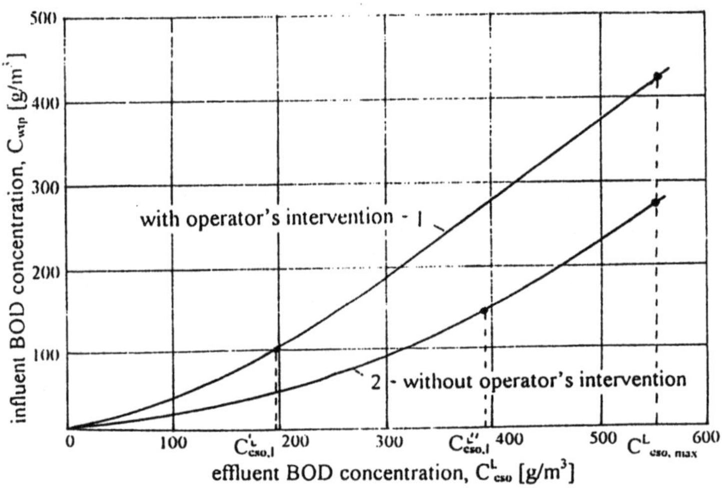

Fig. 4

Dependencies of the WTP effluent BOD concentrations C^L_{WTP} on the WTP influent BOD during a rain - C^L_{CSO} at the WTP biological units parameters: Q_d = 1000 m³/d, F = 500 m², X_u = 6 kg/m³, α = 22, β = 1,7, T_d = 5h, μ_0 = 0,03 h⁻¹, Y =0,6, k_s = 50 g/m³

The fact that line 1 is over line 2 shows, that the operator's intervention only prevents the secondary clarifiers from overloading (and escaping the biomass) at higher dilution ratio, comparing with the case without such intervention, but this in any case results in a higher effluent BOD concentration. Therefore the positive effect of the increasing the WTP capacity in respect of water quantity is compensated at great extend by the decreasing the capacity in respect of effluent quality, especially at higher values at the influent BOD concentrations. This have to be consider in the UDI optimisation as an entity. Taking into account formulas (4), (6) and (11) it appears that the wet weather sewage quality is the main factor, affecting the urban drainage infrastructure performance during a rain. Therefore the knowledge about the pollutant's concentrations values and its variations in the combined sewer network flow during a rain is of primary importance.

IV. Combined Sewer System Impact

The mixed runoff/sewage pollutants concentration - C_{CSO} is a time variable, represented by the relevant pollutograph. The later has to consider the contribution of catchment surface runoff pollutants mass transfer as well as the processes of resuspension of the sediments in the sewer pipe network and the pollutants mass transfer along the pipes between inlets and overflows or waste water treatment plant. Because of the specifics and diversities of the relevant processes on the catchment surface and in the pipe network, they would better to be described separately. For this purpose, it is worthy the urban drainage area to be divided into subcatchments, each drained by one inlet to the sewer pipe network..

a. Subcatchment Pollutograph Determination

The hydraulic processes of the catchment surface runoff movement and associated with them processes of the pollutants mass transfer depend on the land surface topology, soil and unpervious road materials properties, land use and urban planing. Because the influence and interactions of these factors are too complicated, the classic physically based runoff hydraulic models include a lot of simplifications which in general are not acceptable for the runoff quality adequate modelling . In this respect a geographical information systems (GIS) based approaches in runoff hydrographs and pollutographs modelling are more realistic and flexible.

As it is known the GIS allow more precise and realistic description of land topology and use, taking into account the updated terrain parameters, roads and buildings and initial pollutants' spatial distribution, which is a base for more adequate pollutograph modelling. The level of preciseness in such a modelling depends on the adopted grid-cells (pixels) dimensions (at raster based GIS) which in hydrology practice are no less than 10 m (usually - 50m or more for river basins). These dimensions however are too rough for adequate description of the urban areas. The decreasing of the pixels dimensions, from the other hand is restricted by the hardware capacity and in any case increases unproportionally the time of computations, which makes this approach unpractical for larger areas. This problem could be overcome through dividing the global urban drainage area into small subcatchments, drained by one inlet to the sewer network.

For modelling of an urban area runoff pollutograph we used [3] a raster based GIS - IDRISI[1], choosing the pixels' dimensions of 1m and creating overlays for the subcatchment surface properties and pollution's descriptions, which are as follows:

- digital elevation model (DEM) - 1 overlay;
- subcatchment boundaries - 1 overlay;
- land use (roads, parks, buildings, streams, etc.) - 1 overlay;
- Pervious and impervious areas properties (soil permeability parameters by Horton, volumetric water content, Manning coefficient, porosity, pollutants molecular diffusion coefficient, pollutant's soil distribution coefficient, pollution mass transfer coefficient) - 9 overlays;
- initial spatial distribution of a definite pollutant - 1 overlay for every kind of pollutant;
- rain hyetograph -1 overlay.

Every pixel is defined by the DEM with its plane coordinates *(x, y)*, elevation z and slope i. The rain water flowrate $Q_{z,r}$ causes the infiltration flow $Q_{z,f}$, the runoff Q_x, Q_y and the surface retention water volume $W_{x,y}$ on the pixel which in every moment are in dynamic equilibrium with the inflows from the surrounding pixels (Fig. 5). Three parameters - runoff flowrate Q runoff velocity v and column mean height - h are associated by each of the four walls of the water prism on the pixel. At least two of these twelve parameters are in general known by the subcatchment boundary conditions and the pixels slope direction. The rest are defined by a set of ten equations discussed in details in [3]. This set involves the De Saint - Venant's equations, water mass balance equation, flowrate conservation equations and plane water surface definition. Some of them are written for both the axes directions *OX* and *OY*. The infiltration flowrate $Q_{z,f}$ could be defined by Richard's physically based equation, but in [3], the well - known Horton's empirical equation is used for it's simplicity.

---

[1] The IDRISI Project, Clark University, Graduate School of Geography, Worcester, MA 01610, USA.

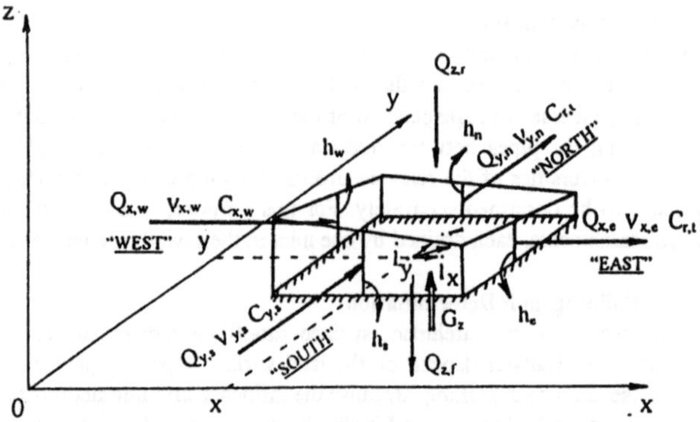

Fig. 5

Flowrates, velocities and pollutants mass and concentrations inflows and outflows at the rain water volume, retained on a pixel at a moment *t* from the beginning of the rain.

In these conditions the initial spatial distributed pollutants on the land surface, defined by the relevant overlay, start moving with the rain water flows. As it indicated in Fig.5 part of them moved into the soil (if the pixel is pervious) with the filtration flow $Q_{z,f}$, but as soon as the concentration gradient is established, the pollutants mass G_z move back to the soil surface, driven by the molecular diffusion factors. On the land surface the pollutant mass $G_{x,y}$ are transported by the runoff $Q_{x,e}$ and $Q_{y,n}$ with the relevant concentration $C_{r,t}$ in them, defined according to the following formula [3]:

$$C_{r,t} = \frac{C_{x,w}.Q_{x,w} + C_{y,s}.Q_{y,s} + C_{r,t-\Delta t}.W_{x,y/t-\Delta t} + \frac{a^2}{\Delta t}\cdot G_z}{Q_{x,e} + Q_{y,s} + W_{x,y/\Delta t} + W_{x,y/t-\Delta t}} \tag{13}$$

where $Q_{x,w}$ and $C_{x,w}$ are the flowrate and pollutant concentrations, respectively coming to the pixel (x, y) from the upper bordering pixel along the axes OX from the "West" direction (Fig.5), $Q_{y,s}$ and $C_{y,s}$ are the flowrate and pollutant concentration, respectively coming to the pixel (x, y) from the upper bordering pixel along the axis OY from the "South" direction; $C_{r,t-\Delta t}$ and $W_{x,y/t-\Delta t}$ are the pollutant runoff concentration and the water volume, retained in the pixel (x, y) at the previous moment $t - \Delta t$, respectively; $W_{x,y/\Delta t}$ is the water volume, retained in the pixel for the current time interval Δt; a is the length of the pixel's sides. All the parameters on the right side of formula (13) are known by the previous time and/or spatial step solution.

Adopting a time step Δt the necessary runoff hydrograph and pollutograph could be obtained for the outlet pixel for the relevant subcatchment. They could be applied to the combined sewer inlet by which the subcatchment is drained, as an input to the combined sewer pipe network pollutographs development. The surface runoff pollutograph modelling based on GIS is described and discussed in details in [3].

b. Determination of the Pollutograph in the Combined Sewer Network

Although the importance of the rainwater, surface runoff and sewage qualities influence to a definite pipe network pollutograph, the effect of the sewer sediments resuspension during the storm events has most significant quantitative impact on the relevant time depending pollution's concentrations. This event is important not only for suspended solids concentration assessment but concerns significantly the BOD concentration. As it was shown by the data, presented by Molokov and Shifrin [6], the BOD concentration in this condition could exceed more than 5 times the dry weather sewage BOD concentration, since the organic content of the combined sewer pipe sediments is about 6%. Therefore, the modelling of combined sewer sediment collection and distribution in time and space is of primary importance.

The physical base and the algorithm for this modelling is described and discussed in [2]. The model describes quantitatively the dry weather sewer sediments mass collection and the suspended solids - Δm_t resuspension by the mixed sewage and rainwater in a definite moment t from the beginning of the storm.

The description of the resuspended mass transportation in the pipe network in [2] is based on the relevant hydrographs obtained by the Figlus' conceptual model [1]. For each pipe the suspended pollutants mass flux - M_t at every moment t is defined by the formula:

$$M_t = \frac{(Q_r + Q_d).\Delta m_t}{R_d} + Q_d.C_d + Q_r.C_{r,t} \quad (14)$$

where Q_r and $C_{r,t}$ are the time depending runoff flowrate and the runoff pollutant concentration, respectively as an input from the relevant subcatchment inlet to the sewer pipe; Q_d and C_d are the dry weather flowrate and pollutant concentration, respectively; R_d is the dry weather flow hydraulic radius of the pipe.

Formula (13) regards both the pipe generated and the surface generated pollutant mass flux and therefore are affected by the relevant two types of pollutants transfer models, described herein.

The cumulative mass flux pollutograph - $M_{t,cum}$ at a given point down the sewer network is obtained by summarising the local mass flux pollutographs, contributed to the total mass flux at this point, considering the time of travelling from each contributing pipe to the point under consideration. The examined pollutograph of mass concentration $C_{CSO} = f(t)$ is obtained by dividing the relevant cumulative mass flux pollutograph - $M_{t,cum}$ to the relevant hydrograph:

$$C_{CSO} = \frac{M_{t,cum}}{Q_r + Q_d} \quad (15)$$

The pollutographs of mass concentrations at the combined sewer overflows and at the wastewater treatment plant affect directly the two other elements of the urban sanitary engineering infrastructure - RW and WTP, respectively.

The design and technological management of the UDI as an entity demands the establishment of an adequate and realistic strategy for admissible pollutants mass transfer into the water environment and development of the relevant tools for it's implementation.

V. A Possible Strategy for UDI Technological Management and Design

The establishment of an strategy for the UDI technological management demands the definition of a goal to be achieved and the possible and adequate remedies for it's achievement.

The goal could be defined as "keeping of the pollutants standards values - C_s in the receiving water during a rain". This goal can be expressed quantitatively in respect to the BOD concentration by the limiting runoff flowrate - $Q_{r,l}$, defined by formula (4). The later can be rearranged, taking into account the formulas (6) and (12) for the case when C^L_{CSO} exceeds the limiting concentration $C^L_{CSO,l}$ and the formulas (8) and (12) for the case with WTP operator's intervention and when C^L_{CSO} is lower then $C^L_{CSO,l}$, respectively:

$$Q_{r,l} = \frac{Q_{RW}(C^L_S - C^L_{RW}) + \dfrac{T_d \mu_0 X_0}{Y(C^L_{CSO} - K_S)}.Q_d(C^L_{CSO} - C^L_{WTP})}{C^L_{CSO} - C^L_S}; \quad (16)$$

$$Q_{r,l} = \frac{Q_{RW}(C^L_S - C^L_{RW}) + \alpha.F.X_0^{-\beta}.0{,}5.\{(C^L_{CSO} - C^L_{CSO,l} + [(C^L_{CSO} - C^L_{CSO,l})^2 + 4C^L_{CSO}K_S]^{0{,}5}\}}{C^L_{CSO} - C^L_S}, \quad (17)$$

where the limiting BOD concentration - $C^L_{CSO,l}$ is a constant, defined by formula (10), and depending entirely on the WTP technological parameters as well as on the mode of it's operation. For the case without WTP operator's intervention, when $C^L_{CSO} < C^L_{CSO,l}$, formula (4) is transformed as follows:

$$Q_{r,l} = \frac{Q_{RW}(C^L_S - C^L_{RW}) + (n_{0,l}+1).Q_d.0{,}5.\{(C^L_{CSO} - C^L_{CSO,l} + [(C^L_{CSO} - C^L_{CSO,l})^2 + 4C^L_{CSO}K_S]^{0{,}5}\}}{C^L_{CSO} - C^L_S}, \quad (18)$$

where the limiting dilution ratio $n_{o,l}$ is defined by the formula (9).

The receiving water parameters $Q_{R,W}$, $C_{R,W}$ and C_S are usually known, as well as the waste water treatment units and processes parameters F, X_0, K_S, α and β and therefore they are constants in formulas (16), (17) and (18). The time depending wet weather sewage BOD concentration - C^L_{CSO} is defined by the relevant pollutographs, associated with the overflows down by the combined sewer pipe network or at the WTP inlet. Formulas (16), (17) and (18) actually defined the limiting hydrographs $Q_{r,l} = f(t)$, associated with the definite point of the combined sewer network and definite WTP mode of operation. They should not be exceeded if the RW BOD concentration standard value - C^L_S has to be kept. If any hydrograph $Q_r = f(t)$ during a rain exceeds the limiting hydrograph (Fig.6), the excess wastewater volume W_r, limited by the two relevant hydrographs have to be retained in the combined sewer system (CSS).

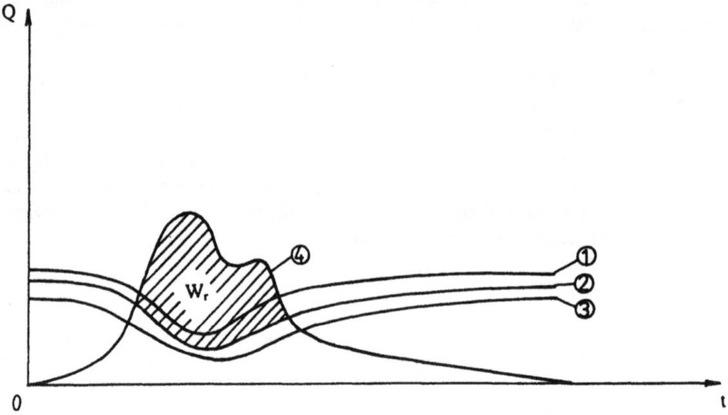

Fig.6

Graphical representation of the wet weather waste water volume W_r which has to be retained at the CSS in order the RW pollution standard concentration to be kept.

1 -limiting hydrograph $Q_{r,l} = f(t)$ at $C^L_{CSO} < C^L_{CSO,l}$, with WTP operator's intervention;

2 - limiting hydrograph $Q_{r,l} = f(t)$ at $C^L_{CSO} < C^L_{CSO,l}$, without WTP operator's intervention;

3 - limiting hydrograph $Q_{r,l} = f(t)$ at $C^L_{CSO} > C^L_{CSO,l}$;

4 - random CSS hydrograph $Q_{r,l} = f(t)$ at the relevant point of the sewer network.

It is clear from Fig.6 that at a given runoff hydrograph the wet weather sewage BOD concentration at the CSS - C^L_{CSO} and the WTP capacity and operation mode define the wastewater volume W_r necessary to be retained. The WTP operator's intervention however could result in decreasing of the volume W_r, when C^L_{CSO} is less then the limiting BOD concentration $C^L_{CSO,l}$, defined by formula (10).

Therefore, the detention (retention) basins appear to be the necessary remedy for the UDI strategy's goal achievement. These facilities have to be envisaged in the CSS initial or rehabilitation design project, considering the UDI as an entity.

The wet weather wastewater retention volume W_r, associated with a random hydrograph, resulted from a random hyetograph can be use mainly for on line UDI technological management, if the systems are equipped accordingly for an operation response in a real time. This is not however such a case with the detention basins design as a tools for the UDI technological management.

The detention basins design volume -W_{db} (adequate to the necessary retention wastewater volume W_r) should be fixed according to the chosen design steady state of the UDI as a whole system. In this case the Rational method is usually use for the CSS pipe's dimensioning, based an a linear (triangular) steady state design hydrograph with a flowrate peak -Q_c corresponding to the time to peak -t_c, equal to the time of concentration.

Accordingly the receiving water flowrate Q_{RW} is choose with 95% probability of occurrence in respect to the mean monthly value for the period of monitoring not less then 25 years. Similarly the RW organic pollutants concentration - C^L_{RW} is obtained as a mean background value for the period of monitoring. The WTP dry weather technological parameter Q_d, F, X_0, X_u, K_s, α and β are also fixed accordingly. The fixed maximum value of the CSS wet weather BOD concentration - C^L_{CSO} is

recommended to be used for the sake of security. It could be obtained by an expert's assessment, based on historical data or could be calculated, using the described above GIS based procedure, considering constant pollutants surface spatial distribution and a constant rainfall intensity which corresponds to a definite rain duration - t_r (Fig.7). The later causes a linear (trapezoidal) hydrograph $Q_r = f(t)$ at the Rational method prepositions if adopted.

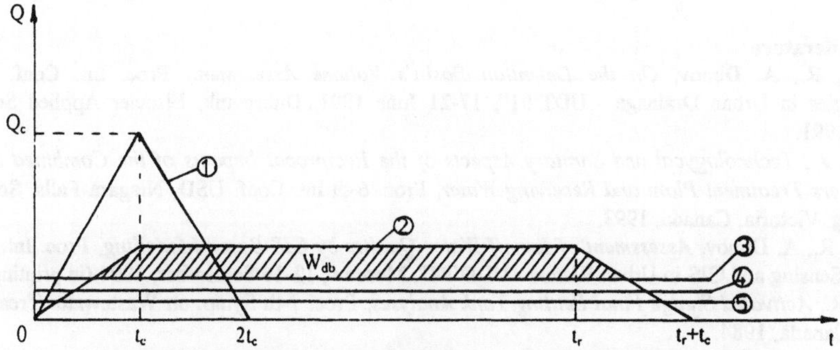

Fig.7
Schemes of the linear design hydrographs for the sewer pipes and detention basins' volume dimensioning, based on the Rational method and defined CSS steady state.
1 - triangular hydrograph for the sewer pipes dimensioning; 2 - trapezoidal hydrograph for the detention basins' volume dimensioning; 3 - limiting hydrograph $Q_{r,l} = f(t)$ at $C^L_{CSO} < C^L_{CSO,l}$, with WTP operator's intervention; 4 - limiting hydrograph $Q_{r,l} = f(t)$ at $C^L_{CSO} < C^L_{CSO,l}$, without WTP operator's intervention; 5 - limiting hydrograph $Q_{r,l} = f(t)$ at $C^L_{CSO} > C^L_{CSO,l}$.

Therefore, the limiting hydrographs $Q_{r,l} = f(t)$, associated with the relevant limiting BOD concentration - $C^L_{CSO,l}$ and the WTP mode of operation become linear (straight lines, parallel to the axes OX - Fig.7), adopting the relevant steady state conditions.

The parameters of the trapezoidal design hydrograph $Q_r = f(t)$, illustrated in Fig.7 are discussed in more details in [1], along with the optimal detention basins volume - W_{db}. The classical formula (19) is always valid for the W_{db} determination, providing the proper value of the parameter K_{db} is used in it:

$$W_{db} = K_{db}.Q_c.t_c. \tag{19}$$

The optimal value of the detention basin's coefficient K_{db} is recommended according to the Table 1 [1], depending on the adopted return period P, and the ratio $Q_{r,l}/Q_c$. As it was discussed above the later depends on the RW design parameters Q_{RW} and C_{RW}, as well as on the WTP technological parameters and the mode of it's operation, which demonstrates again the interactions and reciprocal impacts of the three main UDI elements.

Table 1

$Q_{r,l}/Q_c$	P, years					
	0,5	1,0	2,0	5,0	10,0	15,0
0,1	1,01	1,16	1,28	1,40	1,46	1,49
0,2	0,68	0,74	0,78	0,82	0,84	0,85
0,3	0,50	0,51	0,53	0,55	0,56	0,56
0,4	0,36	0,36	0,37	0,38	0,38	0,38
0,5	0,25	0,25	0,25	0,25	0,25	0,26
0,6	0,16	0,16	0,16	0,16	0,16	0,16
0,7	0,10	0,09	0,09	0,09	0,09	0,09
0,8	0,04	0,04	0,04	0,04	0,04	0,04
0,9	0,01	0,01	0,01	0,01	0,01	0,01

VI. Conclusions

The three main elements of the urban drainage infrastructure - receiving waters, combined sewer network and wastewater treatment plant depend significantly each to the other and can not be considered separately any more as it is the current practice up to now. An attempt is demonstrated here for a possible integrated approach in UDI design and performance assessment, which seems to be more adequate and realistic. It needs however a long way of verification and approval in the practice.

Literature

1. Arsov, R., A. Dimov, *On the Detention Basin's Volume Assessment,* Proc. Int. Conf. "New Technologies in Urban Drainage - UDT'91", 17-21 June 1991, Dubrovnik, Elsevier Applied Science, London, 1991.
2. Arsov, R., *Technologycal and Sanitary Aspects of the Reciprocal Impacts of the Combined Sewer, Wastewaters Treatment Plant and Receiving Water,* Proc. 6-th Int. Conf. USD, Niagara Falls, Seapoint Publishing, Victoria, Canada, 1993.
3. Arsov, R., A. Dimov, *Assessment of Runoff Water Quality by GIS Based Modelling,* Proc. Int. Conf. "Remote Sensing and GIS in Urban Waters - UDD'94", Moscow, 12-16 September, 1994 (in printing).
4. Dick, R., *Activated Sludge Final Settling Tank Analyses,* Proc. 7-th Symp. on Wastewater Treatment, Ottawa, Canada, 1984.
5. Keinath, T. M., *Operational Dynamics and Control of Secondary Clarifiers,* Proc. 57-th Annual Conf. of WPCF, USA, 1985.
6. Molokov, M. V., V. N. Shifrin, *Urban and Industry Territory Runoff Treatment,* Strojizdat, Moskow, 1977 (in Russian).
7. Pitman, A.B., South Africa, 1980.

Waste treatment in Moscow

V.V.Dragalov
Mendeleyev University of
Chemical Technology of Russia
123514, Miusskaya Sq., 9
Moscow, Russia

1. INTRODUCTION

The actual unfavourable ecological situation in the densely populated part of the world is largely due to the malfunctioning of the system for treating or rendering harmless municipal solid waste, as well as industrial and hazardous waste.

The present report is devoted to the analysis of problems of the treatment of such waste materials in Moscow.

Some 6 million tons of *industrial waste (IW)* and about 2,5 million tons of *municipal solid waste (MSW)* form in Moscow every year, not including sludge from sewage purification works. The rough composition of these types of solid waste is represented in the diagrams.

2. STOCKPILING OF WASTE

By now only 59% of IW are subjected to treatment. In principle 90% of IW can be used as secondary raw material and only 10% is eliminated.

The situation with MSW is even worse. Some 10 000 people and 3 000 special machines are engaged in Moscow in operations of collecting, transporting and retreating MSW, nevertheless dumping grounds still remain the main destination for their treatment (close to 90%). At present Moscow is serviced

NATO ASI Series, Partnership Sub-Series, 2.Environment – Vol.45
Environmentally Devastated Areas in River Basins in Eastern Europe
Edited by Alfons Georges Buekens and Vasily Victorovich Dragalov

by only two dumping grounds: "Timokhovo" (the Noginsk region, 113 ha) and "Khmetievo" (Solnechnogorsk, 5 ha), and both are about to be closed. A third dumping ground "Iksha" (the Dmitrov region, 25 ha) was already closed in 1991. A considerable amount of IW and MSW (approximately 1 mln t/year) are brought to non-official ripping sites which number 200-odd in Moscow (take up 1500 ha) and as many in the Moscow region (600 ha). A zone of pollution is usually some 20 - 30 times larger than the territory of the non-official rip site itself.

The non-sanctioned dumping grounds not only spoil the aesthetic appearance of the city, but also present serious epidemic danger. For instance, the microbe SDG number more than 1 000 times exceeds the maximum permissible.

3. TREATMENT OF WASTE

Only 10% of MSW in Moscow (or 4% for the cities of the ex-USSR) is treated in specialized enterprises. On an industrial scale only two methods of MSW treatment are used in Russia: burning in *incinerator plants (IP)* and a biothermal method (composting) applied at *refuse-processing plants (RPP)*. In the states of the ex-USSR there operate 9 IP's and 8 RPP's. Moscow boasts of only one IP (Special plant N 3 in Biryulevo) with a design capacity of 200 000 t/year. The other two plants (RPP №1 in Korovino and IP №2 in Beskudnikovo) are closed for reconstruction.

Equipment for IP №3 was supplied by the firm "Volund" (Denmark) and features energy recovers, in contrast to the two plants built earlier. The specific feature of this construction is an afterburning drum, installed downstream of a cascade of reciprocating grates, which makes it possible to burn out wet waste increased moisture content waste (up to 60%), yet recover energy in a boiler unit. Another structural peculiarity is the division of flue gases into two flows (a forward flow and a counter-flow) which somewhat contributes to the drying waste materials on a fire grate and to completing the combustion in the drum. The operation of the plant has shown, however, that the construction of the boiler unit and gas-treatment does not meet Moscow conditions. At present the flue gases purification system is being up-graded.

In the cities of the ex-USSR refuse incineration is practiced on a wide scale using boiler units with a roller grate manufactured by the firm "ChKD-Dukla",

Czecho-Slovakia. These boiler units with a capacity of 14 t of MSW/h are manufactured under the license of the firm "Deutsche-Babcoch", Germany, obtained in 1973. Moscow also purchased the boiler units of this firm, which are not mounted yet. The firing grate of the firm "Deutsche-Babcoch" comprises 6 rollers having a diameter of about 1.5 m, which provide stable conditions for predrying, ignition, combustion and complete waste outburn.

Moscow-based IP's are equipped only with a single stage of waste gas cleaning (electrostatic precipitator), which cannot ensure the concentrations of pollutants in waste gases, as imposed by sanitary norms (cf. Table 1, 2). The ecological safety of IP's and RPP's can be secured only through use of modern highly effective abatement technology, whose cost generally amounts to 30% of total capital investments in the construction of the plant. Moreover, their construction and selection of operating conditions necessarily follow a design choice and technical-and-economic solutions, which consider the averaged-out composition of waste treated and, accordingly, waste flue gases. The necessity of wet scrubbing must be substantiated with account taken of the cost of neutralizing and purifying the resulting waste water.

Among the existing methods of waste materials treatment, their combustion at IP's calls for the least capital investments and operating expenses (due to the absence of a waste sorting stage and the low level of marketable products obtained from separated utilizable waste fractions), it fulfills the sanitary task of rendering waste harmless and to secure thereat the ecological safety of the treatment process.

IP's can operate according to a no-waste diagram, which is achieved using ash and slag (providing these are non-toxic) in the building materials industry and in highway engineering.

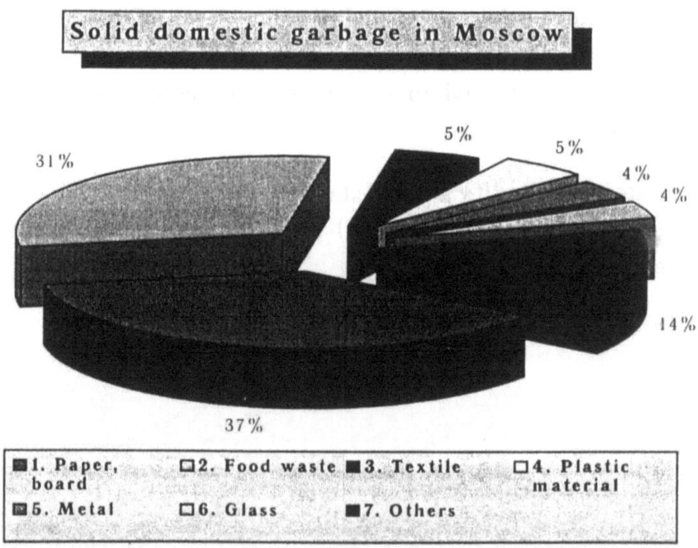

Note also the fact that the composition and properties of solid household garbage is distinguished by instability and changes depending on a degree of the good order of housing resources, season, industrial development, public catering, trade and other conditions

Heat generation, construction products and ferrous metal scrap can partially compensate for the expenses involved in IP construction and operation. The gas cleaning works of modern IP's are more effective than those of heat and electric power plants and, therefore, utilization of the heat generated at IP's has also an additional positive ecological effect. For the consumption of energy to be realized evenly all the year round, there must be selected a most favorable contingent of the users of energy in the case of each and every IP and an economically reasonable balance must be carried out along the lines of utilization of energy: heat, cold and electric energy.

However, given even the competent organization of a burning process and the realization of all the concomitant products, the capital expenditures of and operation expenses production cannot be compensated in full and IP's always remain expensive elimination lines. A great disadvantage of this method is the destruction of waste fraction that could become marketable products. At present active developments are carried out all over the world of the processes of comprehensive waste retreatment. The knotty problem here is represented by MSW treatment as being highly non-uniform and unstable in structure. Of the methods of MSW treatment in Russia today a most popular thus far is a biothermal method, in other words, a biological anaerobic process of the garbage composting, and along with this, MSW treatment includes:

1. Acceptance, sorting and fraction separation with the recovery of ferrous and non-ferrous metals.
2. Specially treated fractions composting.
3. Combustion of non-composted waste materials.

In 1972 the Moscow RPP-1 (in Korovino) was put into operation using this know-how practice, with 8 towers installed pair wise there. The fractions obtained by crushing, metal separating and mixing with drainage water acting at anaerobic process accelerants, are fed to the top of a tower and passed to its bottom within 24 hours. A composting cycle was accomplished upon passage of four towers. Nevertheless, this plant was closed in 1985 because of the low quality of the compost produced which hence was non-marketable, and the imperfection of the manufacturing process as a whole. Apparently, for the purpose of improving the efficiency and convenience in treating waste materials by composting, the latter should be carried out in cities or in regions where there exists a stable source of organic waste devoid of fractions capable of irreversibly worsening the quality of the compost produced.

More universal, though requiring great capital outlay, is a pyrolysis method of waste treatment. At present the Academy of municipal economy named after K.D.Panfilov is engaged in developing the production processes and equipment for the treatment of MSW into gas fuel, oil and activated carbon. MSW non-uniform composition calls for an especially thorough design choice for the fractionation of volatile pyrolysis products.

In selecting a method for rendering harmless MSW in Moscow one should remember that in the critical situation evolved the primary sanitary task resides in scavenging the streets from sweepings, which can be accomplished in a most effective way through the competent organization of dumping grounds and the construction of state-of-the art refuse incinerator plants (IP). However, in the foreseeable future it is necessary to increase the share of MSW retreatment, for only this method will make the rendering-harmless-waste process a more ecological and less cost-based one.

By now, pyrolysis methods should actively force out IW combustion processes. These waste materials are more stable in structure, and what is more, world practice has accumulated no little scientific and practical experience representative of the use of pyrolysis installations of different capacity which are adapted to retreat various types of IW. Just like up-to-date IP's, the pyrolysis installations may represent fully independent power processing plants, totally exclude the effluents of contaminants in amounts exceeding the European Standards and at the same time produce a certain range of marketable products Active in the field of IW and MSW treatment are the Tekhenergokhimprom and the above-mentioned Academy of municipal economy using the pyrolysis methods.

To sum up, a critical situation has long since created in Moscow with rendering harmless and utilizing waste materials that calls for urgent solution.

Apparently, the situation cannot be improved without a comprehensive approach to tackling this problem, which may comprise the steps of:

1. Allotting necessary plots for dumping grounds with observation of state-of-the-art waste-keeping-in-storage know-how practices precluding contamination of the neighboring territories.

2. The analysis of the state of active and non-operative dumping grounds and also non-sanctioned ones; elaboration of a complex of measures to liquidate or isolate them to avoid a further pollution of the environment.

Table 1 Contaminants contained in waste gases (Special plant №3)

Components	Content, мг/м^3	Norms: Bundesrepublik Deutschland 17.BlmSchV/4.90
Fly ash	800.0-3500.0	5.0
NO_x	10.0-160.0	200.0
SO_2	30.0-180.0	40.0
HCl	10.0-210.0	10.0
HF	0.07-3.0	1.0
CO	160.0-250.0	50.0
CO_2, %	4.0-6.0	
Carboxylic acids	25.0-49.0	
Alcohols	11.3-24.8	
Resinous compounds	5.0-10.0	
Pyrene	0.2-98.0	
Triphenylene	0.3-73.0	
1,3-Diphenoxybenzol	0.1-92.0	

Table 2 Heavy metals content in fly ash (Special plant №3)

METAL	CONTENT ;%вес
Mercury	0.0006
Lead	0.16
Chromium	0.06
Nickel	0.011
Tin	0.1
Cadmium	0.004
Copper	0.088
Strontium	0.045
Zinc	0.98
Cobalt	0.0015
Arsenic	0.0007

3. Modernization of existing refuse-processing/burning plants for the purpose of securing the stable work of the main processing equipment and a reduction of noxious substance concentrations in waste gases and ash up to maximum admissible concentrations.

4. The construction of new RPP's and IP's in keeping with the present-day requirements as to reliability in operation, no-waste technology, ecological safety and economic efficiency.

Literature references

A comprehensive territory - branch program of intensification of the socio-economic development of Moscow, "Progress-95", Moscow, Moskovskii rabochii, 1989, 135 pages.

Abramov N.F. Urgent problems of sanitary scavenging in the cities. // Housing and municipal economy. Moscow, 1991, issue 7, pp. 22, 23.

Ben'yanovskii D.N. Thermal methods of rendering harmless solid domestic garbage. Moscow, Stroiizdat, 1979, 192 pages.

Ben'yanovskii D.N., Tarasov N.M. Treatment of solid domestic garbage and some types of trade waste by a high temperature pyrolysis method // Problems of large cities. Review, issue 18, Moscow, GOSINTI, 1981, 30 pages.

Bernadiner M.N., Shourygin A.P. Fire treatment and rendering harmless trade waste. Moscow, Khimiya, 1990, 304 pages.

Ecological vestnik of Moscow. Moscow, 1991, issue 16, pp. 14-16.

Ecological vestnik of Moscow. Moscow, 1992, edition 3 - 5 (25 - 27), pp. 56-70.

Ecological vestnik of Russia. Moscow, 1991, issue 5, p. 60.

Glushkova V.G., Khorev B.S. Socio-ecological development of Moscow. Moscow, Znanie, 1988, 60 pages.

Glushkova V.G., Zholkov A.S. Realization of the planned socio-ecological development of Moscow. Moscow, 1989, 50 pages.

Goulyaev N.F. Removal, rendering harmless and treatment of garbage. Moscow, Stroiizdat, 1973, 130 pages.

Goulyaev N.F., Mirny A.N. Equipment of refuse - processing plants, Moscow, 1973, 75 pages.

Kuzmenkova A.M., Goulyaev N.F., Besfamil'ny I.V. Some problems of the study of properties of garbage in the city of Moscow // Transactions of the Academy of municipal economy named after K.D. Pamfilov.

Mirny A.N., Abramov D.N., Ben'yanovskii D.N., et al. Sanitary scavenging in localities. // Reference-book, Moscow, Stroiizdat, 1990, 413 pages.

Segedinov A.A. Moscow's infrastructure. Moscow, Moskovskii rabochii, 1986, 200 pages.

Villevald R.S., Ben'yanovskii D.N. Designing and operation of refuse - burning plants. Moscow, Stroiizdat, 1982, 52 pages.

Environmental Pollution in Colorado by Nuclear Processing

Thomas G. Sanders[1]
Eric Oppelt[2]
Vujica Yevjevich[3]
Department of Civil Engineering
Colorado State University
Fort Collins, Colorado, USA

Introduction

Many sources of pollution are just beginning to be identified in Colorado, especially since the emphasis on controlling point source pollution for the past twenty years in the USA has not cleaned up the environment. One source, mining, is important due to its volume of waste and its world wide impact, and the other is important not so much for the volume of nuclear wastes but for its toxicity. The production and effects of nuclear wastes has historically suffered from a lack of knowledge due to secrecy. Because of this, the amount and distribution of nuclear wastes are just beginning to be understood. This text reports on the extent of pollutant releases and initial cleanup procedures being conducted at the Rocky Flats Nuclear Processing Plant near Denver, Colorado, USA.

History of Rocky Flats Plant

The Rocky Flats National Defense Facility is located approximately 16 miles (26 km) northwest of Denver, Colorado and 12 miles (19 km) south of Boulder, Colorado (see Figure 1). The

[1]Associate Professor, Department of Civil Engineering, Colorado State University, Fort Collins, Colorado, 80523, USA.
[2]Engineer In Training, Fort Collins, Colorado, 80523, USA.
[3]Emeritus Professor, Department of Civil Engineering, Colorado State University, Fort Collins, Colorado, 80523, USA.

NATO ASI Series, Partnership Sub-Series, 2. Environment – Vol. 45
Environmentally Devastated Areas in River Basins in Eastern Europe
Edited by Alfons Georges Buekens and Vasily Victorovich Dragalov

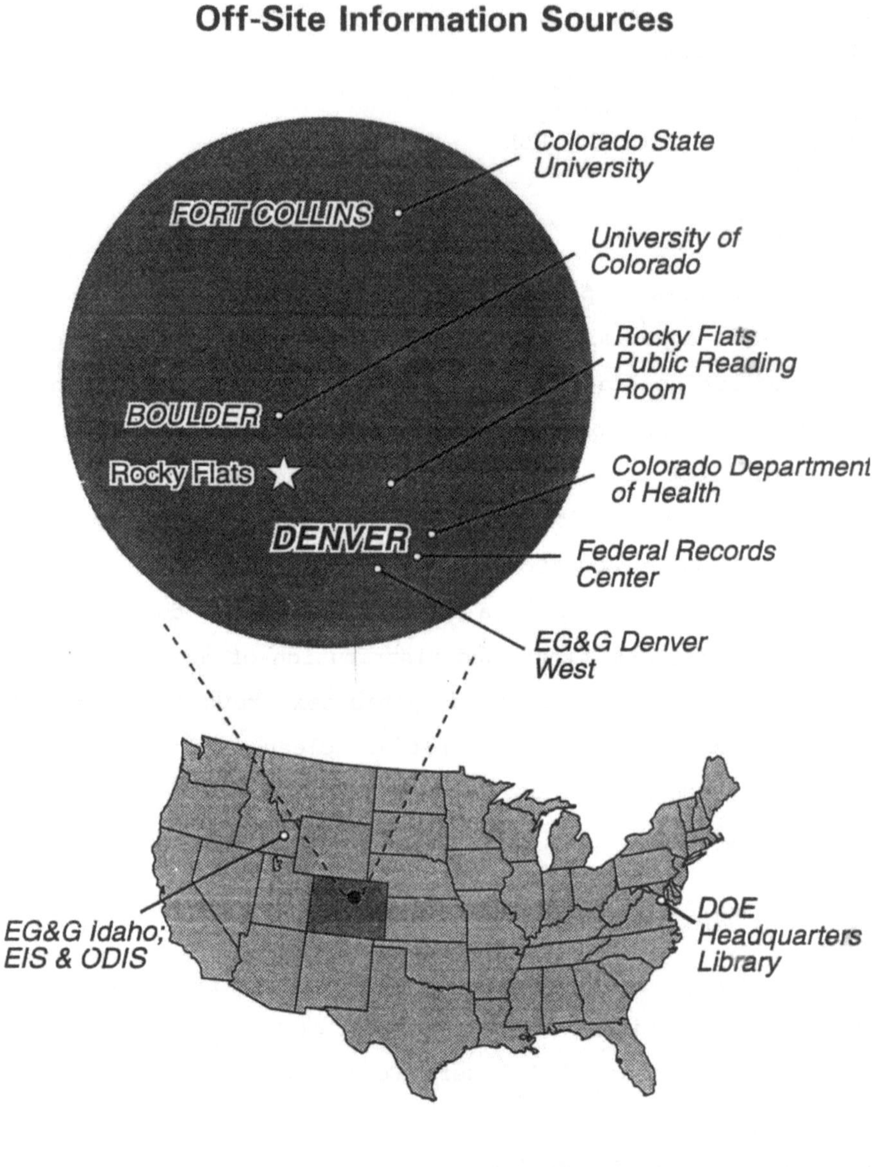

Figure 1. Location of Rocky Flats
(Colorado Council on Rocky Flats, 1993)

primary mission of Rocky Flats from 1951 to 1989 was the production of critical nuclear fission components for the nuclear weapons of the United States of America. The ultra-high security 384 acre (155 ha) plant lies within a 6,550 acre (2,650 ha) restricted preserve (Colorado Council on Rocky Flats, 1993). This land is on a flat, dry, treeless, grass covered plain at an elevation of approximately 6,400 ft (1,950 m) and has a climate similar to the dry steppes of Russia and China. The wind across this plain has achieved sustained velocities of 100 miles/hour (160 km/hour). The steep Foothills Mountain Range rises to the west of the plant area and the surface of the land slopes gently to the east toward the Eastern Plains of Colorado. There are no perennial streams that flow over this land, but there are three intermittent streams called Woman Creek, North Walnut Creek, and South Walnut Creek which collect surface runoff from the plant area. North and South Walnut Creeks flow eastward and drain intermittent flows from northern areas of the Rocky Flats Plant (RFP). Woman Creek flows southeastward and drains intermittent flows from the southern areas of the restricted reserve. The Walnut Creeks drain into Great Western Reservoir which supplies drinking water to the town of Broomfield and Woman Creek drains into Standley Lake which supplies irrigation water and drinking water to the towns of Westminster, Northglenn and Thornton, Colorado. An extensive network of retention ponds and drainage systems is in place at RFP to limit the possible migration of contaminated surface runoff into these natural water courses and drainages (ChemRisk, 1992). There is also a shallow alluvial aquifer below the plant area that contacts the surface downgradient from the plant, and here springs form and produce runoff. Below the shallow aquifer lie two deeper aquifers named the Arapahoe formation and the Upper Laramie formation that may be hydrologically interconnected with the alluvial aquifer. Below these aquifers lies the Laramie-Fox Hills Formation aquifer

which is considered for all purposes to be independent of the upper three aquifers because of an approximately 100 m thick Claystone layer (Colorado Department of Health, Technical Topic, 1993).

The creation of Rocky Flats Plant occurred with the Truman administration's interest in the research and increase in nuclear weapons production during the beginnings of the "Cold War" in the early 1950's. The Atomic Energy Commission announced in March of 1951 their decision to build the Rocky Flats facility. The owner of the plant is the U.S. Government and is under the auspices of the Department of Energy (DOE). The original name for the DOE when the plant came on line in 1951 was the Atomic Energy Commission (AEC), this changed to the Energy Research and Development Administration (ERDA) in 1974, and then to DOE in 1977. Rocky Flats is one of the seventeen facilities that are spread throughout the United States which make up the U.S. "Nuclear Weapons Complex". Two of these plants, one in Hanford Reservation, Washington and the other in Savanaugh River, South Carolina, were the initial major suppliers of the critical mass material to RFP after their reassignment in the late 1950's by the U.S. Department of Defense. This reassignment was to prevent redundancy in the "Nuclear Weapons Complex". This material from Hanford and Savanaugh River consisted chiefly of radioactive isotopes of Plutonium (Pu) and Uranium (U). The other major non-radioactive materials in RFP nuclear weapons components were the light, toxic metal Beryllium (Be), Aluminum (Al) and stainless steel. The suppliers of these non-radioactive materials varied over RFP's history. The DOE Pantex Plant in Texas is where the final assembly and disassembly of a large portion of the nuclear weapons for the U.S. took place. Pantex is where Rocky Flats sent most of its manufactured bomb components. In the later operations of RFP, the Pantex Plant was a major source of Pu through recycled bomb components. RFP received numerous other types of non-radioactive materials (such as chemicals, building

components, machinery, etc.) from outside vendors throughout the plant history (ChemRisk, 1992).

Since July of 1952, the DOE has contracted with private companies to operate the Rocky Flats Plant. In July of 1951, construction at the plant started, and in April of 1952, plant operations began. In July of 1952, plant operation was taken over by the Dow Chemical Corporation. In July of 1975, Rockwell International replaced Dow as primary contractor for RFP, and then in January of 1990, the present day plant operator, Edgerton, Germeshausen, and Grier (EG&G), assumed the position of primary contractor (Colorado Council on Rocky Flats, 1993).

Purpose and Activities of Rocky Flats Plant

The primary purpose for RFP from 1951 to 1990 was to make critical nuclear fission components for nuclear bombs. These primary components were assembled from materials made of purified Pu, U, Be, Al, and stainless steel. At first, these critical components and conventional explosives composed the complete bomb and were made from U^{235} and other isotopes of U and Pu. These were the types of weapons dropped on the Japanese cities of Hiroshima and Nagasaki. Later, around 1957, Pu was found to achieve criticality more efficiently, and with less mass allowing nuclear bombs to be carried as missile payloads or even in artillery shells. When the theory of the "Super", or hydrogen bomb, was explored, Pu was found to be useful as the "trigger" for these much more powerful fusion bombs. From this point on, Rocky Flats utilized Pu as the main critical component of fission "triggers".

The term "trigger" may seem innocuous enough, but to trigger a fusion reaction requires the intense energy involved in a fission bomb. At the highest level of technology at the Rocky Flats Plant, Pu^{239}<10 lbs (4.5 kgs) was machined into hollow spheres of precise geometry known as "pits". Also, with these "pits" were other critical and structural bomb components made from U, Be, Al, and stainless steel. These "pits" and other components were then sent to the Pantex Plant for final bomb assembly. At the Pantex Plant is where Tritium gas was placed within the "pit", and then conventional explosives on the outside (Colorado Council on Rocky Flats, 1993).

When a hydrogen bomb explodes, three steps normally occur; detonators in the conventional explosives ignite the conventional explosives imploding the Pu "pit" into itself, criticality of the Pu and other fissile material takes place causing a fission reaction, and then intense heat of the fission reaction initiates a fusion reaction within the Tritium. These reactions take place in less than one second. Weapons with the explosive force of 10^7 tons (9×10^9 kgs) of Tri-nitro-toluene (TNT) can readily be produced with this procedure (Colorado Council on Rocky Flats, 1993).

The Rocky Flats Plant is basically a complex set of buildings that contain metallurgy laboratories, foundries, and precise mechanical machining facilities. The heart of modern nuclear "triggers" is the Actinide Pu^{239}, and Rocky Flats contains extensive systems for processing, recycling, purifying, and machining this metal. Using processes that were similar for Pu and U, critical bomb components were initially made of enriched U ($U>0.7\% \ U^{235}$), and then later Pu^{239}. Besides the well documented and highly dangerous carcinogenicity of Pu, the elemental metallic form of Pu is highly reactive and pyrophoric. Pu can spontaneously burn to form Pu-oxides in the presence of oxygen. It is reported that Pu oxidizes with little flame or smoke, and that it reduces other materials, such as chlorinated solvents, in contact with which the Pu sponsors dangerous deflagrations. This behavior of the metal was the cause of three major fires within the facility in 1957, 1965, and 1969.

Between January of 1955 and December of 1974, 387 minor fires occurred within the Pu processing areas of the plant (ChemRisk, 1992). Pu is an alpha-particle emitter and can be contact handled inside of sealed glove boxes that have been evacuated with an inert gas such as nitrogen (O_2<5%). This is the primary mode of Pu processing and handling. Sometimes hundreds of glove boxes were linked together in one line, some equipped with conveyor belts between them (ChemRisk, 1992).

Plutonium reached Rocky Flats in several different forms. The first outside source of Pu for RFP was from the breeder reactors at Hanford and the Savanaugh River Plant. This plutonium initially reached RFP in solution as Pu-nitrate, and then later as five pound (2.27 kgs) ingots in need of purification. The second, and after 1960, major outside source was from the Pantex Plant in Texas. This was the plutonium recycled from old dismantled nuclear triggers, that because of fission degradation and bomb design changes, also needed purification (ChemRisk, 1992).

Plutonium was purified by one of three methods that depended on the form of the Pu and type of impurities. The "fast side" used purified Pu-nitrate in a set of precipitation reactions. The "slow side" involved leaching, dissolution and electrolytic stripping of impurities from Pu solutions, and then "fast side" precipitation. The third process, Molten Salt Extraction (MSE), became the preferred method to purify recycled Pu metal in 1967, such as old bomb components. The unusable impurities from the purification processes were stored at Rocky Flats or shipped to other sites for storage. One waste of this process was Americium, which was used in early smoke detectors and anti-static devices. Some money was made by the plant on Americium, but later models of these types of devices do not require Am, so excess Am became a waste again (ChemRisk, 1992). The purification process wastes with high enough Pu or U concentrations were either reprocessed or stored for later purification. It is reported that much of this waste still remains at the plant (Colorado Council on Rocky Flats, 1993).

The operators of RFP were very conscious of recycling and saving all the fissionable materials that they could from their processes. RFP operators historically set low Economic Discard Limits (EDL) on radioactive process waste streams. It would be nice to think that this was because of their concern for the well being of the environment and people within and around the plant. But, serendipitously for all of us, Pu and U and Be are among the most expensive materials be to produced by man. Little of this material could affordably be lost (ChemRisk, 1992).

The manufacturing of nuclear "triggers" is a highly technical, precise process with many steps performed within sealed glove boxes. As with any large metallurgy laboratory, Rocky Flats used extensive amounts of non-radioactive chemicals in the processes. Over 8,000 different kinds of chemicals were reported used at various times at RFP. The list includes both inorganics (such as acids and reagents) and organics (such as solvents and oils). Many of these are hazardous substances by themselves, let alone after they are contaminated with Pu or other radioactive materials. This type of "mixed-waste" poses many complex problems for disposal. On of the most dangerous releases of Pu at the Rocky Flats Plant occurred because of the problems with this type of "mixed-waste" (ChemRisk, 1992).

The present day mission of Rocky Flats was announced in 1992 by the then Secretary of Energy, Admiral James D. Watkins. The Rocky Flats mission was changed from weapons production to environmental restoration, waste management and production readiness in preparation for the future decontamination and decommissioning of the site. This was brought largely about by the changes in world politics and the dissolution of the former Soviet Union. The last proposed weapons mission for the Rocky Flats Plant was to be the production of "pits" for the proposed W-88 warheads for Trident submarines. In 1992, then President George Bush canceled the plans for further production of the W-88 warheads. The name of the Rocky Flats Plant was recently changed by the DOE to reflect the plant's new mission (Colorado Council on Rocky Flats, 1993). The new DOE name for RFP is now the "Rocky Flats Environmental Technology Site".

Concerns over Environmental Risks

Concerns over the environmental risks that RFP presents to the State of Colorado has been present since the proposal to build RFP in the 1950's. But because of strict secrecy during the first two decades of plant operation, little was known about RFP operations. It is reported that security was so strict during the 1950's that even the people who worked in the plant were bussed in and allowed access only to their specific work areas. Some incidences, such as the major fires, could not be hidden because of their magnitude, but even today many records of plant operation are kept classified.

The first real effort to provide public information about RFP came in 1976 with the formation of the Rocky Flats Monitoring Committee. During the 1970's, an increased awareness of environmental problems came to the U.S., and new laws were enacted by Congress that required the Environmental Protection Agency (EPA) to promulgate standards on pollutant emissions. Legislation like the Comprehensive Environmental Response, Compensation and Liability Act (CERCLA) came into being, and this provided funds for expensive environmental cleanups. In 1986, the DOE, EPA and Colorado Department of Health (CDH) signed an agreement to allow regulation of radioactive/hazardous waste at RFP. In 1989, a Technical Assistance Grant (TAG) was provided by the EPA for an independent public review of RFP technical documents. But the biggest change in the environmental policy of RFP operations came in 1989 when the U.S. Government, using helicopters and armed personnel, raided their own facility because of suspected violations of environmental laws. The Federal Bureau of Investigations (FBI) and EPA executed a search warrant on RFP and its then primary contractor, Rockwell International. In 1989, RFP was placed on the National Priorities List (NPL) of hazardous waste sites, allowing CERCLA (Superfund) funds to be used in the cleanup of RFP. Also in 1989, the Governor of the State of Colorado, Roy Romer, and the U.S. Secretary of Energy signed a Principle of Agreement (POA) about environmental issues at RFP which also included funds from the DOE. The POA calls for increased environmental surveillance and oversight, remediation, emergency preparedness measures, accelerated cleanup in high risk areas, and health studies.

In 1992, Rockwell International Inc. accepted a plea bargain and pleaded guilty to breaking some environmental laws, and was fined $18.5 million by a U.S. District Court. No individuals are found guilty. Also in 1992, the Rocky Flats Transition Plan was released to the public; this plan details the initial procedures for changing the RFP mission from weapons production to that of environmental restoration and cleanup. Phase I of the POA health studies has been completed by the CDH contractor ChemRisk, and Phase II of the health studies is being done by the CDH contractor Radiological Assessments Corporation (RAC). Phase II of the POA health studies began in October of 1992 and is scheduled to continue through 1995. The Phase I ChemRisk study of RFP is a toxicologic review and dose reconstruction of pollutant releases, and contains eight tasks. The final draft report of Tasks 3 & 4 (Reconstruction of Historical Rocky Flats Operations & Identification of Release Points) is the most definitive and lengthy study of its kind, and is one of the main sources for this text. Phase II, being done by RAC, is a toxicity assessment and risk characterization of RFP operations. Other agencies and groups also involved in the health studies of RFP include a 12-member Health Advisory Panel, Panel Subcommittees, Colorado State University (CSU), and ENSR Consulting and Engineering (Colorado Council on Rocky Flats, 1993).

Little definitive information is reported on the exact research and breakthroughs made in weapons production at Rocky Flats because of national security reasons, but the general trends in historical plant operation have been acknowledged. In the case of some reviews of RFP operations, classified documents have been examined by individuals to verify other plant reports. And helpfully, many once classified documents are now declassified (Health Advisory Panel, 1993).

RFP construction began in 1951, and facilities additions have been taking place since then. RFP increased from about 700,000 ft^2 (65,000 m^2) in building area in 1954 to more than 2.5 x 10^6 ft^2 (232,300 m^2) by 1990 (see Map 3). More buildings may be added as necessary for the cleanup process of RFP. RFP employed 133 people in 1954, and at its peak of employment in 1984, had almost 6,000 people (Colorado Council on Rocky Flats, 1993). The initial mission of the plant remained essentially the same during all its operation from 1953 to 1989. It is reported that there were only three basic "trigger" designs during all of the production at Rocky Flats, with some variations on these basic designs explored at times. Two of the designs were phased out in the first five years of RFP operations. RFP also did research into the safe handling of radioactive material and of nuclear weapons themselves. There were also many "Special Order" components made for nuclear weapons and other projects, but very little is known of most of these projects (ChemRisk, 1992).

Three Main Ways of Removing Hazardous Materials

There have historically been three main ways of removing hazardous and hazardous/radioactive materials from the RFP process streams, accidental releases, intentional releases, and the removal/storage of process waste streams (ChemRisk, 1992). The accidental releases are the more sensational of the three because of the potential for tragic consequences. Intentional releases have historically appeared to be governed by at least some common sense; either the amounts released were too minuscule to warrant further notice, or so well diluted that exposure to a hazardous off site dose was unlikely. But there is skepticism by some that maybe not all or the true extent of some releases has been reported. This can be settled some by the expected further declassification and public review of some RFP documents. The removal/storage of process waste streams has occurred since the beginning of RFP operations. Most wastes that could be recycled at RFP were, because of economic reasons. Those wastes that could not be recycled, or had been recycled to their EDL, were processed, stored on site, or shipped from RFP for off-site disposal (see Appendix 2). These wastes can be divided into eight categories based on their form, toxicity, and radioactivity:

Sanitary Sewage - This category is similar to that found in most large metropolitan areas and was treated similarly. There is an on-site sewage treatment plant at RFP, and much of the liquid effluent from this plant was evaporated in solar ponds or spray applied to enhance this evaporation. As seen in the case of the chromic acid spill at RFP, this system did occasionally receive dangerous contamination from RFP (ChemRisk, 1992).

Industrial/Municipal Solid Waste - Again, this category is similar to that of a large metropolitan area. In 1992, RFP was generating approximately 2,200 cubic yards (1,680 m³) of solid waste per month. There is the "Old Landfill" at RFP which is closed now, and is handled under the Resource Conservation and Recovery Act (RCRA) because of the disposal of some hazardous materials such as paint thinners. RCRA requires that the landfill be monitored for hazardous material migration, secured, and that appropriate measures be taken to insure the containment of the landfill waste. A new landfill has now been built and will handle only non-hazardous material (Colorado Council on Rocky Flats, 1993).

Hazardous Waste - This category includes such materials as solvents, acids, and alkalis and other chemicals that have been identified as hazardous by federal and state guidelines. Small quantities of this type of material were shipped off-site. The majority of this type of waste generated by RFP was stored on site in specialized waste management units. These waste management units are required to have a five-year interim status permit from the Colorado Department of Health. RCRA and Colorado Hazardous Waste Regulations (CHWR) govern the requirements of such a permit, which are similar to that of those required for landfills (Colorado Council on Rocky Flats, 1993).

Low-Level Radioactive Waste - This category is radioactive waste that does not contain Pu or Pu byproducts (such as Am), or waste that has Pu concentrations below that which is considered transuranic (TRU). The DOE regulates this "pure radioactive waste", and RFP shipped this waste to the Nevada Test Site for disposal. This waste must be stored indefinitely (Colorado Council on Rocky Flats, 1993).

TRU Waste - Transuranic waste includes those elements beyond U in the periodic table, Pu and its isotopes, and U^{233} and its daughter products. This DOE regulated type of waste is divided into two categories based on the danger of exposure in handling it, contact-handled or remote-handled. Contact handling means inside of sealed glove boxes, using lead-lined glove at times. Remote-handling means using robot arms to manipulate a radioactive material in a sealed lead lined environment. All of the TRU generated at RFP is contact handled. Pu^{239} composed the majority of this type of waste generated at RFP, and included Pu contaminated rags, tools, and laboratory glassware. Up until September 1989, TRU for RFP was shipped to the Idaho National Engineering Laboratory (INEL), but is now being stored at RFP pending the opening of the Waste Isolation Pilot Plant (WIPP) in Carlsbad, New Mexico (Colorado Council on Rocky Flats, 1993).

Low-Level Mixed Waste - This category is normally hazardous material that has been contaminated with radioactive material. The CDH regulates this waste under RCRA and Colorado Hazardous Waste Act (CHWA) guidelines. At RFP much of this type of waste has been partially immobilized in concrete blocks in one of two processes. Contaminated sludges from evaporation ponds are mixed with Portland cement to make large blocks of what is known as Pondcrete, and salts from waste process stream evaporators are mixed with Portland cement to make what is known a Saltcrete. Some of the Pondcrete and Saltcrete blocks that have started to crumble from weathering were needed to be placed inside of other containment boxes. The DOE plans to dispose of this waste in the Nevada Test Site (Colorado Council on Rocky Flats, 1993).

TRU Mixed Waste - This category of waste is more radioactive than low-level waste and also mixed with a hazardous waste. The majority of this waste is in solid form and stored in drums at RFP.

The DOE plans to dispose of this type of waste at the WIPP site when it opens (Colorado Council on Rocky Flats, 1993).

Stored Process Waste Streams - This type of mixed waste contains concentrations of Pu that are TRU waste, and possibly at levels above the EDL. This type of waste is stored on site and awaits further processing to remove the valuable Pu metal, and/or off site disposal (Colorado Council on Rocky Flats, 1993).

Releases of Pollutants from RFP into Environment

The types of pollutants released into the environment from RFP have been in all three phases, gas, liquid and solid, and are both radioactive and non-radioactive. It appears that airborne emissions have historically been the major route of exposure to humans outside RFP (Colorado Department of Health, 1993). These airborne emissions include mainly volatilized solvents and particulate borne radionuclides. Research being done at CSU seems to indicate that Actinides, commonly found at RFP, bond well to clay and silt limiting their movement into groundwater (Higley, 1994). Extensive groundwater monitoring being done on and around the plant has shown little or no off-site contamination by RFP. The solid releases of radionuclides has occurred both accidentally and intentionally. The main route of off-plant exposure to radionuclides appears to be the suspension and deposition of wind blown particulates to the east and southeast of RFP. There are also reported limited releases of the radioactive gas tritium. Liquid wastes have been released into the environment through accidental spillage, leaching from evaporation ponds, and a spray evaporation technique used to drain evaporation and retention ponds (ChemRisk, 1992).

Main Accidents - The main accidental releases of hazardous pollutants from RFP are as follows (ChemRisk, 1992):

9/11/57 In building 771, Pu metal shavings ignite in room 180. The fire burns for over 12 hours. Part of the HEPA filter plenum is destroyed leading to Pu release. An estimated 30,000 microcuries of Pu is released. No observed criticality. $818,000 in property losses.

1964-1967 At storage pad 903, 55 gallon (208 liters) drums leak Pu and U contaminated oil and solvents into soil under drum storage area. It has been estimated that approximately 5,000 gallons (18,900 liters) of liquids containing 150 grams of Pu leaked into the soil. To further compound the problem, soil under the pad is excavated for removal on a windy day causing a plume of Pu^{239} contaminated soil to be deposited east and southeast of the 903 pad (Higley, 1994). This plume contains 1.5 curies of Pu^{239} and is the worst known Pu^{239} soil contamination at RFP (Higley, 1994). An asphalt cap is now in place over the old 903 pad.

9/15/65 In building 776/777, a maintenance operation on a glove box drain ignited Pu and then other combustibles in a clogged drain. Twenty-five employees received 1 to 17 times the permissible lung burden of Pu. An estimated 7,000 microcuries of Pu^{239} is released. No criticality observed.

2/68 Tritium was accidentally released from a special project. An estimated 600 curies of Tritium is released through an emissions stack.

5/11/69 In buildings 776 and 777, a plutonium fire burns and spreads by combustibles through several hundred interconnected glove boxes. Fire burns for four hours. An estimated 1,500 microcuries of Pu is released. No criticality observed. There was $45,000,000 in property and decontamination losses.

3/19/73 In building 779-A, shipments of Pu from Lawrence Livermore Laboratories were discovered to have contained 500-2,000 curies of Tritium that was released to the atmosphere by mistakenly handling the Pu through non-tritiated Pu processing.

4/3/74 In building 707-A, maintenance operation on a glove box forced radioactive contamination through an exhaust stack because of a flow reversal in an inert gas recirculation system. An estimated 1,200 microcuries of Pu was released.

8/30/74-9/4/74 In building 777, 1.5 curies of Tritium was released from the building exhaust system.

2/22/89 In building 444, an engineer accidentally left a hose running overnight into a chromic acid plating vat causing the vat to overflow into the RFP sanitary sewer system. Contamination moved through the on-site sanitary sewage system and was spray applied to RFP spray fields. On site pond B-3 and

B-5 were contaminated with chromic acid. No radioactive contamination.

A graph of the historical releases of Pu from RFP compares the relative amounts of Pu released into the environment with each accident (see Figure 2).

Other incidents at RFP involved the spillage of small amounts of PolyChlorinated Biphenyls (PCB) from electrical transformers, and the use of pesticides in and around the plant area. The herbicide, Atrazine, has been detected at high levels in retention ponds at RFP. The historical evidence of off-site migration of PCB's and pesticides suggests little or no migration (ChemRisk, 1992).

There are thousands of minor accidents reported in the 37 years of RFP operations documentation, but only the ones listed in this report are considered a significant enough off-site health risk to warrant special attention (Health Advisory Panel, 1993).

It is important to note that no accidental criticality was ever reported during historical plant operations. Vast amounts of energy can be released during a criticality accident, including acute doses of neutrons, gamma, beta, alpha and X-ray radiation. Many precautions were taken by RFP operators to prevent such an accident, including criticality experiments. There were about 1,600 critical mass experiments performed in building 886. The very small amounts of critical mass used reportedly generated maximum energies of less that one watt/hour, and little radiation was reported lost to the environment (ChemRisk, 1992).

Intentional releases of contaminants from RFP occurred throughout the plants history and include both airborne (see Appendix 1) and liquid (see Appendix 2) emissions. The airborne emissions were both radioactive and non-radioactive. The non-

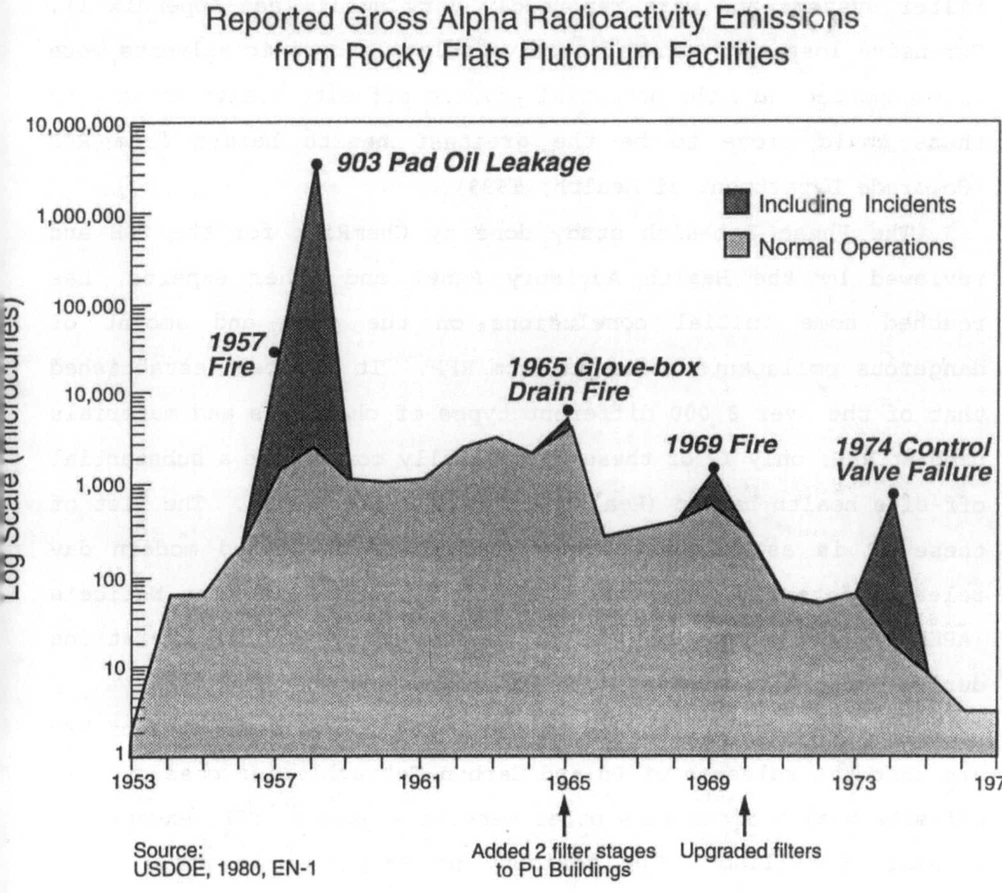

Figure 2. Historical Accidents and Releases at Rocky Flats
(ChemRisk, 1992)

Tritium airborne radioactive losses occurred mostly through HEPA filter systems and were reportedly very small (see Appendix 1). Extensive losses of non-radioactive volatile organic solvents were also reported, and the potential adverse off-site health effects of these could prove to be the greatest health hazard from RFP (Colorado Department of Health, 1993).

The Phase I health study done by ChemRisk for the CDH and reviewed by the Health Advisory Panel and other experts, has reached some initial conclusions on the type and amount of dangerous pollutants released from RFP. It has been established that of the over 8,000 different types of chemicals and materials used at RFP, only 12 of these historically could pose a substantial off-site health hazard (Health Advisory Panel, 1993). The list of these 12 is as follows, along with their estimated modern day releases determined through EG&G Air Pollution Emission Notice's (APEN's), or other EG&G and Rockwell reports of RFP operations during weapons production (see Table 1).

The conclusions reached in the Phase I health studies of RFP are that the releases of Pu and Carbon Tetrachloride pose greater off-site health risks than other materials used at RFP, except for possibly Beryllium. The toxicity of Beryllium is still being established through research, and until more definitive results are obtained about the lung disease Berylliosis, conclusions must be delayed. The major source of Pu off-site of RFP is from wind blown Pu contaminated particles from the old 903 drum storage pad and surrounding area. The plume of Pu contaminated soil lies mainly to the east and southeast of the old 903 pad. The main off-site route of carbon tetrachloride is from the volatilization of this solvent into the air. Carbon tetrachloride is being considered a greater off-site health problem than Pu or other radionuclides because of its potential carcinogenicity, and the great amounts lost into the air during historic plant operations (Colorado Department of Health, 1993).

Table 1. Materials Which Represent a Substantial Health Risk from Releases at RPF

Radioactive Materials of Concern (ChemRisk, 1992)	
Am^{241}	The estimated 1988 emissions from RFP were 2.02 microcuries into the air, and 118 microcuries into water contained on-site.
Pu -238,239,240,241,242	The estimated 1988 releases from RFP were 15.33 microcuries into the air, and 118 microcuries in water contained on-site.
Tritium	The estimated 1988 emissions from RFP were 0.015 microcuries into the air, and 0.23 microcuries into water contained on-site.
U -233,234,235,238	The estimated 1988 emissions from RFP were 11.93 microcuries into the air, and 24,300 microcuries into water contained on site.
Non-radioactive Materials of Concern (ChemRisk, 1992)	
Beryllium	Modern day emission estimate from RFP into air and water is 0.07 ounces (1.98 g)/year.
Carbon Tetrachloride	Modern day emission estimate into air and water is 80,800 pounds (36,700 kgs)/year.
Chloroform	Modern day emission estimate into air and water is 1,688 pounds (765 kgs)/ year.
Methylene Chloride	Modern day emission estimate into air and water is 6,660 pounds (3,021 kgs)/year.
Tetrachloroethylene	Modern day emission estimate into air and water is 2.28 ounces (65 g)/year, but this was probably higher during earlier RFP operations.
1,1,1-Trichloroethane	Modern day emission estimate into air and water is 62,800 pounds (28,485 kgs)/year.
Trichloroethylene	Modern day emission estimates into air and water is 304 pounds (138 kgs)/year.
Note: The majority of these solvents were lost into the air through volatilization, and not lost through water contamination.	

Environmental Cleanup

The environmental cleanup of RFP is now underway, and is expected to cost billions of dollars and take many years. There are 178 Individual Hazardous Substance Sites (IHSSs) that have been identified at RFP, and these have been organized into 16 distinct cleanup areas called "operable units" (OUs). These OUs have grouped the IHSSs according to one or more common features. These features include type and extent of contamination, type of contaminated media, possible remediation technologies, and historic use of the contaminated area. These OUs have also been ordered according to their potential risk to human health, with OU1 being the highest priority area. To initiate the cleanup process of these OUs, RCRA requires that strict remediation and closure plans be written and then approved by the EPA and the CDH before any work can take place. At RFP, these plans are labeled RCRA Facility Investigation/Remedial Investigation (RFI/RI) Reports and final Corrective Measures Study/Feasibility Study (CMS/FS) Reports (U.S. Department of Energy, Rocky Flats Information, 1993). These reports are necessary to prevent such problems as were created by the attempted cleanup of the 903 drum storage pad. The OUs at RFP area are listed below:

OU1 881 Hillside Area

The main concern here is contamination of groundwater by volatile organic compounds and proximity to Woman Creek (U.S. Department of Energy, Rocky Flats Information, 1993). The preliminary interim plan is to construct a French Drain system to intercept contaminated groundwater, and then treat and release this water on-site. Final remediation plans have yet to be completed, and are expected to take years to complete (U.S. Department of Energy, Rocky Flats Information, 1993).

OU2 903 Pad, Mound, and East Trenches Areas

The problems at the 903 pad stem from the contaminated oil that leaked into the soil. The Mound area had similar types of leaking drums stored there, and Trenches contain buried drums of sanitary sewage waste and radioactive material. The drums have all been removed except for some in the Trenches, and a asphalt pad placed over the 903 pad area. The plans for cleanup are still being researched (U.S. Department of Energy, Rocky Flats Information, 1993).

OU3 Off-Site Releases

Some of the land to the east and southeast of RFP has soil contaminated by Pu and Am that was carried there by windblown particles. The sediments of Great Western Reservoir, Standley Lake, and Mower Reservoir are also reported to contain some Pu and Am. Cleanup plans are yet to be finalized (U.S. Department of Energy, Rocky Flats Information, 1993).

OU4 Solar Evaporation Ponds

These solar evaporation ponds were used to store low-level radioactive waste, sanitary treatment plant effluent, and contaminated groundwater collected downgradient for the ponds. Cleanup plans are yet to be finalized (U.S. Department of Energy, Rocky Flats Information, 1993).

OU5 Woman Creek Drainage

Radionuclides, metals, and nitrates for OU1 and OU2 may have found their way into OU5. There is possible soil contamination in OU5 and of surface water, stream sediments, and soil in the Woman Creek

Drainage. OU5 also includes an old landfill, ash pits, and retention ponds. Cleanup plans are yet to be finalized (U.S. Department of Energy, Rocky Flats Information, 1993).

OU6 Walnut Creek Drainage

OU6 has problems very similar to OU5, and may contain elevated levels of radionuclides, metals, and nitrates. Cleanup plans are yet to be finalized (U.S. Department of Energy, Rocky Flats Information, 1993).

OU7 Present Landfill

The soil and water in this area may contain volatile organic compounds and metals, even though the landfill currently accepts only non-hazardous material. Cleanup plans are yet to be finalized (U.S. Department of Energy, Rocky Flats Information, 1993).

OU8 700 Area

This OU contains 38 IHSSs of numerous chemical spills that contain VOCs, radionuclides, metals, nitrates, acids, bases, and solvents. Final cleanup are yet to be finalized.

OU9 Original Process Waste Lines

The soils in this area may have been contaminated by leaking pipes of tanks used for transporting various types of process wastes. These possible contaminants include nitrates, acids, alkalis, and radionuclides. Final cleanup plans are yet to be finalized (U.S. Department of Energy, Rocky Flats Information, 1993).

OU10 Other Outside Closures

These areas consist of RCRA closure sites. These sites include hazardous waste storage facilities, tanks, and areas contaminated by leakage. Final cleanup plans are yet to be finalized (U.S. Department of Energy, Rocky Flats Information, 1993).

OU11 West Spray Field

From 1982 to 1985, this spray field was used to evaporate water from RFP solar ponds. The water from these ponds contained elevated levels of nitrates and other wastes. The vadose zone, and possibly groundwater, may have been contaminated. Final cleanup plans are yet to be finalized (U.S. Department of Energy, Rocky Flats Information, 1993).

OU12 400/800 Area

The soil in this area may have been contaminated from cooling tower ponds, fiberglass operations, leaks, and spills. Contaminants may include VOCs and other organics, metals, and acids. Final cleanup plans are yet to be finalized (U.S. Department of Energy, Rocky Flats Information, 1993).

OU13 100 Area

The soil in this area may have been contaminated from chemical storage areas, waste destruction areas, a valve vault, and other leaks and spills. The contaminants may include VOCs and other organics, depleted uranium, metals, acids, and alkalis. Final cleanup plans are yet to be finalized (U.S. Department of Energy, Rocky Flats Information, 1993).

OU14 Radioactive Sites

This area was/is used to store radioactive soils removed from around radiological operations buildings. Final cleanup plans are yet to be finalized (U.S. Department of Energy, Rocky Flats Information, 1993).

OU15 Inside Building Closures

This OU includes structures within buildings where hazardous materials were handled or processed. Final cleanup plans are yet to be finalized (U.S. Department of Energy, Rocky Flats Information, 1993).

OU16 Low-Priority Sites

This area covers areas that are considered least likely to cause health or environmental problems. Some soil contamination may have occurred by organics, solvents of nickel carbonyl. Final cleanup plans are yet to be finalized (U.S. Department of Energy, Rocky Flats Information, 1993).

There has not yet been much serious cleanup at RFP because of the careful planning needed to minimize problems created by the handling of the hazardous materials involved. Extensive tests are being done on soil, surface water, and groundwater to determine the extent of the environmental problems at RFP, and if these problems are migrating off-site (Higley, 1994). The proper analysis of this data is crucial to identify and prevent health risks, and will take time. Most of the RFI/RI and CMS/FS reports are not due until after 1994. The cleanup plans at different RFP OUs could be as elaborate as complete removal of the hazardous materials and contaminated soil to storage somewhere off-site, or just letting the area remain as it is and not disturb anything. The most health conscious, economical, safe way to handle some of this material may be not to handle it at all. There is some material at RFP though, that must be properly removed for national security reasons. There is reported to be approximately 7,000 - 14,000 pounds (3,175 - 6,350 kgs) of weapons grade Pu still stored at RFP (Colorado Council on Rocky Flats, 1993). There is also reported to be about 60 pounds (27 kgs) of Pu distributed within the ventilation ductwork of some buildings at RFP (Colorado Council on Rocky Flats,

1993). The final removal and storage of this Pu metal has yet to be finalized, and depends on the opening of long-term (indefinite) storage facilities for this radioactive material. Currently, there is much argument on where this material will be stored. Nobody wants highly radioactive material in their backyard.

There is no doubt that the final cleanup plans for RFP will take years to decide, and cost billions of dollars. Some of the most hazardous material will have to be removed, and other less hazardous material will probably be contained and then left alone. Even if the cleanup at RAP takes decades, it is just good to know that the RFP mission has changed to one of peaceful purposes and environmental cleanup.

General Conclusions

(1) The contaminations are the results of complex industrial processes, so that only very systematic and complex procedures, methods, and efforts may be successful in decontamination of nuclear industrial sites.

(2) If the peoples of the major nuclear powers had been informed of the dangers and the very high cost of these activities in the past, it is likely less pollution and health dangers would have been generated.

(3) Countries with science to become nuclearly armed should be well advised to study the problems and the related long-term costs in cleaning up the mess that this status and eventual security may impose.

References

ChemRisk, 1992, Final Draft Report, Reconstruction of Historical Rocky Flats Operations & Identification of Release Points, ChemRisk a Division of McLaren/Hart, 1135 Atlantic Avenue, Alameda, CA, 94501.

Colorado Council on Rocky Flats, 1993, The Handbook on Rocky Flats, Colorado Council on Rocky Flats, Colorado Department of Health, 4300 Cherry Creek Drive South, Denver, CO, 80222-1530.

Colorado Department of Health, 1993, Health Studies on Rocky Flats (Phase I), Toxicologic Review and Dose Reconstruction, Rocky Flats History - Citizens Summary, Colorado Department of Health, 1536 Cole Boulevard, Suite 325, Denver West Office Park, Building 4, Golden, CO, 80402.

Colorado Department of Health, 1993, Technical Topic: Movement of Contaminated Groundwater from the Rocky Flats Plant, Colorado Department of Health, 4300 Cherry Creek Drive South, Denver, CO, 80222-1530.

Health Advisory Panel, 1993, Report to Colorado Citizens on the Phase I Health Studies of Rocky Flats, Colorado Department of Health, 4300 Cherry Creek Drive South, Denver, CO, 80222-1530.

Higley, K.A., 1994, Personnel Interview, Graduate Student studying soil contamination at RFP with the Colorado State University Department of Radiological Health Sciences, Fort Collins, CO, 80523.

U.S. Department of Energy, Rocky Flats Information, 1993, Operable Units: Environmental Restoration Areas of Rocky Flats Plant, EG&G Community Relations, Rocky Flats Plant, P.O. Box 464, Golden, CO, 80402-0464.

Appendix 1

ROCKY FLATS AIRBORNE EMISSION POINTS

BUILDING AND IDENTIFIERS	MAIN INPUTS	STACK OR VENT TYPE	TYPICAL EXHAUST QUALITIES	EXPECTED CONSTITUENTS AND CONTRIBUTIONS TO SITE TOTAL
371C-001 Vent #1; System 1	Dicesium Hexachloroplutonate Ops., Chemical Standard Lab, and Maintenance Ops.	Louvered Penthouse Roof - 37 ft Stack - 4.0 ft	(1)	Nitric acid (NO_x): 0.0030% Beryllium: 11.2% Plutonium: 0.80% Americium: 0.8% Uranium: 2.8%
371C-002 Vent #2; System 2	Plutonium Analytical Support Lab and Maintenance Ops.	Louvered Penthouse Roof - 37 ft Stack - 4.0 ft	(1)	Nitric acid (NO_x): 0.06% Beryllium: 8.1% Plutonium: 0.83% Americium: 1.8% Uranium: 4.2%
374D-002 Vent #3	Waste Receiving and Neutralization	Circular Dia - 16.25" Roof - 37 ft Stack - 8.4 ft above roof	17.537 cfm 3.769 fpm 70° F	Beryllium: 0.67% Plutonium: 0.44% Americium: 1.3% Uranium: 0.58%
374J-001 Vent #'s 7,8,9	Waste Receiving and Neutrilization	Louvered Penthouse Dim. 72" x 54" (3 sides) Roof - 37 ft Stack - 3.6 ft above roof	115.430 cfm 13.22 fpm 70° F	Nitric Acid (NO_x): 5.4% Beryllium: 0.66% Methylene Chloride: 0.74% 1,1,1-TCE: 0.0081% Chloroform: 0.16% PCE: 100% TCE: 0.1% Plutonium: 0.93% Americium: 1.9% Uranium: 3.9%
444 Vent #122	Beryllium Machining Areas	(1)	(1)	Beryllium: 25%
444N-003 Vent #82	Production Plating Laboratory	Rectangular Dim. 30.5" x 24.0" Stack 10.7 ft above ground level	(1)	Uranium: 0.1% Nitric Acid (NO_x): 20%
444N-004 Vent #200 Filter Plenum Building 450	Foundry, Uranium Machining, Titanium Stripping, Assembly, Welding, Brazing Etching, and Coating	Rectangular Roof Height 14.8 ft Stack 2.0 ft above roof Dimensions 108"x96"	(1)	Nitric Acid (NO_x): 47% Uranium: 4.1%

BUILDING AND IDENTIFIERS	MAIN INPUTS	STACK OR VENT TYPE	TYPICAL EXHAUST QUALITIES	EXPECTED CONSTITUENTS AND CONTRIBUTIONS TO SITE TOTAL
447D-001 Filter Plenum 201 Building 451	Electron Beam Welding Maintenance Operations Vacuum Arc Melt Furnace Chip Roaster	Rectangular Roof Height 10.3 ft Stack 2.5 ft above the roof Dimension 60" x ?	Flow Rate: 64,000 ACFM Velocity: 2,133 ft/min Temp: 70° F	Beryllium: 6.9% Methylene Chloride: 0.045% Uranium: 2.4%
Building 460 High-bay Exhaust Vents 2, 4, 5, 6, 35, 36, 38, 39, 40, and 43	Product Inspection, Maintenance Activities	63"x63" vents, 40" above roof	Flow Rate: 34,820 ACFM Velocity: 1,263 FPM Temp: 70° F	TCE: 99.9% 1,1,1-TCE: 0.015% Chloroform: 0.74% TCE = trichlorethane
460-14 Rooms 117 and 118 Exhausts	Material Development Part Cleaning	12" dia stack 10 feet above roof	(1)	Carbon Tetra: 0.0066%
460-23 Hood Exhausts	Aqueous Assembly Cleaning	36"x36" vent 2' above roof	(1)	1,1,1-TCE: 0.011% Methylene Chloride: 0.045%
460-30 Hood Exhausts	Assembly Cleaning- Automated and Internal Cleaning Lines	24" dia stack, 10 ft above roof	(1)	Nitric Acid (NO_x): 3.4%
460-54 Room Exhaust Fan	Nondestructive Testing; Radiographic Testing	24" dia wall vent 13' above grade	(1)	1,1,1-TCE: 0.0009%
559A-001 Exhaust Vent #36 Building 561 Plenum	Gallium Determination, Emissions Spectroscopy, Uranium Analysis, and Plutonium Oxidation	Double Inverted J 57"x48" Roof Height 21 ft Stack 2.0 ft above roof	Flow Rate: 73,600 ACFM Velocity: 3067 ft/min Temp: 70° F	Beryllium: 6.5% Chloroform: 87.7% Nitric Acid (No_x): 0.26% Plutonium: 3.0% Uranium: 2% Americium: 3.7% Tritium: 2.6%
707 Vent 1/2	(1)	Mushroom Diameter 28" Roof Height 37.5 ft Stack Height 8.6 ft above roof	(1)	Beryllium: 0.33% Plutonium: 0.2% Americium: 1.4% Uranium: 0.6%
707 Vent 3/4	(1)	Mushroom Diameter 14" Roof Height 37.5 ft Stack Height 7.8 ft above roof	(1)	Beryllium: 0.33% Plutonium: Same as 707 Vent 1/2 (2) Americium: Same as 707 Vent 1/2 Uranium: Same as 707 Vent 1/2

BUILDING AND IDENTIFIERS	MAIN INPUTS	STACK OR VENT TYPE	TYPICAL EXHAUST QUALITIES	EXPECTED CONSTITUENTS AND CONTRIBUTIONS TO SITE TOTAL
707B-006 Vent 9/10	Pu casting, oxidation, storage, and shearing	2 Inverted J's Dim. 18"x18" Roof Height 37.5 ft Stack Height 3.0 ft above the roof	(1)	Beryllium: 0.31% Carbon Tet.: 8.32% Plutonium: 0.06% Americium: 0.08% Uranium: 0.08%
707B-003 Vent 28	Part assembly, weighing, testing, and inspection	Inverted J Dim. 36"x36" Roof Height 37.5 ft Stack Height 3.0 ft above the roof	(1)	Beryllium: 1.1% 1,1,1-TCE: 5.5% Carbon Tet.: 0.27% Plutonium: 0.80% Americium: 1.1% Uranium: 0.4% Tritium: 2.2%
707B-005 Vent 36	Pu casting ops, rolling and forming, briquetting, machining, and inspection	Inverted J Diameter 14" Roof Height 37.5 ft Stack Height 3.0 ft above the roof	(1)	Beryllium: 0.09% Carbon Tet.: 71.4% 1,1,1-TCE: 0.031% Plutonium: 0.01% Americium: 0.08% Uranium: 0.08%
707 Vent 38/39	(1)	Mushroom Diameter 43.5" Roof Height 37.5 ft Stack Height 10.2 ft above the roof	(1)	Berryllium: 0.33% Plutonium: same as 707 Vent 1/2 (2) Americium: same as 707 Vent 1/2 Uranium: same as 707 Vent 1/2
707 Vent 40/41	(1)	Mushroom Diameter 43.5" Roof Height 37.5 ft Stack Height 10.2 ft above the roof	(1)	Berryllium: 0.33% Plutonium: same as 707 Vent 1/2 (2) Americium: same as 707 Vent 1/2 Uranium: same as 707 Vent 1/2
707 Vent 42/43	(1)	Mushroom Diameter 43.5" Roof Height 37.5 ft Stack Height 10.2 ft above the roof	(1)	Berryllium: 0.33% Plutonium: same as 707 Vent 1/2 (2) Americium: same as 707 Vent 1/2 Uranium: same as 707 Vent 1/2
707 Vent 44/45	radiography	Mushroom Diameter 43.5" Roof Height 37.5 ft Stack Height 10.2 ft above the roof	(1)	Beryllium: 0.32% 1,1,1-TCE: 0.028% Plutonium: same as 707 Vent 1/2 (2) Americium: same as 707 Vent 1/2 Uranium: same as 707 Vent 1/2

BUILDING AND IDENTIFIERS	MAIN INPUTS	STACK OR VENT TYPE	TYPICAL EXHAUST QUALITIES	EXPECTED CONSTITUENTS AND CONTRIBUTIONS TO SITE TOTAL
707B-001 Vent 55	Assembly-superdry	Inverted J Dim. 22"x22" Roof Height 37.5 ft Stack Height 3.0 ft above the roof	(1)	Beryllium: 0.76% 1,1,1-TCE: 0.25% Plutonium: 0.03% Americium: 0.05% Uranium: 0.5% Tritium: 0.5%
707B-004 Vent 65	Pu machining, part degreasing, assembly testing	Inverted J Dim. 36"x36" Roof Height 37.5 ft Stack Height 3.5 ft above the roof	(1)	Beryllium: 1.2% 1,1,1-TCE: 7.92% Plutonium: 0.10% Americium: 0.3% Uranium: 0.7% Tritium: 0.9%
707B-002 Vent 75	(1)	Inverted J Dim. 30"x30" Roof Height 37.5 ft Stack Height 3.0 ft above the roof	(1)	Beryllium: 0.33% Plutonium: 0.21% Americium: 0.4% Uranium: 1.3% Tritium: 0.9%
707 Vent 76/77	(1)	Mushroom Diameter 43.5" Roof Height 37.5 ft Stack Height 10.2 ft above the roof	(1)	Berrylium: 0.33% Plutonium: same as 707 Vent 1/2 (2) Americium: same as 707 Vent 1/2 Uranium: same as 707 Vent 1/2
707 Vent 78/79	Calibration lab, instrument cleaning	Mushroom Diameter 43.5" Roof Height 37.5 ft Stack Height 10.2 ft above the roof	(1)	Berrylium: 0.39% 1,1,1-TCE: <0.0001% Plutonium: same as 707 Vent 1/2 (2) Americium: same as 707 Vent 1/2 Uranium: same as 707 Vent 1/2
707 Vent 80/81	Assembly brazing scanning, testing	Mushroom Diameter 43.5" Roof Height 37.5 ft Stack Height 10.2 ft above the roof	(1)	Berrylium: 0.40% 1,1,1-TCE: 0.49% Plutonium: same as 707 Vent 1/2 (2) Americium: same as 707 Vent 1/2 Uranium: same as 707 Vent 1/2
771C-001 Vent #86 Building 771 Main exhaust	Plutonium Recovery Facility	Stack Height: 145 ft Diameter: 120"	Flow Rate: 184,000 ACFM Velocity: 2,313 ft/min Temp: 70° F	Beryllium: 0.11% Methylene Chloride: 21% Nitric Acid (No_x): 11.1% Plutonium: 70% Uranium: 7.1% Tritium: 20.2% Americium: 64%

BUILDING AND IDENTIFIERS	MAIN INPUTS	STACK OR VENT TYPE	TYPICAL EXHAUST QUALITIES	EXPECTED CONSTITUENTS AND CONTRIBUTIONS TO SITE TOTAL
771C-002 Vent #9 Building 771C Main Plenum	Shipping and Counting Areas	Stack Height 21 ft Diameter 2.0 ft	Flow Rate: 8,279 ACFM Velocity: 2,635 ft/min Temp: 70° F	Beryllium: 0.13% Plutonium: 4% Uranium: 0.33% Americium: 5.8%
771C-005 Building 771C Room Plenum Vent #'s 2 & 8	Shipping and Counting Areas	Stack Height 21 ft Diameter 2.4 ft	Flow Rate: 10,695 ACFM Velocity: 1,168 ft/min Temp: 70° F	Beryllium: 0.12% Plutonium: 8% Americium: 9.1% Uranium: 0.32%
771G-001 Building 771A Main Exhaust Vent #67 Paint Hood Exhaust	Paint Stripping and Paint Applications	Stack Height 16 ft Diameter 1.1 ft	Flow Rate: 1,700 ACFM Velocity: 1,848 ft/min Temp: 70° F	No chemicals of concern listed
774D-001 Filter Plenum 202 Vent #4	Organic and sludge immobilization system (OASIS)	Inverted J Dimensions - 60"x16.3" Roof Height 26 ft Stack 1.5 ft above roof	(1)	1,1,1-TCE: 30.9% Plutonium: 2% Americium: 1.8% Uranium: 0.12%
776E-001 Plenum 250 Vent #24	Baler, Briquetting, Machining, Disassembly and Assembly Operations, and Radiography	Louvered Penthouse Roof Height 38 ft Stack 3.3 ft above roof Four sides (N,S,E,W), all rectangular Dimensions 244"x28"	Flow Rate: 72,800 ACFM Velocity: 150 ft/min Temp: 70° F	Beryllium: 9.0% Carbon Tet: 20% Methylene Chloride: 70% 1,1,1-TCE: 49.7% Plutonium: 1.3% Americium: 1.2% Uranium: 1.4% Tritium: 1.6%
776E-002 Plenum 206 Vent #32	Special Weapons Projects, Plutonium Metallography Lab, Ultrasonic Cleaning System, Foundry Operations, TCE Collection and Filtration System	Louvered Penthouse Roof Height 38 ft Stack 3.3 ft above roof Four sides (N,S,E,W), all rectangular Dimensions 244"x28"	Flow Rate: 19,000 ACFM Velocity: 39 ft/min Temp: 70° F	Beryllium: 7.2% 1,1,1-TCE: 4.8% Plutonium: 0.30% Americium: 0.24% Uranium: 0.36% Tritium: 52%
776E-003 Plenum 201/203 Vent #24	Baler, Briquetting, Machining, Disassembly and Assembly Operations, and Radiography	Louvered Penthouse Roof Height 38 ft Stack 3.3 ft above roof Four sides (N,S,E,W), all rectangular Dimensions 244"x28"	Flow Rate: 72,800 ACFM Velocity: 150 ft/min Temp: 70° F	Beryllium: same as 776E-001 Carbon Tet: same as 776E-001 Methylene Chloride: same as 776E-001 1,1,1-TCE: (same as 776E-001) Plutonium: 0.03% Americium: 0.05% Uranium: 0.15%

BUILDING AND IDENTIFIERS	MAIN INPUTS	STACK OR VENT TYPE	TYPICAL EXHAUST QUALITIES	EXPECTED CONSTITUENTS AND CONTRIBUTIONS TO SITE TOTAL
776E-004 Plenum 205 Vent #32	Special Weapons Projects, Plutonium Metallography Lab, Ultrasonic Cleaning System, Foundry Operations, TCE Collection and Filtration System	Louvered Penthouse Roof Height 38 ft Stack 3.3 ft above roof Four sides (N,S,E,W), all rectangular Dimensions: 224"x28"	Flow Rate: 19,000 ACFM Velocity: 39 ft/min Temp: 70° F	Beryllium: same as 776E-002 1,1,1-TCE: same as 776E-002 Plutonium: 0.16% Americium: 0.25% Uranium: 0.44% Tritium: 7.2%
776E-005 Plenum 204 Vent #24	Baler, Briquetting, Machining, Disassembly and Assembly Operations, and Radiography	Louvered Penthouse Roof Height 38 ft Stack 3.3 ft above roof Four sides (N,S,E,W), all rectangular Dimensions: 224"x28"	Flow Rate: 72,800 ACFM Velocity: 150 ft/min Temp: 70° F	Beryllium: same as 776E-001 Carbon Tet: same as 776E-001 Methylene Chloride: same as 776E-001 1,1,1-TCE: same as 776E-001 Plutonium: 0.36% Americium: 0.41% Uranium: 1.3% Tritium: 6.6%
776E-006 Plenum 251 Vent #45	(1)	Inverted J Roof Height 35.8 ft Stack 6.5 ft above roof Dimensions: 60"x32"	Flow Rate: 4,754 ACFM Velocity: 257 ft/min Temp: 70° F	Beryllium: 0.67% Plutonium: 0.08% Americium: 0.2% Uranium: 0.20%
776E-007 Plenum 252 Vent #44	(1)	Inverted J Roof Height 35.8 ft Stack 7.5 ft above roof Dimensions: 36"x27"	Flow Rate: 5,850 ACFM Velocity: 975 ft/min Temp: 70° F	Beryllium: 0.49% Plutonium: 0.13% Americium: 0.25% Uranium: 0.10%
776E-008 Plenum 202 Vent #17	(1)	Stack to Conical Hat Roof Height 38 ft Stack Height 14.8 ft Diameter 20.25"	Flow Rate: 6,000 ACFM Velocity: 25,000 ft/min Temp: 70° F	Beryllium: 0.85% Plutonium: 0.11% Americium: 0.12% Uranium: 0.10%
776E-009 Plenum 207 Vent #32	Special Weapons Projects, Plutonium Metallography Lab, Ultransonic Cleaning System, Fouridry Operations, TCE Collection and Filtration System	Louvered Penhouse Roof Height 38 ft Stack 3.3 ft above roof Four sides (N,S,E,W), all rectangular Dimensions 224"x28"	Flow Rate: 19,000 ACFM Velocity: 39 ft/min Temp: 70° F	Beryllium: same as 776E-002 1,1,1-TCE: same as 776E-002 Plutonium: 0.26% Americium: 0.6% Uranium: 0.55%

BUILDING AND IDENTIFIERS	MAIN INPUTS	STACK OR VENT TYPE	TYPICAL EXHAUST QUALITIES	EXPECTED CONSTITUENTS AND CONTRIBUTIONS TO SITE TOTAL
778H-001 Laundry	Laundry Facilities	Cylindrical Diameter 48" Roof Height 25.9 ft Stack 5.3 ft above roof	(1)	Plutonium: 0.77% Americium: 0.83% Uranium: 3.6%
779F-001 Vent #71 Building 729 plenum	(1)	Circular Stack Diameter 37.75" Stack Height 93.5 ft	Flow Rate: 17,707 ACFM Velocity: 2,361 ft/min Temp: 70° F	Plutonium: 0.21% Americium: 0.21% Uranium: 0.29% Tritium: 0.92%
779F-002 Vent #70 Building 782 plenum	(1)	Double Inverted J Dimensions 57"x48" Roof Height 20.1 ft Stack 1.8 ft above roof	Flow Rate: 61,506 ACFM Velocity: 4,316 ft/min Temp: 70° F	Plutonium: 4.1% Americium: 0.68% Uranium: 2.2% Tritium: 1.3%
865P-001 Vent Pair 58/59 Building 867	Beryllium Powder work, Research and Development of Metalworking Processes, Metallography Lab, and Grit Blasters	Rectangular Dimensions 56"x56.5" Roof Height 11.0 ft Stack 4.0 ft above roof	(1)	Beryllium: 0.03% Nitric Acid (No_x): 0.64% Uranium: 2.2%
865P-002 Vent Pair 63/64 Building 866	Beryllium Powder work, Research and Development of Metalworking Processes, Metallography Lab, and Grit Blasters	Rectangular Dimensions 60"x56.25" Roof Height 14.0 ft Stack 1.5 ft above roof	(1)	Beryllium: 0.03% Nitric Acid (No_x): 0.64% Uranium: 1.6%
881Q-001 Ducts 1,2,3 and 4. These ducts exit through the same stack as 881Q-002.	Research and Development Activities	Circular (4 outlets) Diameter 96.0" Roof height 32.6 ft Stack 8.0 ft above roof	Flow Rate: 14,258 ACFM Velocity: 2,159 ft/min Temp: 70° F	Benzene: 100% Beryllium: 3.3% Carbon Tet: 0.002% Chloroform: 11.4% Methylene Chloride: 8% Nitric Acid (No_x): 11.8% 1,1,1-TCE: 0.23% Plutonium: 0.45% Americium: 2.1% Uranium: 6.2% Tritium: 2.7%
881Q-002 Ducts 5 and 6. These ducts exit through the same stack as 881Q-001.	Research and Development Activities	Circular (4 outlets) Diameter 96.0" Roof Height 32.6 ft Stack 8.0 ft above roof	Flow Rate: 14,258 ACFM Velocity: 2,159 ft/min Temp: 70° F	Plutonium: 0.24% Americium: 0.19% Uranium: 3.8%

BUILDING AND IDENTIFIERS	MAIN INPUTS	STACK OR VENT TYPE	TYPICAL EXHAUST QUALITIES	EXPECTED CONSTITUENTS AND CONTRIBUTIONS TO SITE TOTAL
883R-001 Vent #44 Duct A Building 879 plenum	Rolling, Shearing, Blanking/Trepanning, and Forming of Depleted Uranium	Rectangular Dimensions 98.25"x52.75" Roof Height 15.2 ft Stack 7.0 ft above roof	Flow Rate: 102 CFM Velocity and Temperature were not available	Beryllium: 4.5% Uranium: 11%
883R-002 Vent #45 Duct B Building 879 plenum	Rolling, Shearing, Blanking/Trepanning, and Forming of Depleted Uranium	Rectangular Dimensions 98.5"x52" Roof Height 15.2 ft Stack 6.7 ft above roof	Flow Rate: 102 CFM Velocity and Temperature were not available	Beryllium: 4.5% Uranium: 17%
883R-003 Vent #34 Room 139 plenum exhaust	Rolling, Shearing, Blanking/Trepanning, and Forming of Depleted Uranium	Circular Stack Diameter 48" Stack height 69 ft	(1)	Uranium: 12%
886S-001 Vent #15 Plenum Building 875	Nitrate Storage Tanks in Rooms 101 and 103	Rectangular Dimensions 48"x24" Roof Height 17.3 ft Stack Height 1.5 ft	Flow Rate: 17,490 CFM Velocity and Temperature not available	Plutonium: 0.03% Americium: 0.04% Uranium: 1.1%
889T-001 Main exhaust plenum	No APEN available Equipment Decontamination	Cylindrical Stack Diameter 28" Stack Height 31.3 ft	(1)	Plutonium: 0.02% Uranium: 0.77%
991U-001 Building 985 plenum	Air handling system for the underground storage vaults 996, 997, and 999.	Rectangle Dimensions 48"x24" Roof Height 18.4 ft Stack 2.0 ft above roof	(1)	Beryllium: 1.4% Plutonium: 0.04% Americium: 0.29% Uranium: 0.80%
991U-002 Main exhaust	Production warehouse and non-destructive testing	Rectangle Dimensions 60"x54" Roof Height: North 2 ft; West 4 ft; South 20 ft Stack 3.7 ft above the roof	(1)	Plutonium: 0.06% Americium: 0.37% Uranium: 0.44%

Notes: (1) Not characterized in information sources identified to-date. Will be investigated further if required for emission modeling.
(2) The percentages for plutonium, americium, and uranium are included in the percentage provided for vent numbers 1/2 in building 707.

Nitric Acid emission percentages do not include emissions from the tank farm (APEN Building 218).
Methylene Chloride emission percentages do not include data from the sludge drying beds.
Chemical emission percentages are based on data from the Air Pollution Emission Notices.
Radi nuclide emission percentages are based on Department of Energy Effluent Information System totals for 1988.

Sources: EG&G Rocky Flats, 1991
Los Alamos, 1991
USDOE, 1991a

Appendix 2

ROCKY FLATS FACILITY WASTE DISPOSAL AREAS

DESCRIPTION	MAP AREA	NATURE OF DISPOSAL ACTIVITY
Spray Fields North of the Plant	H, I-3, 4	These spray fields were used shortly after the present landfill became operational in 1986, to spray water from two ponds over ground surfaces to enhance evaporation. The East Landfill Pond, also known as the existing landfill pond, and the West Landfill Pond were used to intercept groundwater that may have been contaminated by landfill leachate. The South Area Spray field was used first, until runoff was found to be draining into North Walnut Creek. Use of that field was discontinued, and use of the North Area Spray Field was also found to flow into North Walnut Creek. Spraying was then moved to the Pond Area Spray Field, and drainage flowed back into the existing landfill pond. In September of 1973, tritium and strontium were detected in landfill pond water. Several metals and radionuclides have been detected in a downgradient bedrock groundwater monitoring well installed in 1989, but may represent natural background conditions (USDOE, 1991b). In May of 1981, the West Pond was covered over as part of an expansion project for the existing landfill (USDOE, 1991b).
Trenches A, B, C	H, I-4	Trench A appeared to be active from 1964 to about 1974. Trench B was active in 1959, with date of closure unknown. Trenches A and B received uranium- and/or plutonium-contaminated sludge from the sewage treatment plant. Trench C is actually two separate trenches, that apparently were active from 1964 until 1974. Materials placed in Trench C have not been identified, but sewage sludge is most probable. Several metals and radionuclides and TCE have been detected in a groundwater monitoring well in Trench A. Metals and radionuclides may represent background (USDOE, 1991b). The trenches are no longer active. A road was built across Trenches A and C in 1978 (USDOE, 1991b).
Contaminated Concrete Slab Burial Area	I-5	A concrete slab with direct count (non-removable) americium contamination was buried here (Owen and Steward, 1974). The concrete slab was later excavated, and the contaminated portion of the slab was cut off for off-site disposal (Owen and Steward, 1974).

DESCRIPTION	MAP AREA	NATURE OF DISPOSAL ACTIVITY
Nickel Carbonyl Bottle Disposal Area	J-6	Between March and August of 1972, approximately 185 pounds of nickel carbonyl ("X-gas", $Ni(CO_4)$ contained in seven 25-pound cylinders, two 5-pound cylinders, and one lecture bottle were disposed of. A "dry well" hole about fifteen feet deep and three feet in diameter was drilled in a remote area of the plant site, and the cylinders were opened by individuals wearing supplied air packs and suspended in the hole until they were drained. In some cases, the chemical ignited immediately after release to the well. In other cases, the well remained silent for long periods before a muffled ignition occurred. Samples at the lip of the hole indicated concentrations around 10 parts per million during the disposal (Hobbs, 1972). The map location is the approximate location where empty nickel carbonyl bottles were buried after the chemical was destroyed by burning during the 1957 fire in Building 771 or when ready for discard. Explosive charges were used to destructively vent the cylinders and ignite any residual gas (Owen and Steward, 1974). There are reports that an additional 12 cylinders were vented and buried one-half mile north and west of the current sanitary landfill (Smith, 1975).
East Area Spray Field	L, M-5	The East Spray Field became operational in 1989 to provide additional area for spray evaporation of water from Pond B-3, which is sewage treatment plant effluent and local surface runoff (USDOE, 1991b). Use of this area was discontinued shortly after it became operational in late 1989 due to problems with excessive runoff (USDOE, 1991b).
Radioactive Soil Dump Area	K, L-5, 6	The Soil Dump Area received 50 to 75 dump truck loads of soil containing low levels of plutonium. The soil was excavated during construction of Parking Area No. 334 in the middle of the western half of the plant production area, and had been put there after excavation near Building 774, the waste treatment plant (USDOE, 1991b).
Trench T-1	K-7	Approximately 25,000 kg of depleted uranium chips in 125 drums were deposited in the trench during 1952-1962. The drums were covered with about 2 feet of fill dirt (Owen and Steward, 1974). Depleted uranium was put in the trench primarily due to the hazards of transporting the metal. All drums buried were from Building 444 (Putzier, 1970).

DESCRIPTION	MAP AREA	NATURE OF DISPOSAL ACTIVITY
Trenches T-2 through T-8	J-8 K, L, M-7	Approximately 100,000 kg of sanitary sewage sludge and about 275 flattened empty drums contaminated with uranium were disposed of in these trenches. Activities ranged from 800 to 8,000 dpm/g. T-4 also contains some uranium-plutonium contaminated asphalt planking from the 207 solar ponds. Estimated total alpha activity is between 100 and 150 mCi (Owen and Steward, 1974). The first sludge buried on the plant site dates back to July, 1954. Trenches T-2 through T-8 were used for sludge burial up to August 14, 1968, when the sanitary landfill became operational. Concentrations of radioactivity in the dried sludge have not varied much over the years; the maximum reported was 7,900,000 dpm/kg in June 1960, and the minimum was 840,000 dpm/kg in August, 1964. Earlier activity was primarily uranium, with probable increasing plutonium fraction leading up to primarily plutonium composition in later years (Putzier, 1970). Some contaminated asphalt planking discarded from Pond 2A repair work was buried in Trench T-4. Contamination was principally uranium, with minor Pu contamination possible. No quantitative data are available (Putzier, 1970).
Trenches T-9, T-10, and T-11	L, M, N-7	Trenches T-4 through T-11 are all located just east of the East Access Gate outside the security fence. The trenches, approximately 50 by 300 feet in size, were used from 1954 to 1968 for the disposal of flattened drums contaminated with uranium and plutonium. Activity ranges were from 800 to 8,000 dpm/g. Trenches T-4 and T-11 also contain some uranium and plutonium-contaminated asphalt planking from the solar evaporation ponds and quantities of sanitary sewage sludge (USDOE, 1986).

DESCRIPTION	MAP AREA	NATURE OF DISPOSAL ACTIVITY
Mound Area	K-7	A total of 1,045 drums of oil and solid waste were buried. Most contamination was depleted uranium, with some enriched uranium and possibly low-level plutonium (Owen and Steward, 1974). The first mound burial was in April, 1954. Drums were buried here steadily up to March, 1957, at which time uranium contaminated oil from 90 drums was burned. In April, 1957, another 79 were burned. The final burial was in September, 1958, involving 89 plutonium contaminated oil drums from Building 776. The distribution of waste drum sources was as follows: From B-444: 1,298 drums of oils, stillbottoms, sand, perclene From B-776: 90 drums of oils with carbon tetrachloride From B-881: 85 drums of oils From B-991: 79 drums of concentrated dry waste From B-771: 46 drums of oils with carbon tetrachloride From B-441: 9 drums of dry waste, paper, glass Assuming similar concentrations of plutonium as from 903 area drums, the mound contained about 285 grams of plutonium. After September, 1958, oil and coolant drums were moved to the mound area but were not buried. In July of 1959, they were moved across the road to begin accumulation in the Building 903 drum storage area (Putzier, 1970). Complete retrieval and off-site disposal were achieved in May, 1970. No plutonium was detected. Soil samples ranging from 0.8 to 112.5 dpm/g were attributed to 903 Area infiltration (Owen and Steward, 1974).
Pallet Burn Site	J-7	An area southwest of oil burn pit number 2 was used to destroy wooden pallets in 1965. The materials that may have been spilled on the pallets is unknown (USDOE, 1986). A 1974 summation of incidents affecting soils near Rocky Flats indicated on two maps the presence of a "pallet destruction area" south of Building 991. Other than indicating the site was active in 1968, no discussion was provided (Owen and Steward, 1974). There are also indications of pallet disposal activities in a burning pit south of Building 881 in 1965. In May of 1965, a pallet containing 3 sheets (60 kg) of depleted uranium was inadvertently burned in that pit. After discovery of the event, two barrels of contaminated soil were removed for shipment to Arco, Idaho for disposal (Young, 1965).

DESCRIPTION	MAP AREA	NATURE OF DISPOSAL ACTIVITY
Oil Burn Pit #2	J-7	A total of 1,082 drums of oil containing uranium were burned during 1957 and 1961-1965. The resulting approximately 10,000 cub feet of residues and some flattened drums were covered with backfill (Owen and Steward, 1974). A burning pit was cut near the mound, and burning of the contents of 169 drums took place in March and April of 1957. Oil burning area #2 is actually two parallel trenches essentially side by side near the mound. No further burning occurred until June 1961, after which time oils were burned frequently. May 1965 was the last month any burning took place. The total number of drums burned on-site was 1,093, but it is not clear how many were 30-gallon or 55-gallon drums. About 250-300 emptied drums were flattened and probably buried in trenches 3, 4, 5, 6, 7, and 8 or mounded over in the burning pit areas (Putzier, 1970). The pit was cleaned up and removed in the 1970s (USDOE, 1986).
Reactive Metal Destruction Site (the 952 area)	K-8	Approximately 400 to 500 pounds of metallic lithium were destroyed over 1956 to 1970. Residues, primarily non-toxic lithium carbonate, were buried. Smaller quantities of other reactive metals (sodium, calcium, and magnesium) and some solvents were also destroyed in this location (Owen and Steward, 1974).
Gas Detoxification Area	K-8	Building 952, utilized for Toxic Gas Storage, was located in this general area. The gas detoxification area referred to (USDOE, 1987 and Helmstadt, 1988) was most likely associated with the nickel carbonyl cylinders that were stored in Building 952 and later destroyed as described under Nickel Carbonyl Bottle Disposal Area in this table (Hobbs, 1972).
903 Drum Storage Area	J-7, 8	From 1958 through 1967, approximately 5,240 drums of oil containing radioactivity were stored at this location. Of these drums, 3,570 contained plutonium. Corroded drums lead to deposition of plutonium over an area of 98,000 square feet, which was covered with asphalt and fill material in November, 1969 (Owen and Steward, 1974). Over 1959 to 1966, the distribution of drum sources was as follows: B-776 (69%), B-881 (17%), B-444 (8.6%), B-883 (3.5%), B-771 (2.5%). Drums were moved to the area after 1966. Some of the uranium contaminated oils at the 903 area were burned. The contents of 191 drums were processed for Pu recovery at the 903 filter plant. With the transfer of contents into new drums, the equivalent of 4,826 55-gallon drums were transported to Building 774 for solidification. Of these, 3,572 contained Pu contaminated coolant (Putzier, 1970).

DESCRIPTION	MAP AREA	NATURE OF DISPOSAL ACTIVITY
800 Area Radioactive Site Liquid Dumping Area Chemical Burial Area	I, J-8, 9 I-8 I-8	From 1951 until 1972, portions of the "Hillside Area" near Building 881 were used as oil sludge pits, chemical burial sites, liquid disposal sites, solvent drum storage sites, and fire damage refuse disposal sites. As a result, soil and groundwater have been contaminated with volatile organic compounds including carbon tetrachloride, TCE, and PCE. Alluvial groundwater contains 1,1,1-TCE, and chloroform. Uranium was the only radionuclide occurring above estimated background concentrations (USDOE, 1990). One of the Solid Waste Management Units (SWMUs) in the 881 Hillside area, SWMU 130, is sometimes called the 800 Area Radioactive Site #1 (Helmstadt, 1988).
Contaminated Soil Burial	I-8	Plutonium contaminated soil from the periphery of Building 774 waste storage tanks was buried here. The soil averaged 250 dpm/g. The 240 drums of soil were buried under 3 feet of fill dirt (Steward, 1973).
Asphalt and Soil Burial	I-8	Approximately 320 tons of plutonium-infiltrated asphalt and soil from the 1969 Building 776 fire were buried in 1969 under 1 to 2 feet of fill dirt. Less than 1 mCi of plutonium is estimated to be dispersed in about 250 cubic yards of material, with an estimated alpha activity of about 7 dpm/g. About 60 cubic yards of plutonium contaminated soil from the Building 774 waste storage tank area was placed on top of the asphalt disposal area in 1972, and covered with 3 feet of fill dirt. Estimated activity of the soil was less than 250 dpm/g total long-lived alpha (Owen and Steward, 1974). Total contained plutonium is estimated at 0.97 mCi or about 14 milligrams (Putzier, 1970).
Oil Sludge Pit	I-9	Approximately 30 to 50 drums of oil sludge from a storage tank cleanout were emptied into a pit, which was then backfilled. No radioactivity was involved (Owen and Steward, 1974).
Concrete Slab Disposal Area	H-8	An area of several hundred square feet northwest of Building 881 was involved in storage of a contaminated concrete slab in 1958. The slab had been removed from the east side of Building 776 (Owen and Steward, 1974). The slab was later broken up, removed, and the area cleaned (Owen and Steward, 1974).

DESCRIPTION	MAP AREA	NATURE OF DISPOSAL ACTIVITY
Original Landfill	F, G-9	The original plant landfill was used from 1952 to 1968 to dispose of general plant wastes. An estimated 20 kg of depleted uranium ash is buried along with normal plant waste, including small quantities of various chemicals. The 20 kg of depleted uranium resulted when 60 kg was inadvertently burned and only 40 kg were recovered (Owen and Steward, 1974). The landfill may have received nonradioactive hazardous chemical wastes generated at the plant, including solvents. A reported old graphite dump located south of Building 440 that might have received beryllium and uranium was actually the original plant landfill (USDOE, 1986).
Former Incinerator	D-9	From 1952 to August, 1968, all combustible noncontaminated waste from the Rocky Flats Plant was incinerated in Facility 219 along the west access road. All noncombustible noncontaminated trash and ashes from the incinerator were dumped adjacent to the incinerator and covered with dirt (Seastone, 1973). Small quantities of depleted uranium contaminated combustibles were burned along with the general combustible plant refuse over the years 1952 to 1968. It is estimated that less than 100 grams of depleted uranium would be involved (Piltingsrud, 1973). The incinerator burned office-type wastes and some depleted uranium chips. Ashes were put into pits located adjacent to the incinerator or were pushed over the side of the hill into the Woman Creek drainage. Incineration was discontinued and the incinerator demolished in the early 1960s (USDOE, 1986).
Incinerator Ash Pits I-1 through I-4	D, E-9	An estimated 100 grams of depleted uranium was burned with general combustible waste in the nearby incinerator from 1952 through 1968. Ashes from the incinerator were buried in these trenches (Owen and Steward, 1974). Some unknown quantity of depleted uranium contaminated incinerator ashes were dumped in an area south of West Road and within a few hundred feet southeast and southwest of the incinerator (Putzier, 1970). Ashes from operation of the incinerator were put into pits located adjacent to the incinerator or were pushed over the side of the hill into the Woman Creek drainage. Incineration was discontinued and the incinerator demolished in the early 1960s. The ash pits were covered with fill (USDOE, 1986).
Concrete Wash Pad	D-9	There have been reports that material from Buildings 444 and 881 was placed between the original sanitary landfill and the incinerator ash pits. More recently, cement trucks were washed in that area (Smith, 1975). It appears that the area was used to dispose of waste concrete from plant construction activities. It is also likely that concrete trucks were washed down in this area after delivering concrete (USDOE, 1991e).

DESCRIPTION	MAP AREA	NATURE OF DISPOSAL ACTIVITY
West Spray Field	A, B, C, D-7, 8, 9	From 1982 to 1985, the West Spray Field was spray irrigated with water from solar evaporation ponds that contained elevated levels of nitrates and other wastes. The practice may have contaminated the ground water and the water in the soil lying just above the ground water (USDOE, 1991d).
Lithium Metal Destruction Areas	F-7	Approximately 400 to 500 pounds of metallic lithium were destroyed over 1956 to 1970. Residues, primarily non-toxic lithium carbonate, were buried. Smaller quantities of other reactive metals (sodium, calcium, and magnesium) and some solvents were also destroyed in this location (Owen and Steward, 1974). Building 335 is located over an old lithium metal destruction site. Lithium metal was disposed of at this location by placing it in trenches and reacting it with water. Residues were covered with soil (USDOE, 1986).
Soil Burial Area, Building 334 Parking Lot Area	G-7	Soil containing low levels of plutonium was placed near Building 334 after excavation near Building 774, the waste treatment plant. The volume of the soil containing plutonium and the associated concentrations are not known (USDOE, 1991b). Between 50 to 75 dump truck loads of soil were removed during construction of Parking Area No. 334 and placed in the Soil Dump Area on the northeast side of the plant (USDOE, 1991b).
Oil Burn Pit #1	F-7	Ten drums of oil containing depleted uranium were burned in August, 1956. The residue was covered with backfill. The area is now located under Building 335 and involves approximately 70 cubic feet of depleted uranium residue (Owen and Steward, 1974). The first oil was burned in August, 1956 in what was referred to as the garage oil burning pit (Putzier, 1970). Building 335 was constructed over burn pit number 1 and a lithium metal destruction site (USDOE, 1986).
Solvent Burning Ground	G-7	A "solvent burning ground", designated at solid waste management unit number 171, is listed in various documents (Helmstadt, 1988; EN-589 and USDOE, 1987). Building 335 has been used in the past, and still is to some degree, for training of fire department personnel. The original, preconstructed building was placed in an area north of Building 331 after the 1969 fire (PAC 700-157.7). Experiments took place to test heat and water effects on different types of materials, for example, filter plenums. When this area was first used for training purposes, magnesium chips coated with a water soluble material were burned. Diesel fuel was the main material that was used. Gasoline was utilized to ignite the diesel fuel. The fire fighters may have also used waste solvents.

DESCRIPTION	MAP AREA	NATURE OF DISPOSAL ACTIVITY
Scrap Metal Disposal Area	H-6	Scrap metal components, mostly from original construction, were buried in this area. Although no detectable radioactive or chemical contamination was observed, some pieces came from process areas and low level contamination of a small percentage is possible (Owen and Steward, 1974).
Former Cooling Tower Blowdown Retention Ponds	F, G-8	These small ponds were used to contain water from cooling towers. Hexavalent chromium is present. Some quantity of lithium was also destroyed in the two eastern-most ponds. These ponds were covered with fill (Owen and Steward, 1974) and may have been used to bury small amounts of depleted uranium (USDOE, 1986).
Present Landfill	G, H-3, 4	Started operation in 1968, and currently in use. Materials with less than Minimum Detectable Activity radioactivity (500 dpm/60 square cm direct or 50 dpm/square ft smear) are accepted for burial. From August 1968 to February 1970, approximately 1,000 kg of sanitary sewage sludge, with 800 to 8,000 dpm/g of alpha activity, were buried. Estimated total activity was 1 to 1.5 mCi. Recent surveys have detected other radionuclides, including tritium, in small quantities (Owen and Steward, 1974).

Environmental Pollution by Mining in Colorado

Thomas G. Sanders[1]
Herman Feissner[2]
Vujica Yevjevich[3]
Department of Civil Engineering
Colorado State University
Fort Collins, Colorado, USA

Part I. Historical Pollution

Introduction

Among the largest sources of pollution in Colorado are mining and nuclear processing. One source, mining, is important due to its volume of waste and its world-wide impact, and the other is important not so much for its volume of wastes but for the toxicity and lack of knowledge, due to secrecy, of amounts and distributions. This text reports on the pollution problems associated with mines abandoned many years ago, and more recent mining initiated at a time when it was thought that the environment would be protected. The problems of the nuclear processing industry is addressed in a previous text.

Clear Creek Mining

Edward Blair writes, "The spring of 1849 saw the greatest western migration in American history. California gold lured men from all over the world. Colorado, in 1849, did not exist, even in men's imaginations. It was merely a mountainous barrier that lay

[1]Associate Professor, Department of Civil Engineering, Colorado State University, Fort Collins, Colorado, 80523, USA.
[2]Engineer In Training, Fort Collins, Colorado, 80523, USA.
[3]Emeritus Professor, Department of Civil Engineering, Colorado State University, Fort Collins, Colorado, 80523, USA.

NATO ASI Series, Partnership Sub-Series, 2. Environment – Vol. 45
Environmentally Devastated Areas in River Basins in Eastern Europe
Edited by Alfons Georges Buekens and Vasily Victorovich Dragalov

between the cities and farms of the East and the waiting wealth of California" (Blair,1980). Chicago Creek, near present-day Idaho Springs, was the site where George A. Jackson located a "rich find" in January of 1859 (Blair, 1980). Only a few miles away and a short time later, John H. Gregory found "paydirt" on Clear Creek (Blair, 1980). By the fall of 1859, optimistic prospectors were coming from all over in search of Colorado gold. Figure 1 illustrates the locations of the many mining tunnels and tailings/waste rock piles in the Clear Creek basin.

Estimates suggest that over 800 inactive mines and tunnels exist within the Clear Creek and Gilpin Counties. Therefore, it was not surprising that in September, 1983, the 400 square mile Clear Creek drainage basin was selected for addition to the Superfund National Priorities List (EPA, 1991).

Several present-day environmental problems are the result of the mining activity from over a century ago, including mine tunnels that were used to haul ore and drain mine water and a large number of small-volume mine waste rock/tailings piles. Drainage tunnels became necessary as mining shifted from the oxidized surface zones of veins (CDM, 1991) to the deeper and more difficult to mine sulfide ores, which lay below the oxidized zone (EPA, 1991). Ore taken from the oxidized surface zone was easily mined and gold was removed by crushing in a stamp mill and sluicing and amalgamation (CDM, 1991). Today, many of the mine drainage tunnels are still functioning and discharge acid mine drainage is one of the primary problems associated with abandoned mines where ore- or coal-mining activities have exposed rocks that are rich in pyrite (FeS_2) (Gough, 1993).

Pyrite (FeS_2) is an iron-sulfide mineral that, when exposed to atmospheric oxygen and moisture, "results in one of the most acidic of all known weathering reactions" (Förstner and Wittman, 1979).

Stoichiometry of Pyrite Oxidation

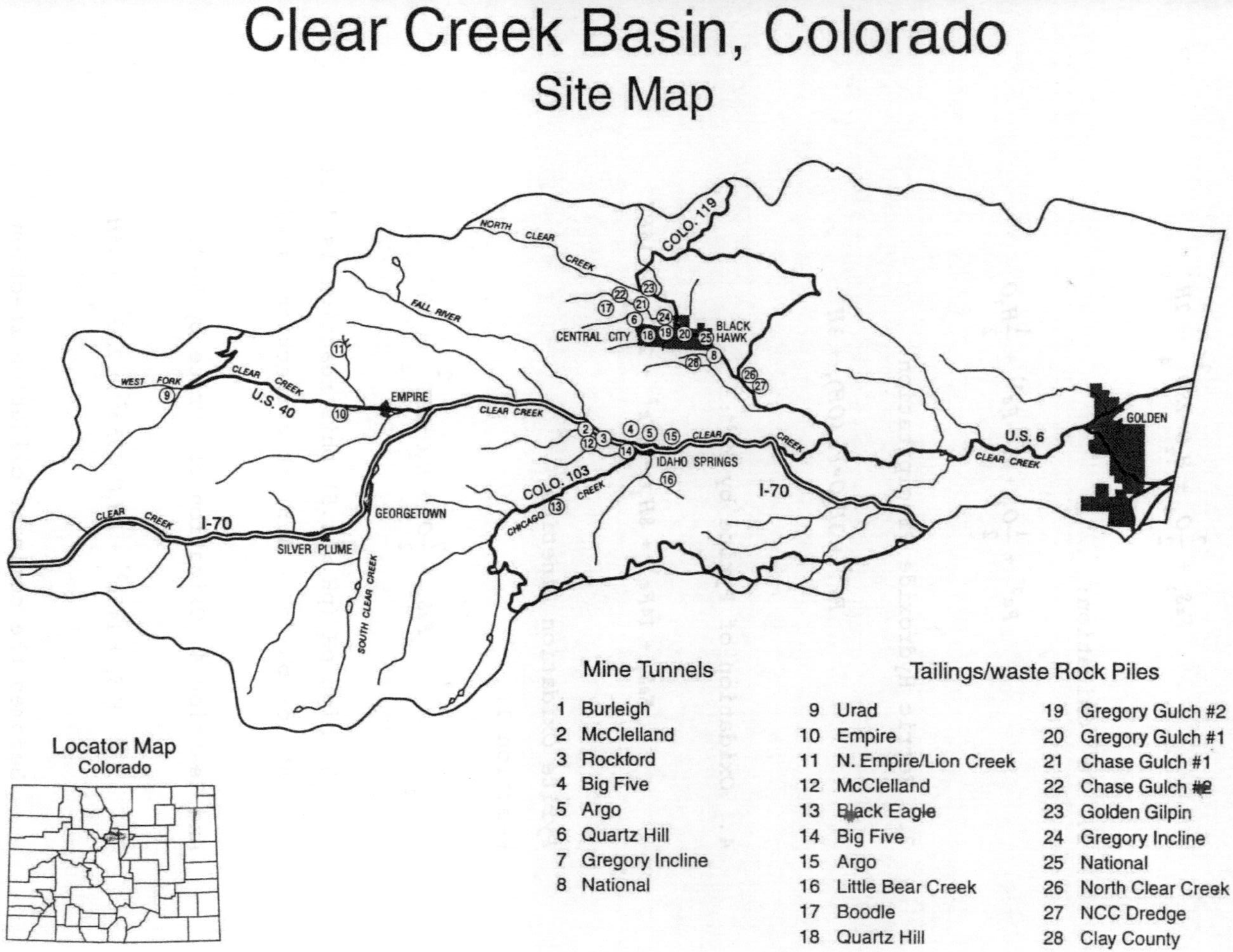

Figure 1. Mine Tunnels and Tailings/Waste Sites in Clear Creek Basin

1. Pyrite Oxidation:

$$FeS_2 + \frac{7}{2}O_2 + H_2O \rightarrow 2SO_4^{2-} + 2H^+$$

2. Fe^{2-} Oxidation:

$$Fe^{2+} + \frac{1}{2}O_2 + H^+ \rightarrow Fe^{3+} + \frac{1}{2}H_2O$$

3. Ferric Hydroxide Precipitation:

$$Fe^{3+} + 3H_2O \rightleftharpoons Fe(OH)_3 + 3H^+$$

4. Oxidation of Pyrite by FE^{3+}:

$$FeS_2 + 14Fe^{3+} + 8H_2O \rightarrow 15Fe^{2+} + 2SO_4^{2-} + 16H^+$$

Pyrite Oxidation Kinetics

Reaction 1:

$$FeS_2 + \frac{7}{2}O_2 + H_2O \rightarrow 2SO_4^{2-} + 2H^+$$

At high pH (pH > 4.5) this reaction is *fast*

At low pH (pH < 4.5) this reaction is *slow*

And Reaction 4: Oxidation of Pyrite by Fe^{3+}:

$$FeS_2 + 14Fe^{3+} + 8H_2O \rightarrow 15Fe^{2+} + 2SO_4^{2-} + 16H^+$$

Becomes the dominant oxidation reaction

But Reaction 2: Fe^{2+} Oxidation

$$Fe^{2+} + \frac{1}{2}O_2 + H^{+} \rightarrow Fe^{3+} + \frac{1}{2}H_2O$$

is also *slow* and becomes *the rate limiting step*

Reaction 2 is Catalyzed by Bacteria:

examples: Ferrobacillus ferrooxidans

Thiobacillus ferrooxidans

(Robertson, 1992)

This section of the text will serve to explain the chemical reactions responsible for acid mine drainage from the mining of sulfide ores, investigate some of the present-day effects of acid drainage occurring in the Clear Creek drainage basin, and look at several remedial actions being used/suggested to mitigate the problem.

In many cases, "acid" is not the primary problem in acid mine drainage. Heavy metals dissolved into solution, as the "acid waters contact exposed mineral zones", constitute an even greater risk to the environment. "Many of these metals are toxic to aquatic and terrestrial life, if the concentrations are high enough". The variety and concentration of the dissolved metals are controlled by the "mineralogy of the ore body; lead and zinc mines may produce metal migrations of lead and arsenic ... gold mines may produce flows containing arsenic" (USDA/FS, 1993). Acid drainage water entering streams usually depletes the fish and stream organisms in a relatively short period of time. Take, for example, the following conditions: a cold water fishery and soft water conditions. The copper concentration lethal to trout is one part per million (USDA/FS, 1993).

As of 1982, the Clear Creek/Central City site was ranked no. 174 of 400 on the National Priorities List, thereby making it eligible for action under the Superfund Program. Choosing/developing remedial actions based upon feasibility, effectiveness, implementability, cost effectiveness, and which ultimately "eliminate or substantially reduce the release or threat of release of hazardous substances" appears to be a primary goal of the Superfund Program. Pursuant to this goal, the EPA carries out a series of steps beginning with discovery and a preliminary evaluation of the "site". Once qualified under the Superfund Program, a "site" is required by the EPA to conduct a Remedial Investigation/Feasibility Study (RI/FS) to determine the remedial actions required to satisfy the objectives of the Superfund Program. Upon completion of the RI/FS, primarily an evaluation of remedial action alternatives developed during the Feasibility Study (FS), a Proposed Plan for "site" remediation is developed by the EPA. The next step provides an opportunity for the public to comment on the FS and Proposed Plan. "Based on the FS, the Proposed Plan, and the comments received from the public, EPA develops a Final Plan for inclusion in the Record of Decision (ROD)", (CDM, 1991).

As for the Clear Creek/Central City Superfund site, Camp Dresser & McKee, Inc. Of Denver, Colorado was chosen to perform the RI/FS for the Colorado Department of Health. In brief, the RI/FS was performed in two phases. Phase I "focused on the discharges and milling and mining wastes from the following five mines/tunnels: the Big Five and Argo Tunnels in Idaho Springs, the Quartz Hill Tunnel west of Central City, and the Gregory Incline and National Tunnel near Blackhawk". Phase II RI/FS was a basin-wide study which included "approximately 400 square miles of the Clear Creek drainage above the Golden gaging station and extending to the Continental Divide" (Figure 1), (CDM, 1991). According to the EPA, "the reason for the expansion was to obtain a better understanding of the various contamination sources in the basin; to enable the State, EPA, and the local communities to make informed decisions as to what could and could not be done to control these sources; and to assess what effect these controls would have on Clear Creek" (1991).

Highlights of a Recent Study Conducted by the Colorado Division of Wildlife (DOW)

A recent study by the Colorado Division of Wildlife addressed "the impacts of heavy metals and acid mine drainage on the aquatic biota of Clear Creek" (DOW, 1991). Point source pollution from active mine drainage, and non-point source pollution, which includes heavy metals loading from mine waste piles and mine tailings, were considered.

Heavy metals carried in riverine systems are potentially lethal to aquatic ecosystems if concentrations reach high enough levels. According to the Colorado Division of Wildlife, several factors influence the toxicity of metals including: pH, alkalinity, hardness, the form in which metal is found, and organism specific tolerance (DOW, 1991). Currently, three "effect levels" or "categories for metals toxicity" exist: no-effect level, the chronic toxicity level, the acute toxicity level (DOW, 1991). The no-effect level, or no-effect concentration (NEC), is one at which aquatic life will not suffer toxic effects. Primarily, the NEC functions as a baseline at which all aquatic species are protected. Exceedance of the NEC will not cause all aquatic species to be effected. The chronic toxicity level provides information on long-term, low concentration exposure to metals. Several effects are possible from this type of exposure: stunted growth, reduced reproduction, or physical deformities. The acute toxicity level is a high concentration usually resulting in direct mortality (DOW, 1991). Refer to Table 1 for a complete listing of the respective "effect levels" for rainbow, brown, and brook trouts.

Table 1
NO EFFECT CONCENTRATION (NEC)
(µg/*l*)

Cadmium	0.4
Copper	5.0
Iron	1000.0
Lead	4.0
Manganese	1000.0
Zinc	47.0

CHRONIC TOXICITY VALUES
(µg/*l*)

METAL	RAINBOW	BROWN	BROOK
Cadmium	0.7 - 1.5	2*	1.7 - 3.4
Copper	11.4 - 31.7	22.0 - 43.2	9.5 - 17.4
Iron	1000 - 2200	unk	unk
Lead	4.1 - 7.6	unk	58 - 119
Manganese	770 - 1530	unk	unk
Zinc	47	225.0	532 - 1368

ACUTE TOXICITY VALUES
(µg/*l*)

METAL	RAINBOW	BROWN	BROOK
Cadmium	3	1.4	3.6 - 60.0
Copper	52 - 56	unk	99 - 110
Iron	unk	unk	unk
Lead	1170	unk	4100
Manganese	14000 - 30000	unk	unk
Zinc	240 - 800	640	2000

(DOW 1991)

According to the Colorado Division of Wildlife, the smaller tributaries "carry elevated concentrations of metals to the mainstream" (1991). The water quality of some tributaries, such as the North Fork and West Fork, have declined to the point they have no fish populations. Mainstream metals concentrations often reached acutely toxic levels. "In mainstream Clear Creek, each metal has occurred at acute toxicity levels from Idaho Springs to Golden" (DOW, 1991).

In March 1991, Colorado Division of Wildlife reports detailing the impacts of mining on water quality and the aquatic ecosystem in Clear Creek Basin. Chicago Creek was described as follows: (Refer to Table 2 for water chemistry data.)

> "Chicago Creek, south of Idaho Springs, is toxic in its lower reaches due to elevated metals concentrations. With a maximum zinc concentration of 253 mg/l just upstream of the Clear Creek confluence, the NEC of 47 mg/l is exceeded. Copper reached 25 mg/l, exceeding the 5 mg/l NEC. At these values, zinc can be acutely toxic to rainbow trout and chronically toxic to brook and brown trout. This copper concentration can be chronically toxic to all three salmonids. In the upper reaches of the Chicago Creek drainage, there is no indication of elevated heavy metals" (DOW, 1991).

According to the DOW, 1991, "North Clear Creek water quality is severely degraded by heavy metals pollution". North Clear Creek, which encompasses the Central City and Blackhawk areas, was mined extensively during the late 19th century. Signs of the past activities such as tailings piles and abandoned adits (mine tunnels) are still visible today. Stream reaches below Blackhawk contain "huge piles of rock that were removed from the stream bed during placer mining operations" (DOW, 1991). Minerals contained in these piles that would have otherwise been concealed are exposed

Table 2
WATER CHEMISTRY DATA
CHICAGO CREEK

Source:
*Camp Dresser & McKee
!Colorado Department of Health

DATE	STN LCTN	ALKAL mg/*l* CaCO3	HARDNESS mg/*l* CaCO3	pH	CADMIUM (µg/*l*)	COPPER (µg/*l*)	LEAD (µg/*l*)	ZINC (µg/*l*)
06-76	1	-	-	-	0	6.0	0	0!
06-76	2	-	-	-	0	4.0	0	20!
06-76	3	-	-	-	0	0.0	0	0!
06-76	4	-	-	-	0	9.0	0	90!
11-01-85	5	-	-	-	-	11.0	-	82*
02-05-86	5	-	-	-	-	23.0	-	253*
05-20-86	5	-	-	-	-	25.0	-	202*
04-89	5	-	-	6.8	-	-	-	<20*
NEC					0.4	5.0		47
Acute Toxicity Values					1.4-60	52-110		240-2000

STATION LOCATIONS
1 - 150 yards above Idaho Springs Reservoir
2 - 150 yards below Idaho Springs Reservoir
3 - Above West Chicago Creek Confluence
4 - 0.4 miles above Clear Creek Confluence
5 - Above Clear Creek Confluence

to the processes of leaching and erosion. Concluding remarks by the Colorado Division of Wildlife described the water quality situation on North Clear Creek: (Refer to Table 3 and Table 4 for water chemistry data.)

> "The aquatic ecosystem of North Clear Creek has been almost destroyed by mining activities. From the surrounding riparian zone to the in-stream water quality, the entire drainage has been effected. Specifically, in the area surrounding Central City and Blackhawk, mining activities have resulted in a stream barren of trout life. Many of the tributaries to North Clear Creek carry elevated concentrations of heavy metals into the mainstream. The proximity of several of theses waterways to mine tunnels and tailings allows them to collect runoff and seepage which contain highly concentrated metals" (DOW, 1991).

Mainstream Clear Creek (refer to Table 5 for water chemistry data) water quality displays the following metals loading concentration pattern: "low metals concentrations are present in the headwaters and increasing levels as loading occurs from a variety of downstream sources". Around the area of Idaho Springs, the Argo and Big Five are two tunnel complexes which contribute to elevated metals concentrations. Zinc concentrations below Argo tunnel have reached 16,900 mg/l, a level acutely toxic to trout (DOW, 1991).

In general, seasonal variations in flow contribute to the metals concentrations in mainstream Clear Creek. According to the Colorado Division of Wildlife, "there is (also) a discernable pattern of increased zinc concentrations during lower seasonal flows" (see Figure 2) (1991). During spring runoff, zinc concentrations are shown to decrease as stream flow increases. Dilution is the probable cause. Likewise, as the stream flow decreases, zinc concentrations are shown to increase in most cases.

Table 3
WATER CHEMISTRY DATA
NORTH CLEAR CREEK
(Source: Colorado Department of Health)

DATE	STN LCTN	ALKAL mg/*l* CaCO3	HARDNESS mg/*l* CaCO3	pH	CADMIUM (μg/*l*)	COPPER (μg/*l*)	IRON (μg/*l*)	LEAD (μg/*l*)	MANGANESE (μg/*l*)	ZINC (μg/*l*)
06-09-76	1	-	-	-	-	5.0	260.0	-	140.0	20.0
06-15-76	1	-	-	-	-	0.0	60.0	-	0.0	0.0
06-15-76	2	-	-	-	-	0.0	210.0	-	60.0	60.0
07-20-76	2	-	-	-	-	2.0	840.0	-	60.0	30.0
07-20-76	3	-	-	-	-	15.0	8400.0	-	1300.0	580.0
06-09-76	4	-	-	-	-	68.0	3200.0	-	670.0	270.0
05-15-76	4	-	-	-	-	54.0	2700.0	-	580.0	260.0
07-20-76	4	-	-	-	-	95.0	9300.0	-	1300.0	1000.0
08-01-76	4	-	-	-	-	210.0	13000.0	-	1500.0	1000.0
08-11-76	4	-	-	-	-	7.0	110.0	-	630.0	30.0

STATION LOCATIONS

1 - Headwaters
2 - 7 miles above Clear Creek
3 - Above Russel Gulch
4 - 0.1 miles above Clear Creek

Table 4
WATER CHEMISTRY DATA
NORTH CLEAR CREEK
(Source: Camp Dresser & McKee)

DATE	STN LCTN	ALKAL mg/*l* CaCO3	HARDNESS mg/*l* CaCO3	pH	CADMIUM (μg/*l*)	COPPER (μg/*l*)	IRON (μg/*l*)	LEAD (μg/*l*)	MANGANESE (μg/*l*)	ZINC (μg/*l*)
07-17-85	1	-	-	-	U*	12.0	328.0	3.0	149.0	259.0
10-31-85	1	-	-	-	2.0	19.0	242.0	7.0	359.0	556.0
02-06-86	1	-	-	-	U*	41.0	322.0	7.0	600.0	778.0
05-27-86	1	-	-	-	U*	U*	659.0	U	77.0	98d
07-17-85	2	-	-	-	4.4	40.0	6760.0	5.6	1300.0	507.0
10-31-85	2	-	-	-	3.0	106.0	16700.0	U	3240.0	1180.0
02-19-86	2	-	-	-	U*	104.0	21900.0	7.3	4710.0	1680.0
05-22-86	2	-	-	-	U*	15.0	1340.0	U	204.0	14.0
07-17-85	3	-	-	-	1.6	44.0	6430.0	8.8	1370.0	563.0
10-31-85	3	-	-	-	3.0	106.0	15800.0	7.0	3190.0	1260.0
02-05-86	3	-	-	-	5.4	183.0	25900.0	21.0	13200.0	4860.0
05-20-86	3	-	-	-	U*	44.0	2200.0	3.0	412.0	396.0
07-18-85	4	-	-	-	14	41	6360	14	1260	533
10-31-85	4	-	-	-	4	102	15900	7	3460	1390
02-05-86	4	-	-	-	4.3	166	20600	15	4660	1870
05-20-86	4	-	-	-	U*	34.0	2120.0	4.0	423.0	403.0
07-18-85	5	-	-	-	1.4	60.0	6030.0	24.0	1250.0	552.0
10-29-85	5	-	-	-	2.0	104.0	5270.0	U*	2530.0	1400.0
02-05-86	5	-	-	-	U*	99.0	3870.0	U*	2010.0	1440.0
05-02-86	5	-	-	-	U*	61.0	2140.0	4.0	521.0	475.0

STATION LOCATIONS
1 - Above Gregory Tailings
2 - Below Gregory Tailings
3 - Above National Tunnel
4 - Below National Tunnel
5 - Above Clear Creek

U = below detection limit
U* = detection limit above NEC
d = dissolved measurement

Table 5
WATER CHEMISTRY DATA
CLEAR CREEK
(Source: Camp Dresser & McKee)

DATE	STN LCTN	ALKAL mg/l CaCO3	HARDNESS mg/l CaCO3	pH	CADMIUM (µg/l)	COPPER (µg/l)	IRON (µg/l)	LEAD (µg/l)	MANGANESE (µg/l)	ZINC (µg/l)
02-06-86	1	-	-	-	U*	9.9	228.0	U*	1900.0	238.0
02-24-86	2	-	-	-	13.0	1240.0	4120.0	U*	9580.0	3570.0
02-24-86	3	-	-	-	18.0	960.0	17800.0	88.0	13800.0	5750.0
05-22-86	4	-	-	-	25.0	527.0	1970.0	6.0	10700.0	3890.0
02-05-86	5	-	-	-	U*	100.0	311.0	5.1	2160.0	1260.0
02-20-86	5	-	-	-	U*	20.0	243.0	U*	1420.0	356.0
05-21-86	5	-	-	-	U*	13.0	411.0	3.0	1070.0	174.0
07-19-85	6	-	-	-	U*	15.0	450.0	8.5	354.0	167.0
10-30-86	6	-	-	-	2.0	19.0	202.0	U*	852.0	258.0
02-24-86	6	-	-	-	U*	26.0	261.0	U*	1120.0	368.0
05-22-86	6	-	-	-	U*	14.0	698.0	3.0	1090.0	226.0
07-19-85	7	-	-	-	U*	47.0	1400.0	3.7	929.0	639.0
10-30-86	7	-	-	-	3.0	230.0	6620.0	U*	4690.0	2090.0
02-25-86	7	-	-	-	15.0	567.0	17000.0	5.8	12200.0	5570.0
02-25-86	7	-	-	-	51.0	1800.0	54700.0	15.0	36800.0	16900.0
02-25-86	7	-	-	-	7.0	285.0	7850.0	U*	6220.0	2710.0
02-25-86	7	-	-	-	23.0	785.0	23700.0	14.0	16400.0	7440.0
06-03-86	7	-	-	-	U*	69.0	4140.0	15.0	2030.0	801.0
07-18-85	8	-	-	-	U*	22.0	544.0	5.6	474.0	214.0
10-29-85	8	-	-	-	2.0	37.0	883.0	U*	1430.0	449.0
02-05-86	8	-	-	-	U*	62.0	1340.0	5.5	2200.0	849.0
05-21-86	8	-	-	-	9.0	281.0	2500.0	4.0	1560.0	1680.0
07-18-85	9	-	-	-	U*	22.0	1100.0	8.0	555.0	244.0
10-29-85	9	-	-	-	2.0	48.0	1410.0	U*	1480.0	547.0
02-05-86	9	-	-	-	U*	51.0	912.0	U*	2300.0	793.0
05-20-86	9	-	-	-	U*	51.0	1100.0	3.0	1310.0	525.0

STATION LOCATIONS
1 - Lawson @ Gauge
2 - Above Rockford Tunnel
3 -
4 - Ditch above Idaho Springs
5 - Above Big Five
6 -
7 - Below Argo
8 - Above North Clear Creek
9 - Below North Clear Creek

U* = detection limit above NEC

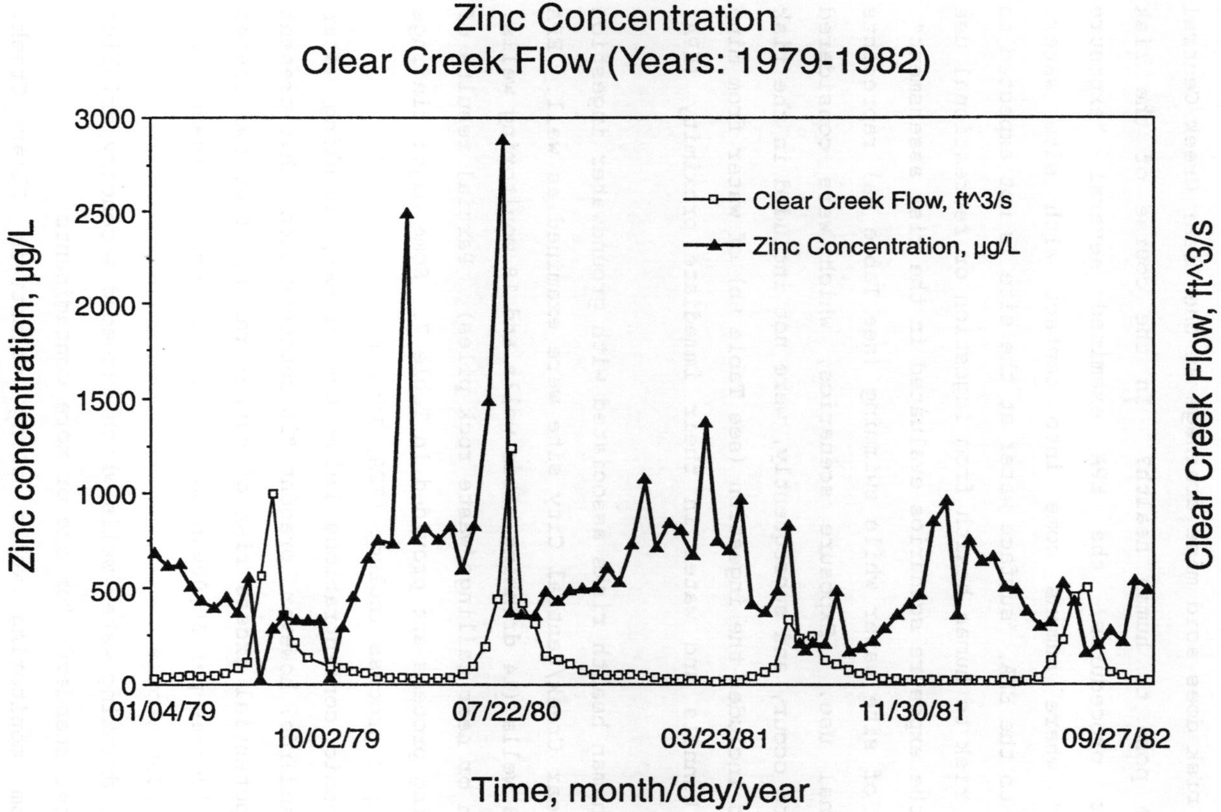

Figure 2. Zinc Concentrations in Clear Creek (1979-1982)
(Colorado Division of Wildlife, 1991)

What risk does acid mine drainage at the Clear Creek/Central City site pose to human health? In the course of the risk assessment procedures, the EPA examined several "exposure scenarios" where humans come into contact with site water. According to the EPA, "surface water at the site is not expected to present a risk to human health from ingestion or recreational use based on the exposure scenarios evaluated in the risk assessment". Ingestion of site water while swimming (see Table 6a) represents recreational use. Exposure scenarios, which were considered unlikely to occur, and subsequently, were not included in the risk assessment included the ingestion (see Table 6b) of water from mine drainage tunnels and water in their immediate proximity (EPA, 1991).

The human health risks associated with groundwater ingestion at the Clear Creek/Central City site were examined as well. EPA sampled 33 wells (14 drinking water wells and 19 monitoring wells, located in or near tailings/waste rock piles). Partial results of the sampling process are provided in Table 7. Some major findings of the sample process include (EPA, 1991):

- Arsenic concentrations below the primary drinking water standards; however, present "in concentrations that present a potential excess risk of cancer ranging from two cancer incidences per 10,000 people to seven cancer incidences per 100,000 people".
- Ten drinking water wells which exceeded secondary drinking water standards for one or more contaminants.
- From monitoring wells along mainstream Clear Creek, "alluvial ground water in or near tailings and waste rock piles had concentrations of cadmium, copper, fluoride, manganese, and zinc, that could pose non-carcinogenic risks to human health if used as a drinking water supply".

Table 6a

COMPARISON OF RISK-BASED SURFACE WATER TARGET CONCENTRATIONS TO MAXIMUM SURFACE WATER CONCENTRATIONS: INGESTION OF SURFACE WATER WHILE SWIMMING

(Concentrations reported in μg/*l*)

Chemical (c)	Risk-Based Target Concentrations	CCI	CC2	CC3	CC4	South Fork Clear Creek	WF1	WF2	NF1	ChC1	FR1
Arsenic	9100 37 (b)	ND	ND	ND	ND	ND	ND	ND	3	ND	ND
Cadmium	4600	1	ND	0.9	1.2	0.5	3.7	0.2	19	ND	ND
Chromium	46000	ND	ND	ND	ND	ND	ND	ND	59	ND	ND
Copper	370000	30	42	3.2	5	4	7	ND	203	5	ND
Fluoride	550000	600	630	900	400	300	4800	500	300	300	ND
Manganese	1800000	714	895	1040	18	22	10000	114	4580	11	30
Nickel	180000	42	50	ND	30	ND	ND	ND	87	ND	27
Zinc	1800000	364	760	215	408	81	998	27	2220	5	356

(a) Except as noted, all target concentrations derived based on non-carcinogenic effects.
(b) Target concentrations derived based on carcinogenic effects
(c) Only detected chemicals are evaluated for the contaminants of concern; aluminum and iron are not evaluated because no health effects criteria are available.

ND = Not detected

(EPA 1991)

Table 6b

COMPARISON OF RISK-BASED SURFACE WATER TARGET CONCENTRATIONS TO MAXIMUM SURFACE WATER CONCENTRATION AT MINICIPAL DRINKING WATER DIVERSION POINTS: INGESTION OF DRINKING WATER

(Concentration reported in μg/*l*)

Chemical (f)	Risk-Based Target Concentration (a)	Idaho Springs (b)	Black Hawk (c)	Georgetown (d)	Empire (e)
Cadmium	18	ND	0.5	0.5	ND
Copper	1400	5	12	4	ND
Fluoride	2100	300	ND	300	ND
Manganese	7000	11	135	22	7
Zinc	7000	5	177	81	24

(a) Based on non-carcinogenic effects.
(b) Sample SW-08
(c) Sample SW-48
(d) Sample SW-25
(e) Sample SW-32
(f) Only detected chemicals are evaluated for the contaminants of concern; aluminum and iron are not evaluated because health effects criteria are not available.

ND = Not detected

(EPA 1991)

Table 7

COMPARISON OF RISK-BASED GROUNDWATER CONCENTRATIONS TO GROUNDWATER CONCENTRATIONS IN DOMESTIC WELLS: INGESTION

(Concentrations reported in μg/*l*)

Chemical (c)	Risk-Based Target Concentrations (a)	DW-04	DW-05	DW-06	DW-08	DW-09	DW-10	DW-12
Arsenic	35 0.047 (b)	ND	2.7	7.5	ND	1.3	ND	1.6
Cadmium	18	ND	ND	ND	ND	ND	NA	ND
Copper	1400	5.6	ND	ND	ND	ND	6.6	1.2
Fluroide	2100	600	500	1800	700	1800	700	1600
Manganese	7000	ND	3.5	293	118	27	2	1510
Nickel	700	ND	ND	ND	ND	ND	ND	29
Silver	110	ND	ND	0.2	ND	ND	ND	ND
Zinc	7000	7	4	14	37	77	77	50

Table 7 (cont.)

Chemical (c)	Risk-Based Target Concentrations (a)	West Clear Creek DW-01	West Clear Creek DW-02	North Clear Creek DW-03	Virginia Canyon DW-07	Spring Gulch DW-13	Fall River DW-11	DW-14
Arsenic	35 0.047 (b)	ND	ND	ND	ND	ND	ND	ND
Cadmium	18	5.7	4.6	ND	28	ND	ND	ND
Copper	1400	781	900	1.9	2.7	ND	8.5	ND
Fluroide	2100	1800	ND	300	900	200	200	1200
Manganese	7000	6550	ND	ND	18500	153	3	240
Nickel	700	68	ND	ND	160	ND	ND	ND
Silver	110	ND	15	ND	ND	ND	ND	ND
Zinc	7000	516		183	18	-18	22	8

(a) Except as noted, target concentrations derived based on non-carcinogenic effects.

(b) Target concentration derived based on carcinogenic effects.

(c) Only detected chemicals are evaluated for the contaminants of concern; aluminum and iron are not evaluated because health effects criteria are not available.

ND = Not detected

NA = Not analyzed

(EPA 1991)

Remedial Action Alternatives

The following represents a brief summary of the remedial action alternatives technologies developed for the Clear Creek/Central City Superfund site.

> "One of the primary objectives of the Superfund Amendment and Reauthorization Act of 1986 (SARA), as specified in Section 121, is to select a remedial action that is protective of human health and the environment, that is cost effective, attains Federal and State requirements that are legally applicable or relevant and appropriate (ARAR's), and that utilizes permanent solutions and alternative treatment technologies or resource recovery technologies to the maximum extent practicable. Additionally, SARA Section 121 requires EPA to give preference to remedies that employ, as their principle element, treatment which permanently and significantly reduces the toxicity, mobility, or volume of hazardous substances" (CDM, 1991).

According to the FS prepared by CDM, "the remedial action objectives specify the COCs (Chemicals of Concern), the exposure scenario(s), and acceptable chemical level or range of levels for each exposure scenario" (CDM, 1991). Site-specific cleanup objectives for three contaminated media follow:

- *Surface Water* - "reduce metals loading to streams from point discharges in order to reduce in-stream metals concentrations to levels protective of aquatic life".
- *Tailings/Waste Rock* - "reduce exposure to arsenic concentrations in surface tailings/waste rock to a level such that cancer risk is less than 10^{-6} ".
- *Groundwater* - "ensure that ground water quality is protective of human health for all present and future uses".

What options exist for response actions that will meet the stated remedial action objectives? The following is a list of general response actions that "will meet the remedial action objectives for the site or will provide a baseline against which actions may be compared" (CDM, 1991):

- *No Action* - required by the National Contingency Plan (NCP, 1990) to serve as a baseline against which other alternatives can be compared. Under "No Action", no remedial activities are conducted which "reduce or clean up the hazards on the site" (CDM, 1991).
- *Institutional Controls* - serve to "control human contact with the contamination rather than remediating the contamination itself...actions may be physical actions such as fences or barriers...legal actions including relocation, zoning or security restricted access".
- *Containment* - serves to "limit the mobility of a containment" through the use of technologies such as slope stabilization and capping.
- *Removal/Collection* - the contaminated media is physically removed or collected from the site, thereby restoring full use of the area.
- *Treatment* - involves complete removal of the contaminant from the media or an alteration of the contaminant. Results include: reduction in mobility, volume, or toxicity of the contaminant.
- *Disposal/Discharge* - contaminated media, concentrated contaminants, and related or treated materials are transferred to a site designed for long-term storage of said materials.

Within a specific category, several treatment options/ technologies exist for the treatment of a particular contaminated media. Refer to Table 8a, Table 8b, and Table 8c for a summary of this information. Each treatment option/technology is subject to a multi-step evaluation process. The first step involves determining the feasibility of a technology. Site conditions and

Table 8a
IDENITIFICATION AND SCREENING
REMEDIAL ACTION TECHNOLOGIES AND PROCESS OPTIONS

Media	Response Action	Remedial Technology	Process Option	Screening Status
Tailings/Waste Rock	No Action	Inspection	Inspection	Technically Feasible
	Institutional Controls	Resident Relocation	Temporary	Technically Feasible
			Permanent	Technically Feasible
		Access Restrictions	Fencing	Technically Feasible
		Use Restrictions	Deed Restrictions/ Zoning Ordinances	Technically Feasible
			Regulatory Agencies	Technically Feasible
	Containment	Slope Stabilization	Retaining Walls	Technically Feasible
			Recontouring	Technically Feasible
			Grout Stabilization	Not Feasible
		Capping	Synthetic Liners	Technically Feasible
			Bituminous Pavement	Technically Feasible
			Soil Cement	Technically Feasible
			Soil Cover	Technically Feasible
			Clay Barrier/Soil Cover	Technically Feasible
			Mod RCRA Cap	Technically Feasible
			Vegetative Mat	Technically Feasible
	Removal	Excavation	Hydraulic Mining	Not Feasible
			Excavator	Technically Feasible
	Treatment	Fixation	Neutralization	Technically Feasible
			Soil Washing	Technically Feasible
			Soil Flushing	Technically Feasible
		Reprocessing	Pyrometallurgical	Technically Feasible
			Hydrometallurgical	Technically Feasible
			(flotation)	Technically Feasible
		Biological	Bacterial Treatment	Not Feasible
		Thermal	Pyrolysis	Technically Feasible
			In-situ Vitrification	Technically Feasible
			Incinceration	Not Feasible
	Disposal	On-Site	Consolidation w/ Other Tailings	Technically Feasible
			Mine Shafts	Technically Feasible
		Off-Site	RCRA Subtitle D (Municiple)	Technically Feasible
			RCRA Subtitle C (Hazardous Waste)	Technically Feasible

(CDM 1991)

Table 8b
IDENITIFICATION AND SCREENING
REMEDIAL ACTION TECHNOLOGIES AND PROCESS OPTIONS

Media	Response Action	Remedial Technology	Process Option	Screening Status
Groundwater	No Action	Monitoring	Monitoring	Technically Feasible
	Institutional Controls	Use Restrictions	Well Restrictions	Technically Feasible
			Banning Wells	Technically Feasible
		Alternative Water Supply	Wellhead Treatment	Not Feasible
			Bottled Water	Technically Feasible
			Minicipal Supply	Technically Feasible
	Containment	Vertical Barrier	Slurry Wall	Not Feasible
			Grout Curtain	Not Feasible
			Sheet Piling	Not Feasible
			Hydraulic Barriers	Not Feasible
		Infiltration	Wellhead Protection	Not Feasible
		Extractions Wells	Well Field	Technically Feasible
			Existing Wells	Technically Feasible
	Treatment	Biological	Active	Technically Feasible
			Passive	Technically Feasible
		Physicochemical	Ion Exchange	Technically Feasible
			Reverse Osmosis	Technically Feasible
			Air Stripping	Not Feasible
			Activated Carbon	Not Feasible
			Precipitation	Technically Feasible
			Oxidation/Precipitation	Technically Feasible
			Electrodialysis	Not Feasible
		Physical	Freezing	Not Feasible
			Distillation	Technically Feasible
			Evaporation	Not Feasible
		POTW	Blackhawk/Central City	Not Feasible
			Empire	Not Feasible
			Clear Creek	Not Feasible
			Georgetown	Not Feasible
			Idaho Springs	Not Feasible
	Discharge	On-Site	Surface Waters	Technically Feasible
			Reinjection	Technically Feasible
			POTW	Technically Feasible

(CDM 1991)

Table 8c
IDENITIFICATION AND SCREENING
REMEDIAL ACTION TECHNOLOGIES AND PROCESS OPTIONS

Media	Response Action	Remedial Technology	Process Option	Screening Status
Surface Water	No Action	Monitoring	Monitoring	Technically Feasible
	Institutional Controls	Access Restrictions	Fencing	Technically Feasible
	Containment	Source Control	Air/Water Seals	Technically Feasible
			Backfilling Mines	Technically Feasible
			Flooding the Mines	Technically Feasible
		Discharge Control	Controlled Release	Not Feasible
			Water Tight Bulkhead	Technically Feasible
			Control Bulkhead	Technically Feasible
			Culverts	Technically Feasible
			Buffer Zones	Technically Feasible
			Sediment	Technically Feasible
	Collection	Channelize Flow	Inlet Structures	Technically Feasible
	Treatment	Biological	Active	Technically Feasible
			Passive	Technically Feasible
			Ion Exchange	Technically Feasible
			Reverse Osmosis	Technically Feasible
			Air Stripping	Not Feasible
			Activated Carbon	Not Feasible
			Precipitation	Technically Feasible
			Oxidation/Precipitation	Technically Feasible
			Electrodialysis	Not Feasible
		Physical	Freezing	Not Feasible
			Distillation	Technically Feasible
			Recycling	Not Feasible
			Evaporation	Not Feasible
		POTW	Blackhaw/Central City	Not Feasible
			Clear Creek/Empire	Not Feasible
			Georgetown	Not Feasible
			Idaho Springs	Not Feasible
		On-Site	Surface Waters	Technically Feasible
			POTW	Technically Feasible

(CDM 1991)

waste characteristics are the primary criteria used in determining the feasibility of a technology. Any treatment option determined to be feasible is further evaluated on its effectiveness, implementability, and cost (CDM, 1991).

Acid drainage is a problem that has been around since the late 1800's and one that will continue to persist with present and future mining operations, creating "acid drainage for years or decades after the mines cease operation" (USDA/FS, 1993). A simple remedial solution is quite elusive at this time due to several factors. For example, current technology lacks the ability to predict acid mine drainage from present and future mining activities. Information on the "acid-forming potential of ore deposits" is currently needed by state and federal permitting agencies in order to "provide for the development of necessary environmental controls". Acquiring such information seems unlikely if the industry continues to grow at such a brisk pace. Take for instance the gold mining industry, "gold and other precious metal operations have experienced a 30-35 percent growth in domestic production in each of the last five years" (USDA/FS, 1993). In the case of recent mining activities, "Federal regulatory controls on some modern mines has, in some instances, been able to limit the development of acid mine drainage, and consequently, reduce the long-term environmental effects" (USDA/FS, 1993). On the other hand, the time has passed to subject abandoned mines to any regulatory controls. New technology is needed in order to remediate these sites.

How widespread is the problem? Estimates for the number of streams affected by acid and metal mine drainage, in the Western United States, stand at approximately 20,000 to 50,000 and affecting 5,000 to 10,000 miles of streams (USDA/FS, 1993). How serious can one incident be? Consider Iron Mountain Mines in California. Iron Mountain Mines is a 100-year-old iron, zinc, copper, silver, gold, and pyrite mine. "Mining caused fractures which exposed the mineralized zone to air and water, thus allowing mine drainage bearing sulfuric acid, cadmium, copper, and zinc to be discharged into watersheds along with additional contamination from heavy metals leached from abandoned tailings" (39 Rocky Mt. Min. L. Inst., 1993). Reports estimate as much as 500 pounds of copper wash into the Sacramento River per day. The discharge threatens "habitat for spawning salmon and aquatic life as far away as San Francisco Bay" (39 Rocky Mt. Min. L. Inst., 1993). What has been done to remediate the problem? The State of California has spent $10 million cleaning up the site. Cost estimates for full cleanup are estimated at approximately $50 million (30 Rocky Mt. Min. L. Inst., 1993).

Part II. Modern Cases of Pollution

The first half of this text focused on the environmental impacts of mineral mining operations, primarily gold, within the historic mining districts of Clear Creek/Gilpin Counties, Colorado, USA. Many of the environmental problems of this area are the result of several hundred abandoned mining operations from over a century ago. The next section of this text looks at the environmental problems currently being faced from the Summitville mine of the San Luis Valley in southwestern Colorado. Unlike Clear Creek, mining operations at Summitville were carried out at a single "Site" during the years of 1986 to 1992, at which time the mine was abandoned, and subsequently, declared as a Superfund site. Historically, mining activity within the Summitville area dates back to the late 1800's. However, the period of critical activity occurred during the specified years.

Summitville

According to Plater and others, mining on public lands represents "an extreme form of dominant use", which is governed primarily by the General Mining Law of 1872 (Plater, 1992). This law is best described s follows, "Under this statute, a prospector locating valuable minerals can stake his or her claim without notifying or obtaining permission from the federal land management agency, exclude other prospectors and interfering recreationists, and use whatever natural resources present on the claim that he or she needs for mining purposes" (Plater et al., 1992). Presently, the opposition seems to maintain the upper hand in the fight to institute new and tougher standards for companies/persons desiring to begin a hard-rock mining operation. For example, Crown Butte Mines, a subsidiary of a subsidiary of Noranda, Inc., a Canadian company, is proposing to open up a round-the-clock gold mine and processing mill just a few miles from the far northeastern corner of Yellowstone National Park, in Cooke City, Montana (Skow, 1993). Recent exploration shows ore deposits valued at approximately $1 billion. According to Skow, 1993, "The mine threatens the environment, social and economic stability of Yellowstone Park, and nearby Wyoming. Exploratory drilling has already scared away many of the area's elk, moose, bighorn sheep, and grizzly bears". To gain some perspective on the magnitude of the proposed mine, one should look to the now abandoned Summitville mine in southwestern Colorado, USA.

History of Summitville Mining

During the years 1986 to 1992, Summitville Consolidated Mining Company, Inc. (SCMCI hereafter), a subsidiary of the Canadian company Galactic Resources, operated a low-grade/high tonnage open-pit gold mine in Summitville, Colorado (Gray et al., 1993). SCMCI removed approximately 9,250,000 metric tons of ore from the mine and produced 9,400 kg of gold (302,000 oz) valued at approximately $98 million (Skow, 1993); thereby, making Summitville the largest producer of gold in Colorado for the given six-year period (Gray et al., 1993).

Mining operations at Summitville stopped in October, 1991 and gold recovery ceased in October/November, 1992 (Posey, 1993). The following month, on December 4, 1992, SCMCI announced bankruptcy, and on December 15, 1992, Galactic/SCMCI abandoned the Summitville mine (Posey, 1993). On January 27, 1993, Galactic Resources filed for bankruptcy in Canada (Posey, 1993). What triggered this sudden and abrupt declaration of bankruptcy from both companies? Possibly an increase from $2.2 million to $7.2 million for a bond required for reclamation costs (Skow, 1993). Although cleanup had already begun during the later half of 1992, future remedial efforts at Summitville were placed in serious jeopardy when SCMCI declared bankruptcy. Justifiable concern existed over the short-term and long-term effects associated with adverse environmental impacts from an abandoned mining site, such as Summitville. The "presence of cyanide- and metal-bearing solutions in the mine's heap-leach pond" represent short-term environmental concerns, while "acidic, metal-rich mine waters emanating from several sources on the mine site" represent several long-term concerns (Gray et al., 1993). Currently, the Summitville mine is a Superfund site. Cleanup costs are estimated to cost taxpayers approximately $100 million (Skow, 1993).

Like the proposed Noranda mine in Montana, where mining activity occurred prior to the 1950's, Summitville enjoyed a rich history of mining activity before the opening and subsequent closing of the current site. Summitville mining activity dates back to 1870, when James L. Wightman first discovered placer gold (Posey, 1993). A placer is defined as a waterborne or glacial deposit of gravel containing heavy ore minerals, as gold, platinum, etc., that have been eroded from their original bedrock and concentrated as small particles that can be washed out (Webster's New World Dictionary, 1988). Within the next 15 years, most of the underground workings were constructed except for the Reynolds adit which was not built until later, in 1903 (Posey, 1993). The year 1934 saw the first cyanidation circuit being built and tested (Posey, 1993). Cyanidation or the cyanide process is a method of extracting gold or silver from low grade ores by treating them with a solution of sodium cyanide or potassium cyanide and then recovering the gold or silver by electrolysis (Webster's New World Dictionary, 1988). During the late 1960's, several companies carried out mining and mining related activities at the Summitville mine site. In 1966, the "W.S. Moore Company and Union Pacific Railroad conducted exploration, partly underground...(and) rehabilitated the Reynolds adit" (Posey, 1993). Just prior to the arrival of SCMCI, ASARCO conducted exploration during the 1970's and 80's (Posey, 1993).

Effects of Mine-Drain Water on Irrigation and Irrigated Crops

The Summitville mine is located along the Wightman Fork of the Alamosa River in the San Juan Mountains of southwestern Colorado (Figure 3). The Wightman Fork is just one of several tributaries that drains directly into the Alamosa River that together serves to drain many highly mineralized and sulfide-rich areas of the San Juan Mountains (Plumlee and Severson, 1994). The nearness of the Summitville open-pit mine to the Wightman Fork, and ultimately the Alamosa River, brings attention to several matters of environmental concern, including: "(1) recent increases in concentrations of metals and decreases in pH of the Alamosa River waters, San Luis Valley irrigation waters, and irrigated soils; (2) increased corrosion and fouling of irrigation equipment due to the lower pH

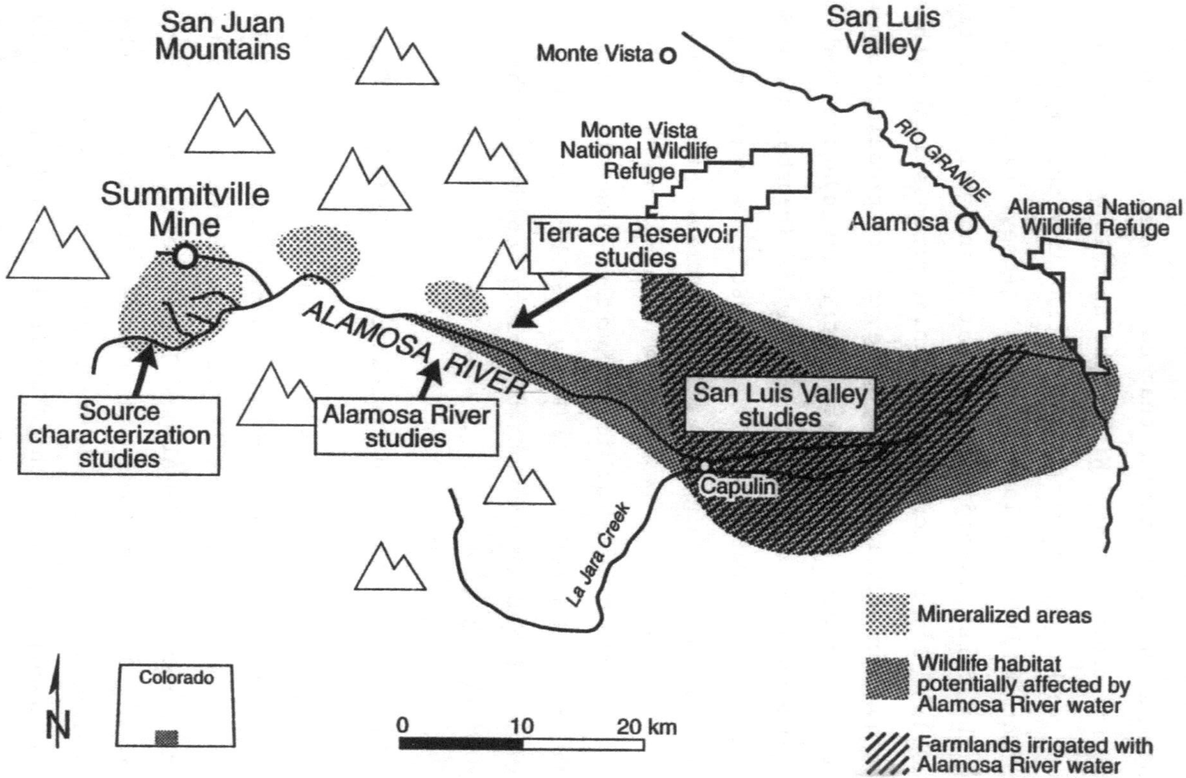

Figure 3. Location of Summitville Mine

(Plumiee and Severson, 1994)

of the irrigation waters; and (3) reduction of aquatic life in the Alamosa River" (Plumlee and Severson, 1994). Downstream, Alamosa River water is used extensively for "agricultural irrigation in the San Luis Valley for domestic purposes and for water supply to the Alamosa National Wildlife Refuge and nearby wetlands (seasonal hosts to endangered migratory species such as the whooping crane)" (Plumlee and Severson, 1994). Currently, several projects/studies are being conducted to investigate the downstream environmental impacts of the Summitville mine. Once finished, these projects/studies will offer a broader understanding of the efforts that can be undertaken to remediate the Summitville mining site and future mining sites (Plumlee and Severson, 1994).

The highly acidic and metalliferous waters draining from the Summitville mine are a predictable consequence of the now abandoned mine site by virtue of the geologic characteristics. Currently, the Summitville mine generates highly acidic, metalliferous mine drainage waters from several areas on-site including the Reynolds Adit (Reynolds Tunnel, the main drainage tunnel at the site), heap leach pad, and the open-pit. Waters sampled in late 1993 showed drainage water pH levels from 1.7 to 3.2 (Plumlee et al., 1994). In addition to the extremely low pH levels, the drainage waters carry high concentrations of metals and toxic elements, "including Cu and Zn (hundreds of parts per million [ppm]); As, Cd, Cr, Co, Li, Ni, rare-earth elements, Se, U, Th, V (hundreds of parts per billion [ppb] to tens of ppm); Be, Ga, Ge, Te (ppb to hundreds of ppb)" (Plumlee et al., 1994). Based on the geologic characteristics of the mine deposit, the highly acid, metalliferous drainage waters were a predictable consequence. For example, "Many of the ore minerals, including pyrite, marcassite, chalcopyrite, and enargite, generate large amounts of acid during aqueous weathering; the highly altered wallrock, coupled with the lack of carbonates in the ores, resulted in very little means for buffering the acid generated during sulfide oxidation" (Plumlee et al., 1994). In addition to predicting characteristics of the drainage waters by virtue of the geologic history of the deposit, several geochemical factors prove to be significant. "Important geochemical controls on the drainage chemistry, in addition to sulfide oxidation and acid generation, mineral dissolution/ precipitation, evaporation, and to only a limited extent, sorption" (Plumlee et al., 1994). For example, Reynolds Tunnel waters show an increase in metals during spring runoff which reflects the "dissolution of soluble metal-sulfate salts in the old mine workings and fracture systems". Such geologic and geochemical information will make significant contributions to the direction of future remediation efforts at the Summitville mine site (Plumlee et al., 1994).

Located upstream of the Terrace Reservoir, along the Alamosa River, is the abandoned Summitville mine (Figure 3). The Terrace Reservoir is located along the Alamosa River some distance downstream from the point the Wightman Fork intersects the Alamosa River. Presently, Terrace Reservoir water is used for irrigation on approximately 45,000 acres of farmland in the San Luis Valley (Stout et al., 1994). Recently, concerns have been raised regarding the use of Terrace Reservoir water for irrigation due to possible hazards associated with runoff into the Wightman Fork from the Summitville mine. Low pH and metal laden surface water runoff from the mine travel to the Terrace Reservoir via the Wightman Fork of the Alamosa River (Erdman and Smith, 1993). Recent concerns by the agricultural community question the safety of this water for irrigation uses. Currently, the Terrace Reservoir supplies irrigation water to the southwestern San Luis Valley (Erdman and Smith, 1993). During June 1993, a study was conducted to understand the effects of irrigating with Terrace Reservoir water. The study comprised eight fields: four sprinkler irrigated fields using Terrace Reservoir water, and four "control" fields utilizing a similar irrigation system. Four sites from within each of the eight fields were chosen to sample irrigation water, soils, and alfalfa. The results of the study show that fields receiving irrigation waters from the Terrace Reservoir are more acidic and contain higher concentrations of dissolved metals than fields receiving irrigation water from other sources (Erdman and Smith, 1993). For example, the Terrace Reservoir water's pH levels ranged between 5.6 and 6.8, while those irrigation waters from other sources ranged between 7.6 and 9.2 (Erdman and Smith, 1993). Likewise, dissolved metal concentrations for copper, zinc, and manganese were 60-350 μg/l, 150-190 μg/l, and 360-520 μg/l, respectively for the Terrace Reservoir irrigation waters (Erdman and Smith, 1993). Irrigation waters from other sources contained the following concentrations of dissolved metals: copper (less than 25 μg/l), zinc (less than 25 μg/l), and manganese (less than 20 μg/l) (Erdman and Smith, 1993). Soil and alfalfa stem and leaf samples were taken as well.

Concentrations in Soils

Soils irrigated with Terrace Reservoir water showed no significant differences in element concentrations at the 0.01 significance level; however, significant concentration differences occurred at the 0.05 significance level for copper and phosphorus (Erdman and Smith, 1993). Likewise, element concentrations were compared in alfalfa stem and leaf samples. When tested at the 0.01 significance level, no significant differences occurred between alfalfa samples watered from the different sources; however, significant differences were recorded for copper and manganese at the 0.05 significance level (Erdman and Smith, 1994). In summary, metal concentrations in the alfalfa stem and leaf samples, "affected by both water source (1) meet published nutritive requirements for ruminants, (2) are far below concentrations reported to be toxic to cattle, and (3) are comparable to concentrations in alfalfa found in other parts of the country" (Erdman and Smith, 1993). According to the National Academy of Sciences, toxic levels in cattle for Cu, Mn, and Zn are: 115 μg Cu/g, 150 μg Mn/g, and 900 μg Zn/g, respectively (1976).

The objective of the following study was to determine if "the use of Terrace Reservoir water significantly increased metal uptake in malt barley and to determine relationships between soil chemistry and plant metal concentrations" (Stout et al., 1994). For the study, six sample farm sites were chosen. Of the six, three were irrigated with water from the Terrace Reservoir and three were irrigated with water determined to be unaffected by the Summitville mine. Test site soil composition was similar and classified as "well-drained sandy loam". Soil and barley samples were taken from each test site at three different times during the summer, while irrigation water was sampled at each site two times during the summer (Stout et al., 1994).

According to test results, the pH levels of water samples collected from the Terrace irrigated fields ranged between 5.3 and 6.7 with an average of 5.95, while the pH range for the control field ranged between 7.3 and 9.2 with an average of 8.25 (Stout et al., 1994). Although exact pH values were not given, the pH values of the Terrace Reservoir soils were significantly lower than the pH values for the control fields (Stout et al., 1994).

Soil analyses showed that fields irrigated with the Terrace Reservoir water contained significantly higher concentrations of Co, Cu, Fe, Mn, P, V, and Zn. Soils from the control fields showed higher concentrations of Ca, Li, Mg, Na, and Sr than soils from the Terrace Reservoir irrigated fields (See Table 9 and Table 10) (Stout et al., 1994).

Table 9

CONCENTRATIONS OF SELECTED METALS IN ALFALFA COLLECTED FROM FIELDS IRRIGATED BY TERRACE RESERVOIR (ALAMOSA RIVER) WATER AND FIELDS IRRIGATED BY WATER FROM OTHER SOURCES

μg Cu/g Dry Matter Mean	μg Cu/g Dry Matter Range	μg Mn/g Dry Matter Mean	μg Mn/g Dry Matter Range	μg Zn/g Dry Matter Mean	μg Zn/g Dry Matter Range
			4 Terrace-water fields		
11	9.0 - 14	50	42 - 63	29	19 - 36
			4 "Control" fields		
8.4	4.2 - 11	32	22 - 43	20	11 - 29
			Probable dietary requirement for cattle:		
			Church (1971)		
	10		20		30 - 40+
			National Academy of Sciences (1976)		
	4		1 - 10		20 - 30
			Agricultural Research Council (1965)		
	10		40		50
			Possible toxic levels in cattle (National Academy of Sciences)		
	115		150		900

Table 10

Western U.S. Soil^			Southwestern San Luis Valley Agricultural Soil	
Element	Baseline	GM*	Observed Range	GM*
Aluminum (%)	1.5 - 2.3	5.8	7.0 - 8.1	7.4
Barium (ppm)	200 - 1700	580	820-910	859
Beryllium (ppm)	0.13 - 3.6	0.68	2 - 2	--
Calcium (%)	0.19 - 17	1.8	1.0 - 3.1	1.5
Cerium (ppm)	22 - 190	65	57 - 87	70
Chromium (ppm)	8.5 - 200	41	21 - 33	26
Cobalt (ppm)	1.8 - 28	7.1	12 - 17	15
Copper (ppm)	4.9 - 90	21	26 - 51	34
Gallium (ppm)	5.7 - 45	16	15 - 19	17
Iron (%)	0.55 - 8.0	2.1	3.9 - 5.6	4.7
Lanthanum (ppm)	8.4 - 110	30	32 - 47	37
Lead (ppm)	5.2 - 55	17	13 - 32	19
Lithium (ppm)	8.8 - 55	22	12 - 21	16
Magnesium (%)	0.15 - 3.6	0.74	0.62 - 0.84	0.72
Manganese (ppm)	97 - 1500	380	620 - 1000	787
Neodymium (ppm)	12 - 110	36	25 - 39	30
Nickel (ppm)	3.4 - 66	15	11 - 15	12
Phosphorus (%)	0.0059 - 0.17	0.032	0.09 - 0.14	0.12
Potassium (%)	0.38 - 3.2	#1.8	2.3 - 2.6	2.4
Scandium (ppm)	2.7 - 25	8.2	9 - 10	10
Sodium (%)	0.26 - 3.7	0.97	1.5 - 2.1	1.8
Strontium (ppm)	43 - 930	200	400 - 510	450
Thorium (ppm)	4.1 - 20	9.1	7 - 10	9
Titanium (%)	0.069 - 0.70	0.22	0.44 - 0.53	0.48
Vanadium (ppm)	18 - 270	70	87 - 140	120
Ytterbium (ppm)	0.98 - 6.9	2.6	1 - 2	2
Yttrium (ppm)	8.0 - 60	22	13 - 23	18
Zinc (ppm)	17 - 180	55	83 - 110	93

^Data from Shacklette, H.T., and Boerngen, J.G., 1984, Element Concentrations in Soils and Other Surficial Materials of the Conterminous United States, U.S. Geological Survey Professional Paper 1270, 105 p., Table 2.

*GM is the geometric mean except as noted.

#Arithmetic mean.

Other soil tests included DTPA (diethylene triamine pentaacetic acid) extractions, which determine/estimate the availability of minerals contained in the soil to plants. DTPA test results showed that "extractable concentrations of La, Ni, Cu, Ce, Mn, Fe, and Zn were significantly greater in soils from the Terrace-irrigated fields" (Stout et al., 1994).

At the time this article was written, "analytical data was still pending"; therefore, the effects of the irrigation waters with higher metal values was unknown (Stout et al., 1994).

Effects of Acidity of River Waters

A recent study by the Colorado Division of Minerals and Geology and the U.S. Environmental Protection Agency was done "to assess the source and fate of acid, metal-rich water in the river system" (Walton-Day et al., 1994). The study included the headwaters of the Alamosa River to the outlet of the Terrace Reservoir. Of interest were the pH, specific conductance, streamflow, and total copper loads (Walton-Day et al., 1994). Data was collected between the months of April and August, 1993 approximately every two to four weeks. Samples were collected at eight sites along the Alamosa River and three sites along tributaries. The three tributary sites included: Pipeline Creek, which is a tributary to the Wightman Fork upstream of the Summitville mine, the Wightman Fork directly downstream of the Summitville mine, and 5.5 miles (8.8 km) downstream of the Summitville mine along the Wightman tributary (Walton-Day et al., 1994). Therefore, the study included Summitville mine acid-rock drainage and some drainage from cyanide heap-leach piles. According to some preliminary results of the study, "Values of pH, specific conductance, and discharge exhibit large seasonal and spatial variation ... sources of acid water to the upper Alamosa

River in addition to the Summitville (exist)...(an) increase in pH and decrease in specific-conductance values occur as water moves down the Wightman Fork from the Summitville mine to the confluence with the Alamosa River during and following runoff (late May to late June)" (Walton-Day et al., 1994).

The heap-leach pad at the Summitville mine comprises 55 acres of the total 1,231 acre permit area. When the Summitville mine was in operation, sodium cyanide was used to dissolve gold contained in the pulverized rock. The heap-leach pad was used to catch the sodium cyanide solution as it was flushed through the rocks. Recently, several environmental concerns have been raised regarding the past and present condition of the heap-leach pad, including: past accidental releases of "cyanide bearing heap-leach solutions into the Wightman Fork of the Alamosa River, and the ongoing efforts to remove cyanide from solutions remaining in the heap-leach pad" (Plumlee et al., 1994). A recent research project sought to "understand the geochemical processes (1) that are currently active in the Summitville heap-leach pad, and (2) that would occur during an accidental release of the heap-leach solutions into the Wightman Fork" (Plumlee et al., 1994). By virtue of the copper-rich ores and copper complexes, the predominant metal in the heap-leach solutions is copper (210 ppm). Other metals in the heap-leach solution include: Fe, Co, and Ni (1-2 ppm); Al (0.5 ppm); and Ag, Mo, and Zn (0.1-0.2 ppm). Current heap-leach solutions undergoing remediation have pH levels around 9.4 and cyanide concentrations of 120-150 ppm (Plumlee et al., 1994).

Laboratory experiments were designed to learn about the mixing of heap-leach solutions with the Wightman Fork waters of the Alamosa River (Figure 4). With reference to the graphs, mixing of the heap-leach solution and the Wightman Fork waters, the WAD cyanide concentration decreased to less than 20 mg/l.

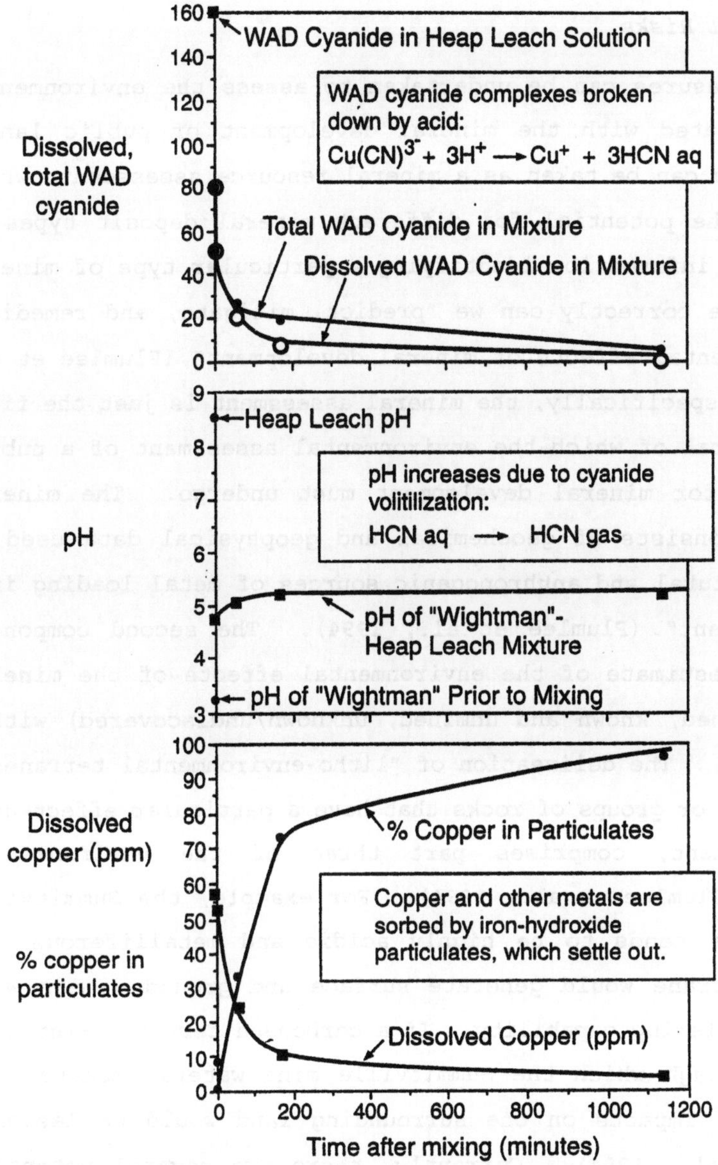

Figure 4. Graphs of WAD Cyanide, pH, Dissolved Copper, and Particulates Versus Time After Mixing (Plumlee, Ficklin, Montour et al., 1994)

Environmental Risks

What measures can be undertaken to assess the environmental risks associated with the mineral development of public lands? Such measures can be taken as a mineral resource assessment, which identifies the potential for different mineral deposit types to occur. With information identifying a particular type of mineral deposit, more correctly can we "predict, mitigate, and remediate the environmental effects of mineral development" (Plumlee et al, 1994). More specifically, the mineral assessment is just the first part of several of which the environmental assessment of a public land chosen for mineral development must undergo. The mineral assessment consists of geochemical and geophysical data used to "identify natural and anthropogenic sources of metal loading into the environment" (Plumlee et al., 1994). The second component involves an estimate of the environmental effects of the mineral deposits (mined, known and unmined, unknown/undiscovered) within the land area. The delineation of "litho-environmental terranes", or rock units or groups of rocks that have a particular effect upon the environment, comprises part three of the environmental assessment (Plumlee et al., 1994). For example, the Summitville mine drainage tends to be highly acidic and metalliferous. A carbonate terrane would generate surface and ground waters with high acid-buffering capability. If a carbonate terrane existed in the area through which the Summitville mine waters emanate, the environmental impacts on the surrounding land would be lessened (Plumlee et al., 1994). Currently, there are several potential uses for environmental assessments based upon a mineral assessment. For example, "the potential environmental effects resulting from the development of different types of mineral deposits can be better predicted and planned for by industry, regulators, and land-use planners ...Finally, land managers can better estimate the potential environmental effects resulting from the development of undiscovered mineral resources contained in public lands" (Plumlee et al., 1994).

Conclusions

The effect of mining on the environment depends highly on the mineral ores being mined, mining technology, and the lack of carbonates to neutralize the highly acid water. Sometimes the cost of cleanup of abandoned mines and their tailings exceeds the benefits created by the mining. Although there has been a tremendous effort to clean up pollution in this country and to prevent future environmental disasters, it is apparent with the Summitville mine that there is a lot to be done before the Summitville mines of the future do not occur.

References

39 Rocky Mountain Mineral Law Institute, 1993, Closing a Hardrock Mine - Can you Ever Walk Away?, Gregory P. Williams, Denver, CO, Rocky Mountain Mineral Law Foundation.

Agricultural Research Council, 1965, The Nutrient Requirements of Farm Livestock, No. 2 - Ruminants, Technical Reviews of Summaries, Agricultural Research Council, London.

Blair, Edward, 1980, Leadville: Colorado's Magic City, Boulder, CO, Pruett Publishing Company.

Camp Dresser & McKee, Inc. (CDM), 1991, Final Clear Creek Phase II Feasibility Study Report, March.

Church, D.C., 1971, Digestive Physiology and Nutrition of Ruminants, Volume 2 - Nutrition, Albany Printing Co., Albany, OR.

Colorado Division of Wildlife (DOW), 1991, Clear Creek Basin - The Effects of Mining on Water Quality and the Aquatic Ecosystem, March.

Erdman, J.A., and K.S. Smith, 1994, See U.S. Department of the Interior, 1994.

Förstner, Ulrich, and Gottfried T.W. Wittman, 1979, Metal Pollution in the Aquatic Environment, Springer-Verlag, Berlin, Heidelberg, NY.

Gough, Larry P., 1993, See U.S. Department of the Interior, 1993.

Gray, J.E., M.F. Coolbaugh, and G.S. Plumlee, 1993, See U.S. Department of the Interior, 1994.

Hilton, Joanne, 1984, Heavy Metal Concentrations from Abandoned Mine Drainage in Coal Creek, Colorado, M.S. Thesis, Colorado State University.

National Academy of Sciences, 1976, Nutrient Requirements of Domestic Animals, Number 4 - Nutrient Requirements of Beef Cattle, National Academy of Sciences, Washington, D.C.

Plater, J.B., Zygmunt, Robert H. Abrams, and William Goldfarb, 1992, Environmental Law and Policy: Nature, Land and Society, St. Paul, MN, West Publishing Company.

Plumlee, Geoffrey S., Walter H. Ficklin, Kathleen S. Smith, Maria Montour, John Gray, Philip Hageman, Paul H. Briggs, and Allen L. Meier, 1994, See U.S. Department of the Interior, 1994.

Plumlee, Geoffrey S., Walter H. Ficklin, Maria Montour, Kathleen S. Smith, Allen L. Meier, and Paul H. Briggs, 1994, See U.S. Department of the Interior, 1994.

Plumlee, Geoffrey S., and R.C. Severson, 1994, See U.S. Department of the Interior, 1994.

Plumlee, Geoffrey S., Steven M. Smith, Margo Toth, and Sherman S. Marsh, 1994, See U.S. Department of the Interior, 1994.

Posey, Harry, 1993, Update on Summitville Activities, Denver, CO, Department of Natural Resources, Division of Minerals and Geology.

Robertson, Dr. A. MacG. Of Steffen, Robertson and Kirsten (Canada), Inc., 1992, Prediction, Prevention, and Control of Acid Mine Drainage in the Western United States Workshop, Breckenridge, CO, August 18-21, 1992.

Skow, John, 1993, Mother Lode vs. Mother Nature, Time Magazine, 22 November, 58-60.

Stout, P.R., J.A. Erdman, and J.C. Emerick, 1994, See U.S. Department of the Interior, 1994.

U.S. Department of Agriculture (USDA), Forest Service (FS), 1993, Acid Mine Drainage from Mines on the National Forests: A Management Challenge, Program Aid 1505, March.

U.S. Department of the Interior, 1993, Understanding Our Fragile Environment: Lessons from Geochemical Studies, compiled by Larry P. Gough with contributions from Sigrid Asher-Bolinder et al., edited by Carole A. Bales, Sonja K. Runberg, and Robert K. Wells, U.S. Geological Survey Circular 1105.

U.S. Department of the Interior, 1994, Geologic Framework and Environmental Geology of the Summitville, Colorado Acid-Sulfate Mineral Deposit, by J.E. Gray, M.F. Coolbaugh, and G.S. Plumlee, Open-file Report, U.S. Geological Survey, 1-2, Denver, CO, U.S. Geological Survey.

U.S. Department of the Interior, 1994, USGS Environmental Geoscience Studies of the Summitville Mine and its Effects on Agriculture and Wildlife Ecosystems in the San Luis Valley, Colorado - Project Overview, by Geoffrey S. Plumlee and R.C. Severson, U.S. Geological Survey Circular 1103-A, 78-80, Denver, CO, U.S. Geological Survey.

U.S. Department of the Interior, 1994, Geologic and Geochemical Controls on the Composition of Acid Waters Draining the Summitville Mine, Colorado, by Geoffrey S. Plumlee, Walter H. Ficklin, Kathleen S. Smith, Maria Montour, John Gray, Philip Hageman, H. Briggs, and Allen L. Meier, U.S. Geological Survey Circular 1103-A, 78, Denver, CO, U.S. Geological Survey.

U.S. Department of the Interior, 1994, Impact of the Summitville Mine on Irrigation Water, Agricultural Soils, and Alfalfa in the Southwestern San Luis Valley, Colorado, by J.A. Erdman and K.S. Smith, U.S. Geological Survey Circular 1103-A, 32-33, Denver, CO, U.S. Geological Survey.

U.S. Department of the Interior, 1994, The Geochemistry and Environmental Degradation of Cyanide at the Summitville Mine, Colorado, by Geoffrey S. Plumlee, Walter H. Ficklin, Maria Montour, Kathleen S. Smith, Allen L. Meir, and Paul H. Briggs, U.S. Geological Survey Circular 1103-A, 77-78, Denver, CO, U.S. Geological Survey.

U.S. Department of the Interior, 1994, Integrated Mineral Resource and Environmental Assessments of Public Lands - Applications in Land-Use Management, by Geoffrey S. Plumlee, Steven S. Smith, Margo Toth, and Sherman P. Marsh, U.S. Geological Survey Circular 1103-A, Denver, CO, U.S. Geological Survey.

U.S. Department of the Interior, 1994, Impact of Acid Mine Drainage from the Summitville Mine on Barley Fields in the San Luis Valley, South - Central Colorado, by P.R. Stout, J.A. Erdman, and J.C. Emerick, U.S. Geological Survey Circular 1103-A, 101-102, Denver, CO, U.S. Geological Survey.

U.S. Department of the Interior, 1994, Preliminary Results of a Synoptic Water-Quality Study of the Upper Alamosa River to the Outlet of Terrace Reservoir, by Katherine Walton-Day, Roderick F. Ortiz, and Paul von Guerard, U.S. Geological Survey Circular 1103-A, 107, Denver, CO, U.S. Geological Survey.

U.S. Environmental Protection Agency (EPA), 1991, Record of Decision, Clear Creek/Central City Superfund Site Operable Unit #3, Gilpin and Clear Creek Counties, CO, September 30.

Walton-Day, Katherine, Roderick F. Oritz, Paul von Guerard, 1994, See U.S. Department of the Interior, 1994.

Webster's New World Dictionary, 1988 ed. s.v., "Placer" and "Cyanidation".

REMOVAL OF HEAVY METALS IONS FROM INDUSTRIAL WASTE WATERS

Vladimir A. Kolesnikov
Mendeleev University of Chemical Technology
Department of Elecrochemical Engineering
Miusskaya sq. 9
Moscow
Russia

Ecological problms of electroplating production in the past 10-15 years have called wide attention of scientists and experts in industry. Due to unaccomplishment of technology in Russia and the CIS, about 1 billion cubic m of waste waters and about 50 thousand tons of metals : Cu, Ni, Zn, Cr, Al, Fe, Sn, etc. are being discharged annually by industrial enterprises.

Selection of process flow diagrams for treating waste waters of electroplating production is essentially a complex task caused by many forms of contaminating ingredients and by high requirements for quality of treated water. But, without doubt, such methods are preferable. By using these methods minimum reagents consumption, minimum permissible concentration, water and valuable components return for reuse are achieved. Prospective methods for solving the stated problems in Russia can be membrane and electrochemical, particularly electroflotation with insoluble anodes, membrane electrolysis, electrodialysis.

Though electroflotation has been known since the 1970-ies, but the method was not used for solving ecological and resources conservation problems in electroplating and printed boards productions. At the same time, the method is essentially universal, highly effective, economical, ecologically safe.

In this report work experience of specialists of the University and *Impuls* Company (Ship-building Industry) on solving problems of removing heavy

NATO ASI Series, Partnership Sub-Series, 2. Environment – Vol. 45
Environmentally Devastated Areas in River Basins in Eastern Europe
Edited by Alfons Georges Buekens and Vasily Victorovich Dragalov

metal ions from waste waters of electroplating enterprises in the Moscow Industrial Region and other cities of the Central Russia is considered.

Along with the widely used methods such as reagents with mechanical separation of dispersed phase, ion-exchange as a method for additional water purification, modern electrochemical technologies : electrolysis, electrodialysis, electroflotation, pH electrocorrection, electrocoagulation are analyzed. Prospect of applying electroflotation technology as a universal method of removing dispersed phase, containing ions of heavy metals such as Cu, Ni, Zn, Cr, Cd, Fe, Al, Sn, Pb in individual form as well as in form of mixtures of two, three or more components, is illustrated.

According to assessment by specialist, wide use of the developed electroflotation technology is connected in that the technology is ecologically safe (no secondary contamination of liquid wastes), high effectiveness (purification efficiency - 98-99.9 %, residual concentration of heavy metal ions - 0.05-0.01 mg/l), characterized by low energy consumptions (0.3-0.5 kW.hr/m^3) and easy operation.

Problems of realizing new technologies and equipment for solving problems of removing heavy non-ferrous metals ions from effluents in Russia and CIS countries as well as possibility of joint use of new technologies for water treamtment with foreign partners, are discussed. More details on this new technology will be discussed below.

The prospect of applying electroflotation process, is connected with the formation of finely dispersed gas (hydrogen and oxygen) bubbles during electrolysis of water, which are evenly distributed in the bulk of the solution. The method is characterized by effectiveness, easy operation and small equipment size.

In addition to metal hydroxide removal, electroflotation method of water treatment is successfully used for the removal of dispersed particles, petroleum products, surfactants, oils, greases, etc. In order to avoid pollution and oxidation of the foam product during electroflotation, cathode and anode spaces are separated by electrochemically active (ion-exchange) or inactive membranes. When membrane is used for electroflotation, cathode space is leached which may have positive effect on the ...

In the past years, we are able to carry out a number of work on development and application of titanium-based CO_3O_4-, MnO_2-coated electrodes, which are readily stable in area of anodic polarization at a pH value of the solution in the interval of 1-11.

Based on the studies carried out, it was established that solution content, unit's design, operational treatment condition and other physico-chemical parameters of flotation systems have influence on electroflotation process.

Process of removing a large group of ions Cr^3, Ca^2, Mg^{2+}, Pb^{2+}, Al^{3+}, Cu^{2+}, Ni^{2+}, Zn^{2+}, Fe^{2+}, Fe^{3+} from rinse and waste waters, individually as well as in mixture, was experimentally investigated.

We consider the influence of few factors such as bulk current density, pH scale of the solution, concentration of anions and cations on the process of electroflotation removal of metal hydroxides.

At low current densities, gas saturation of the liquid is low and as a result of this, complete removal of insoluble compounds of metal to form foam product is not achieved. However, at higher current densities, solution purification efficiency falls, which may be connected with turbulence of the treated solution. This makes formation of the foam layer difficult. The optimum current densities are between 125 and 180 A/m^2 during metal hdyroxides removal, while it is between 5 and 10 A/m^2 for phosphates and sulfides.

Electroflotation removal of phosphates and sulfides of Ni, Cu, Zn occurs in a wide range of pH values from 3 to 12. More complete removal of phosphates occurs at pH 9-10, while sulfide at pH 7-8. At these pH values, complete precipitation of phosphates and sulfides of group III is observed.

Effect of solution content on physico-chemical properties of floating particles and their removal was confirmed.

Relationship between optium current passing through the apparatus and the size of the removed phosphate, sulfide and hydroxide particles of Ni, Cu, Zn is established. Thus, for the sulfide particle of Ni, Cu, whose mean diameter is not higher than 5 µm, optimum current density value, $i_{opt.}$, at the highest achievable removal efficiency is 0.5 A/m^2. For Ni, Cu, Zn phosphates and Zn sulfide with particles' diameter between 18-20 µm, $i_{opt.} = 0.1$ A/m^2, diameter of hydroxides particles is 35 nn, while $i_{opt.} = 2\text{-}2.5$ A/m^2.

The change in optimum current density as a dependence on size of the removed particles is suggested to be connected with relationship of sizes of floating gas bubbles and particles.

The studies carried out showed that electrolytic gas bubbles' diameter ranges from 20 to 40 µm, but for effective trapping of the particles by the bubble, definite relationship of their geometrical sizes is necessary. that is why only small quantity of bubbles low bouyance force can attach themselves to particles with small surface area. In addition to that, minute flotocomplexes are well removed in turbulent flow, i;e., probability of their self destruction increases.

One of the methods of increasing particles trapping effectiveness by the bubble, as a result of increasing rate of flotation removal at finely dispersed particles is the use of flocculants, under the action of which sizes of floating particles increase. Using radioisotope method, it was confirmed that on all hydroxides studied adsorption of flocculant was observed. Here, the mean diameter of hydroxides particles in the presence of flocculant increases up to 70-75 mm. Increase in size of the floating particles makes it possible to intensify the process of electroflotation removal of hydroxide particles due to decrease in process time. As the experiments showed, removal efficiency under this condition practically doesn't change. Adsorption of organic flocculants on the surface of the phosphates and sulfides was observed.

It was found that the dependence of removal efficiency on current has a maximum for all compounds studied. At a low current density gas saturation of the liquid is small, as a result of which the process is not all that effective. At high current density considerable turbulence of the solution occurs, which is accompanied by the destruction of the already formed foam layer.

Dependence of size of gas bubbles on current density was established. At a value of $i>20$ A/m^2 increase in size of the bubble was observed.

The morphology of cathode materials (Zn, steel-3, Ni) with different hydrogen over-potentials, does not have any significant effect on electroflotation process.

It was established that the content of the solution affects the dispersed content of the gas bubbles. As a result of coalescence of the later, their mean diameter changes along the height of the flotation chamber.

It was found that addition of surfactants like polyelectrolyte A, KMN, into solution raises the effectiveness of the electroflotation not only because of the provision of particle attachment to the bubbles, but also due to stability of gas-water dispersion and decrease in size of the gas bubbles.

It was established that chemical morphology of the electrolytic bubble affect effectiveness of electroflotation removal of particles. Process of electroflotation removal of finely dispersed particles of flotoconcentrate of bubbles of hydrogen, oxygen and their mixture was studied. Studies carried out showed difference in parameters of flotation process as a dependence on the type of floating gas. Thus, electroflotation removal of hydroxide precipitates by oxygen is worse than that of phosphates. Hydroxides removal using hydrogen is better than that of phosphates.

It was also established that most efficient electroflotation removal in the pH interval from 9 to 10 of positively charged surface of copper and zinc hydroxides by hydrogen bubbles, which carry negative charge. In this cas electrostatic force of attraction increases the effectiveness of particles trapping by the bubbles. At a negatively charged surface of particles (phosphates and sulfides) removal process will be suppressed.

Some parameters of electroflotation removal of ions di- and trivalent of metals Cu, Ni, Zn, Cr, Al, Fe, etc. are presented in tables 1 and 2.

It was established that during electroflotation removal of mixture of phosphates, sulphides, hydroxides better result are achieved than when they are removed individually. Obviously, this is in connection with the formation of solid phase of mixed contents with their own physico-chemical properties. Addition of flocculant enhances intensification of removal process.

It was shown that the addition of flocculant lowers the residual ions concentration in the solution by 5-10 times.

Optimum treatment regime for mixed waters was found, at which high non-ferrous heavy ions removal efficiency down to a concentration of 0.01-0.05 mg/l is achieved.

Table 1 : Effectiveness electroflotation process for some divalent metals hydroxides pH = pH_{opt} : C_{Me} = 100 mg/l

Hydroxide	pH min. solutn.	pH opt α_{max}	π min $\alpha = 90$ %	C_{res} mg/l el. flot.	C_{res} mg/l filtr.	C_{res}* mg/l calc'd
$Cu(OH)_2$	9.8	9-9.5	3	0.5-1.8	0.3	0.02
$Zn(OH)_2$	9.9	9.5	2	0.5-1.0	0.4	0.16
$Ni(OH)_2$	10.8	9.5-10	3	0.4-0.5	0.1	0.08
$Cd(OH)_2$	12.0	10-10.5	2	1-2	1-2	1.4
$Mn(OH)_2$	11.9	10	3	1-2	1-0.5	0.03
$Sn(OH)_2$	7.4	6-6.5	12	10	1.0	0.66

The first module (M1) consists of double-chamber pH electrocorrector with an electroflotator, tanks for rinse and treated water and pumps.

The module enhances removal of heavy metal ions from waste waters with an initial concentration from 50-200 mg/l down to 0.5-5 mg/l. The module operates on the principle that in the cathode chamber of the pH electrocorrector occurs leaching of the waste water, oxygen evolution takes place in the anode chamber of the pH electrocorrector and concentration of acidic solution.

Flotocomplexes formation takes place in the first chamber of the electroflotator at a high current density, while in the second chamber with low current density, bubbles' drift closed to laminar regime. In order to increase the rate of flotation process, the solution must be passed directly into electrode zone, because in thsi zone the mean bubble radius is minimal. Out-flow of the treated solution from the second chamber is better to be done below the electrode block, under this condition additional purification effect is connected with solutin filtration through compact filter-like electrolytic gas bubbles.

pH electrocorrector carries out 4 functions : pH adjustment of catholyte and anolyte to required values; necessary gas saturation of the solution for carring out the process, demineralization of the catholyte due to anions migration to the anode chamber and formation of heavy metal hydroxides. The most important function is the acidity change of treated solution in the cathode chamber, that is why required current load in the pH corrector must be determined starting from this function [1,2].

Optimum conditions for the purification of mixed content : Cu^{2+}-Sn^{2+}-Pb^{2+} and Cu^{2+}-Zn^{2+}-Cd^{2+}-Cr^{3+}-Ni^{2+}, which are characterized for the production of printed circuit board and electroplating, are determined. Optimum conditions for individual components are different. It is better to use maximum preventive loss at operation of the unit as a criterion of optimization condition for mixed effluent treatment. Results of water treatment in the first module are presented in table 3.

Table 2 : Electroflotation process for some divalent metals hydroxides at optimum pH

Hydroxide	pH min. solutn.	π min α = 90 %	pH opt	C_{res} mg/l el. flot.	C_{res} mg/l filtr.	C_{res}* mg/l calc'd
Cu $(OH)_3$	8.0	15	7.0	1-3	0.5	0.05
Al $(OH)_3$	6.2	4	6-7	1-2	0.5	0.28
Fe $(OH)_3$*	9.0	2	5-11	0.1	0.05	0.01
Mn $(OH)_3$*	10-11	2	10	0.1	0.01	-
Ni $(OH)_3$*	10-11	3	11.5	0.2	0.01	-
CO $(OH)_3$*	10-11	2	10-12	0.1	0.01	-
* = Flotoconcentrate contains mixture of oxides and hydroxides						

Based on the studies carried out, electroflotation module for additional purification (MAP) of waste waters was developed, whose main elements are mixing chamber and electroflotation unit. The module is used for additional removal of waste waters of electroplating products for non-ferrous heavy metals ions, petroleum products, greases, oils, flotoreagents and other impurities after the reagent method of purification, electrocoagulation, electrolysis, electroflotation in the pH corrector with initial metal concentration in waste water from 1 to 20 mg/l. The big advantage of applying electroflotation is that it makes provision for creating universal treatment unit, to be intergrated with other methods for the purpose of creating universal equipment.

After purification, residual cations concentration is between 0.005 and 0.01 mg/l, the dispersed substances - between 0.5 and 1 mg/l.

Electroflotation module effectively removes greases, oils, dispersed particles of organic origin, lowering chemical oxygen demand COD down to 80-100 mgO/l, also as a result of adsorption of anions like SO_4^{2-}, CO_3^{2-}, Cl^- etc. by 15-20 %.

Electric energy consumption is 0.5-1 kW*hr/m^3, reagent consumption - 5-10 g/m^2. One module with a capacity up to 10 m^3/hr, occupies an area of up to 30 m^2.

The suggested scheme of purification makes it possible to considerably cut energy consumption, reduce hardness of water and remove non-ferrous heavy metal ions down to the lowest permissible concentration from waste waters.

Positive development of technical solutions, connected with the reduction of contents of Ni, Zn, Cr, Cu, Ca, Fe, Mg, as well as greases, oils allowed for the use of MAP in water preparation stage for demineralization with the aim of recycling it to process line. One of the prospective methods of demineralization is electrodialysis. For practical realization of this, series of laboratory and pilot-scale testing on electrodialysis demineralization of applying module for additional demineralization (MAD) in water preparation stage and electrodialysis for demineralizing electroplating production effluents, showing its operational condition were conducted.

After the electrodialysis desalination, total salt content in diluate compared with the initial solution is reduced averagely by 6 times, while concentrate increases averagely by 25 times.

The volume of the demineralized water is 97 %, concentrate - 3 %. pH of the initial solution may be lowered in the diluate by 3-4 units. It is possible to use titanium-based oxide-coated electrode as anode during the process of electrodialysis.

Presently, technology and unit for demineralization of wastewaters with preliminary purifications in module (M2), are developed. Process flow diagram for purification of rinse and waste waters in the electrochemical engineering process are developed by the author, original units' design are developed directed towards water recycling and utilization of valuable components - metal hydroxides, which is expressed in the practical application of three modules of wastewater treatment.

The advantages of using electroflotation in combination with pH electrocorrector are shown, if comparative analysis of the existing methods of wastewaters treatment is done.

The second module of tertiary treatment of waste waters (M2) consists of electroflotator, tank for reagents, pumps, foam-collecting device. This operates on the principle of adhension of pollutant particles to highly dispersed electrolytic gas bubbles, which are formed durng electrolysis. Density of the flotocomplexes is considerably lower than that of the solution, that is why they are drifted to upper part of the apparatus forming stable foam layer. Electroflotation method enhances removal of only the insoluble metal compounds. Solubility of metal hydroxide dependson the solution content and is between 0.2 and 0.5 mg/l of metal ion at pH = 10. After electroflotator, sodium phosphate is added into the solution for additional purification. As a result of this, more insoluble metal phosphate is formed.

Table 3 : Basic parameters of electrochemical removal of Cu, Ni, Zn, Cr, Fe ions from rinse waters

Parameter	Cu	Ni	Zn	Cr	Fe
Initial concentration, mg/l	50-100	50-100	50-100	25-75	10-50
Removal efficiency, %	98-99.3	98-99.4	93-99.4	93-99.3	99.3-99
Residual concentration, mg/l	0.5-1.6	0.5-1.5	0.4-1.5	0.3-5.0	0.05-0.5
Specific power consumption, kW*hr/m^3	1.4-4.5	1.8-4.8	2.6-5.0	0.9-4.1	0.8-3.2

Solubility of metal phosphates is about 0.01-0.05 mg/l of metal ions enhancing complete ions removal from the solution.

Metal phosphates particles have considerably smaller size than hydroxides causing decrease in flotation rate. Because of this coagulant is added into the solution for the intensification of flotation process. This causes increase in particle size and thereby raising the flotation rate. Excess phosphate ions combine with iron, magnessium and calcium ions to form insoluble phosphates. Analysis showed that electroflotability depends on nature of floating material, some data are given in table 4.

Electroflotation removal of insoluble compounds of heavy non-ferrous metals (HNM) in form of hydroxides, phosphates, sulfides individually or in mixtures, was carried out at pH 7-10 due to their flotation by bubbles of electrolytic gases - hydrogen and oxygen. The use of insoluble anodes made from titanium with oxide depassivating coating enhances high quality purification and does not cause secondary water pollution.

Table 4 : Effectiveness of electroflotation of insoluble compounds of Cu, Ni, Zn, Cd, Mn, Fe, Al, Cr (without flocculant)

Me^{+n}	$Me(OH)_n$	$Me_x(CO_3)_x$	Me_xS_x	$Me_x(PO_4)_x$
Cu+2	high	average	average	high
Ni+2	high	average	low	average
Zn+2	very high	average	average	high
Cd+2	very high	average	low	high
Mn+2	high	average	-	average
Fe+3	very high	average	-	average
Al+3	high	average	average	average
Cr+3	low	-	-	-

Electrochemical module provides for tertiary treatment after reagent method, electrocoagulation, electrolysis, pH electrocorrector at initial metals concentration of 1-10 mg/l in waste waters.

The unit makes it possible to remove Cu(II), Ni(II), Zn(II), Cd(II), Cr(III), Fe(II), Fe(III), Al(III) ions etc. from waste waters of electroplating and printed circuit board productions at any components ration in the presence of anions such as Cl^-, SO_4^{2-}, CH_3COO^-, CO_3^{2-}, PO_4^{3-}, F^-, BF_4^- etc.

Residual concentration of HNM ions is :

Cu(II), Ni(II), Zn(II)	not more than 0.01 mg/l
Cd(II), Pb(II)	not more than 0.005 mg/l
Cr(III), Sn(II)	not more than 0.01 mg/l
Fe(II), Fe(III), Al(III) etc.	not more than 0.05 mg/l
Dispersed substances	not more than 0.5-1 mg/l

The unit enhances additional removal of Ca(II), Mg(II) ions and SO_4^{2-}, CO_3^{2-}, Cl^- anions by 15-20 % due to their adsorption on insoluble compounds of HNM. PO_4^{3-} and S^{2-} are removed down to residual concentration - 0.01 mg/l.

In recent years, many foreign firms from Italy, Canada, China, Germany have indicated their interest in this electroflotation technology and equipment. It is being developed at the Mendeleev University of Chemical Technology (MUChT) and on operation at many plants. In 1992, contracts on sales of the equipment, technology and know-how were signed with foreign partners.

In order to develop, realize highly effective ecologically safe equipment in Russia, USA and other countries, in MUChT Technical Centre *Econhimbiznes-2000* is being organized.

The main aims of this Centre are the following :

- Complex solution to large research-technical problems in areas of ecology and resources conservation of industrial productions, provision of ecological safety of people living in big cities of th Russian Federation and Moscow City.
- Co-ordination of efforts of scientists of the University, industry experts, financio-industrial capital, investments of foreign partners for introduction of new modern technologies at industrial enterprises of the Russian Federation.

- Creation of latest highly effective, ecologically safe technologies and equipment for solving problems of environmental protection and rational use of natural resources at industrial enterprises of Russia.
- Personnel training and in-service training for solving ecological tasks and developing business in area of high technologies.

The major areas of activities of the Technical Centre for the period 1994-1993 include the following :
- Development of new technologies and units for the detoxification of hazardous liquid wastes of industrial production with utilization of valuable components.
- Development of new technologies and units for the detoxification of hazardous solid wastes of industrial production with utilization of valuable components.
- Development of new technologies and units for cleaning atmosphere of industrial hazardous wastes. Manufacturing of small series of highly effective, ecologically safe equipment for the detoxification of solid, liquid and gaseous wastes. Preparation of series of productions.
- Realization of new highly effective, ecologically safe technologies and equipment in Russia and the CIS.
- Analysis of technological solutions on areas of activities of the Centre.
- Research-technical cooperation with foreign partners.
- Realization of technologies developed by the Centre in form of contracts to foreign companies. Working on grants provided by foreign organizations.
- Personnel training and in-service training for solving ecological tasks and developing business in area of high technologies in the Russian Federation.
- Ecology and business : market analysis of high technologies, equipment for solving ecological tasks in Russia and the CIS. Advertizing activity.
- Certification of new and existing technologies and units created for solving problems of ecology and resources conservation of industrial production.
- Analysis of investments projects connected with the creation and reorganization of industrial enterprises in Russia.

REFERENCES

Kolesnikov V.A., Ilin V.I., Aarinola P.K., Shalyt E.A., Development of Electrofloation Technology and Equipment for Industrial Effluents Treatment - Intern. Flotation Conference, 26-28 April 1994, Orlando, Florida, USA

Kudryavtsev V.N., Kolesnikov V.A., Majorov V.B., Electroflotation removal of aluminium ions from electroplating waste waters - Surface Treatment of Aluminium and Other Light Metal. Berichtsband uber den 3. EAST-Kongress von 12 bis 13 November 1992 in Schwabisch Gmund., p. 120-122

IV

CONCLUSIONS

REHABILITATION OF ENVIRONMENTALLY DEVASTATED RIVER BASINS

MAJOR WORKSHOP PRESENTATIONS AND DISCUSSIONS

Vujica Yevjevich
Civil Engineering Department
Colorado State University
Fort Collins, Colorado, 80523
USA

INTRODUCTORY STATEMENTS

The presentations and related discussions by participants of the subject matter of the Moscow NATO Advanced Research Workshop were held on a ship in Russia, October 11 through 17, 1994. Nearly all the major aspects of pollution of air, water, soil, land and biomass were treated in presentations at this workshop. The principal sources of contamination were discussed from various standpoints, however in different depths of coverage. The human-induced pollution of the environment was investigated as being produced by the major seven sources of environmental degradation.

Seven major sources of gaseous, liquid and solid contaminants were discussed at the workshop:

(1) The urban and rural land drainage and sewage waste water and sludge evacuation and disposal;

(2) The industry-generated effluents and rejects of the related production;

(3) The remnants and contaminants in the generation of energy;

NATO ASI Series, Partnership Sub-Series, 2. Environment – Vol. 45
Environmentally Devastated Areas in River Basins in Eastern Europe
Edited by Alfons Georges Buekens and Vasily Victorovich Dragalov

(4) The chemical non-point pollutants in advanced agriculture, forestry and food processing;
(5) The mining and related smelting pollutants as the waste gases, liquids and solids;
(6) The nuclear energy processing and the related radioactive pollution; and
(7) The rejects and pollutants produced by constructing and use of the transportation lines.

As another NATO Advanced Research Workshop was held recently on the subject of pollution by the nuclear processing and radioactive contamination, these sources of pollution were not covered as much and in depth and details as the others of the above seven sources of pollutants and contaminants. The major contribution of the workshop has been on the problems with urban and industrial pollution of the air, water, soil, land and biomass.

PROBLEMS OF WATER AND SOIL POLLUTION AS RELATED TO THE GENERAL WATER AND LAND RESOURCE MANAGEMENTS

The presentations at the Moscow workshop mainly treated the subject of pollution of rivers and related river basins. This was the case even when specific examples of polluted rivers were discussed. Because most often and in these cases "the devil of pollution is in details", the treated subjects did not always realistically, sufficiently and broadly portray the high degrees to which most of the presented example river basins and rivers have been polluted in Eastern Europe. Neither did the examples, cited for the central and western European countries and for the U.S.A. have demonstrated in depth and in details all the seriousness of degradation of European and American environments during the past 200 years of the industrial and post-industrial eras.

The examples of pollution covered the rivers Danube, Volga, Rhine, Dnieper, Oder, Emsher River of Germany, Aren River of Slovakia, Ton, Upper Ob and Oka rivers of Siberia, as well as others. The major emphasis was on the character and degree of pollution, rather than on the type and quantity of environmental degradation and the necessary solutions.

The contaminations discussed were either those from a particular single source, or those being the product of the combined consequences of multiple sources of pollution. What has been missing in presentations and discussions at this NATO-sponsored Advanced Research Workshop was the in-depth classification of sources of pollution and for each source the description of the characteristics of these contaminants as well as the repercussions generated by their pollution. While the past and present aspects of environmental degradation were broadly covered and looked at from a variety of the points of view, the projection and anticipation of new forthcoming sources of contamination and the expected problems of pollution have been almost completely absent.

Some of the recent scientific and technological developments will likely become the significant future sources of the yet undocumented pollution characteristics. The repercussion of the potential pollution by the products and rejects of bioengineering, electronic technology, medicinal drug manufacturing, and similar contemporaneous activities now under very active research and development, may become also the important sources of pollution. Likely they will be also very expensive to manage as are several present sources of contamination. Some of them are already well understood as the sources of undesirable pollution. Piles of the old and obsolete computers and their accessories may be already competing for the waste storing space with the hills of accumulated, used automobile tires or the other industrial project rejects, or the remnants of their recycling.

POLLUTION AS A COMPONENT OF COMPREHENSIVE AND INTEGRATED WATER AND LAND RESOURCES MANAGEMENT

One of the major aspects of the Moscow NATO-sponsored Advanced Research Workshop was the position of the majority of participants for the need that the river and river basin pollution be treated within the context of the comprehensive and integrated management of water and land resources. The environmental degradations of the entire river basins were thus demonstrated, monitored and eventually resolved. In other words, the pollution abatement should be best accomplished if it is treated as an important, integral aspect and crucial component of the water and land resources development, conservation, control and protection.

The comprehensive management is conceive here as the treatment of pollution problems in water and land resources management from the points of view of the general and specific goals, interests and constraints which are imposed on the management by the economic, social and political requirements in the pollution abatement. They may be considered as the external influences on the pollution control systems. The integrated management is conceive here as taking into account all the purposes, interests and aspects that management of a resource may require. It can be thus considered as the internal influences of all those factors which may affect defense of the environment from contamination in one way or another .

Many regions of the world are already so much polluted that they have become hazardous to humans as well as to the other biologic species of the Earth in many ways. As a consequence, the comprehensive and integrated rehabilitation and protection of various kinds, which depend on the type of resource and the kind of contaminants, have become unavoidable for the near and forthcoming future. If one defines the civilization as a composition of the sets and varieties of infrastructures, needed for a society to function as expected by the standards of living criteria of that civilization, those infrastructures must be protected, further improved, and renovated when they have become obsolete. Protecting and maintaining the civilizational infrastructures at a proper level of performance have become thus the major tasks of the comprehensive and integrated managements of water and land resources for the best protection of the environment.

SOURCES OF POLLUTANTS IN MANAGEMENT OF WASTE WATERS AND IN TREATMENT OF CONTAMINATED SOILS

To manage any complex system, such as the system of defenses against the contaminants and general pollution, the scientific classification and characterization of the basic subject matter is a proper way to approach the related complex problems. Here the seven major and specific contamination sources are described. This approach does not assume that there are not some other minor sources of supply of pollutants which may be important in specific cases, such as the non-agricultural rural sources of pollution.

Urban wastes

The urban gaseous, liquid and solid wastes are composed of both the standard waste materials (urban areas produce often the same wastes, say like the automobile exhaust gases) and the specific wastes. These latter wastes come from the particular urban facilities and activities (such as general and specialized hospitals, research laboratories, particular maintenance and sport facilities, etc.). Specific wastes may require also special treatment technologies. Therefore, it is not surprizing that several presentations at the Moscow NATO Advanced Research Workshop were related to either the special or the standard waste monitoring and treatment technologies.

Industrial wastes

The industrial wastes are composed most of the time from all the three types of pollutants, namely gaseous wastes carried via the atmosphere, liquid wastes reaching the rivers, lakes and aquifers, and solid wastes deposited in the soil and/or on the land. They result from both the point and non-point sources of contaminants. The composition of industrial wastes depends on the type of industry and the degree of investment into the recycling and cleaning up of wastes from these sources of pollution. The general trend in waste management in the most recent types of industrial waste treatments is to filter pollutants from the gaseous emission, to recycle water of these wastes and to treat the liquid solid wastes as much as the appropriate technologies are available and economical to use.

Mining and smelting wastes

The mining in many regions of the world acts also as a potent source of pollution. It does it in three major ways. First, the drained water from mines, either during their operation or after they have been abandoned for any reason, usually leaches the heavy metals and other contaminants from the rocks, which are disposed into the streams, most often as the toxic compounds. Second, all mines produce tailings which are most often piled up onto the land, also exposed to water leaching contaminants as well as supplying the undesirable sediments to rivers. Third, smelting of ores to extract major marketable compounds and elements produces the air-borne materials, deposited in the environment either into the river beds or on the land. Fumes of smelting copper ores in Arizona are carried by air hundreds of miles, which are partially deposited also in the Rocky Mountains region, along with the automobile exhausts from Southern California. The Moscow NATO-sponsored Workshop has also covered this source of contaminants, with discussion of major aspects of these important pollutants.

Wastes generated in energy production.

The wastes which result from electricity or heat producing facilities (thermal power plants run by burning coal, natural gas or lignite; heat producing facilities; nuclear energy production) are specific sources of air, water and soil contamination. Scrubbing the effluent gases and treating or burying the resulting solid wastes is the method of how one can resolve, at least temporarily, the economic dilemmas between the need for energy and the resulting degradation of the environment. The solid wastes are then only temporally stored on the land and in soils and rocks until they migrate into rivers, lakes, aquifers and seas. These sources are discussed at the Moscow Research Workshop only marginally, along with the other sources of pollution.

Agriculture and forestry produced wastes.

Although the agricultural and forestry non-point sources of pollution were touched on in some presentations and discussions at the Workshop, their long-range effects and ways of remedying the consequences have not received a sufficient, necessary coverage. It was likely considered as the contamination which needs special attention and treatment as a particular overall pollution and problems of the economics, to be solved by new approaches and special economic methods and cleaning up technologies. The pollutants come from both, the use of chemical fertilizers as well as the use of pesticides, herbicides and insecticides.

Radioactive wastes.

Nuclear contamination was covered at the Workshop only as the radioactive pollution from the point of view of nuclear material processing and the resulting cleaning up of the nuclear industry processing facilities. The polluted Rocky Flats of Colorado were discussed in details as an example of the variety of pollution aspects and needs for their cleaning up, with the huge costs involved.

Transportation generated wastes.

Similarly to the agriculture and forestry related non-point sources of pollution, the various kinds of transportation produce contaminations and wastes. They have been also only marginally covered and discussed at the Moscow Workshop. Likely, this source of pollution is less covered because it is also the non-point source, especially the automobile-produced air-borne contaminants.

Cost of managing wastes.

The estimate or computation of the cost of defense from the environmental pollution for each of the above seven sources of contamination, or their combinations, but especially of the nuclear and agricultural sources of pollution, show the cost to be enormous. Even the richest industrial countries are not in the position to finance all the projects which should be undertaken in order to clean up and then protect the most dangerously polluted areas. The typical estimates and examples are those which require the cleaning up and work from a duration of 10 or more years, with the annual cost in hundreds of millions of dollars, to the work duration of 30 years or more, with the total cost counted in billions of dollars.

COMPREHENSIVE AND INTEGRATING MANAGEMENTS OF CONTAMINATED AREAS

There is no inhabited area in the world which has not been subject to changes with some kind of pollution by human activities. Those pollutions now require the cleaning up and after it a rigorous protection. Living with pollutions of various kinds, by monitoring and fighting them, has become the basic consequence of the modern industrial, advanced agricultural and new technology activities. The basic dilemma is whether the pollutions should be fought from their beginnings, or authorities should allow them to develop, and when they become intolerable to fight them. The rationale for the former is the economic reality (not enough capital for the defense from pollution at the same time as for the development of land and water resources). The latter assumption is to first undertake the development in order to produce sufficient wealth with time in order to finance the cleaning up.

The experience of the second half of 20th century clearly shows that the pollution abatement should parallel the resource development. First, the defense from pollution will be cheaper in the long run. Second, the real standard of living, especially its component of the healthy lives of the populace, will be attained. Third, a large number of issues, controversies and conflicts may be avoided. Here, the approach of avoiding pollution at the same time as undertaking the resource development with pollution potential is described as the most rational approach, especially if it is done by a comprehensive and integrated development of resources.

The complex planning of the future activities in a comprehensive and integrated rehabilitation of the already polluted environment is being used in many developed industrial countries at present. The Moscow NATO-sponsored Research Workshop has shown examples, particularly of the Emsher River in Germany and the Oder River in Poland, with versatile problems of pollution approached and solved. The treatment of many contaminated areas in the Netherlands also may serve as pilot projects and examples for planning rehabilitations of already polluted environments. These examples have shown that planning the protection of the environment does not stop with cleaning up, but it should be included in the follow-up phases of protection.

The comprehensive and integrated treatment of polluted environments should mean looking at the problems from many aspects as well as from today's to the future solutions, with the clear recommendations on how to do it by the most efficient and the least costly approach. It is clear that economics and environmental protection are interwoven in many intricate ways, especially in comparing the problems generated by the water and land resources development and the undesirable environmental impacts. The entire field of environmental pollution and the defense from it create oppositions in the society with various issues raised, many controversies generated and often serious conflicts made unavoidable.

REMEDIES TO VARIOUS TYPES OF POLLUTED ENVIRONMENTS

The treatment of already polluted environments was widely discussed at the Workshop. New technologies were reviewed and often recommended as the best solutions for the contaminated environment discussed. Not only the scientific reasons were behind the presentation of those new methods of treating the polluted air, water or land by solid wastes, but also the advertising of services and equipments stimulated those presentations and discussions. The physical, biological and chemical aspects of new treatment technologies are manyfold, however looked at from very simple to very complex ways of implementation.

It was also demonstrated at the Workshop that several new technologies may be needed in the future, especially those to be used for cleaning up of the already devastated areas by the large sources of contamination. The future NATO-sponsored advanced research workshops will be needed to further push the research and treatment technologies in creating the efficient economical technologies for fighting various types of contamination and pollution of the environment. Also, the future NATO-sponsored advanced study institutes may be desirable to show the states-of-the art in the ever-changing status of these new technologies, reviewed from time to time.

FUTURE RESEARCH AND DEVELOPMENT OF TECHNOLOGIES ON RENOVATION OF CONTAMINATED RIVERS, RIVER BASINS, AQUIFERS, LAKES AND OTHER ENVIRONMENTS

Because the total cost of cleaning up and the further protection of polluted rivers, aquifers, lakes and other water environments around the world is likely to be measured in trillions of dollars, the versatile research for discovery and development of new technologies for treating wastes and polluted environments are already justified. The workshop topics, likely to be found needing research and development, could be:

(1) The identification of materials and elements which in their defined quantities and the level of concentration represent the bulk of ongoing contamination and permanent pollution;

(2) The development of new methods and the appropriate instrumentation for monitoring the dynamic processes in creation and further evolution of contamination;

(3) The development and/or further improvement of the cleaning-up technologies, especially in the liquid, already polluted environments;

(4) The assessment and definition of the new criteria of where, when and how much a particular environment is polluted;

(5) The evaluation and systemization of the seven basic sources of pollution, and several other minor sources. with their major characteristics, as well as the prediction of the likely future sources of contamination of air, water, soil, land and biomass; and

(6) Making scientific distinction between the cleaning-up approaches of contaminated liquid environments and the related polluted solids and solid environments.

POTENTIAL FUTURE ADVANCED STUDIES FOR CLEANING UP POLLUTED ENVIRONMENTS AND PROTECTING THEM FROM FURTHER CONTAMINATION

The advanced study institutes are conceived in the NATO scientific and environmental activities as the vehicle to arrive for any important subject at the state-of-the art at selected time intervals. From the presentation and discussion at the Moscow NATO-sponsored Research Workshop on large area pollutions the following study institutes may be of future interest in the field of significant pollution cleaning-up and protection of the environment:

(1)-(7) The new technologies for the treatment of each individual source of pollution of the above-described major seven sources of contamination of the environment (urban drainage and sewage, industry, energy, agriculture, mining and smelting, nuclear processing, transportation);

(8) The knew methods and techniques for treating wastes generated by two or more sources in producing the specific wastes composition and problems;

(9) The anticipation of the potential new sources of contamination of the environment, with characteristics and forthcoming approaches and methods of treatment;

(10) The methods and sources of financing the cleaning up and further protection of the environment from pollution;

(11) The ecological factors guiding the cleaning-up and protection of the environment from pollution.

(12) New technologies for monitoring and/or avoiding pollution.

FINANCING CLEANING UP AND PROTECTION OF ENVIRONMENT FROM POLLUTION

Each solution in the treatment of wastes and the disposal of remnants of this process involves a relatively non-negligible cost. Financing any activity related to wastes is a major problem in most cases. The simple reason is that the society and the governing authorities have not been conditioned to the need that a large part of their budget be allocated to waste treatment and its temporary or final disposal.

It would be highly educational for all involved if the realistic and detailed analyses and estimations of the past, present and future costs of waste processing and disposal under various conditions would become widely available. The interested parties may be people, professionals of waste treatment and disposal, decision makers, those threatened by contamination, and all the other interested individuals, groups and institutions. That information would help in anticipating the problems and in the search for the most attractive solutions of the ever-increasing gravity and social concern for the pollution of the environment

CONCLUSIONS

The Moscow NATO ARW has shown clearly that the cleaning up of the already polluted environments, especially of the large rivers and their river basins, represents very important economic, social and political problems, especially in the advanced industrial regions. This then leads to the conclusion that it is better to prevent future contaminations than to fight them after they have reached the expensive levels of needed clanging up.

The Partnership Sub-Series incorporates activities undertaken in collaboration with NATO's Cooperation Partners, the countries of the CIS and Central and Eastern Europe, in Priority Areas of concern to those countries.

The volumes published as a result of these activities are:

Vol. 1: **Clean-up of Former Soviet Military Installations.** Edited by R. C. Herndon, P. I. Richter, J. E. Moerlins, J. M. Kuperberg, and I. L. Biczó. 1995

Vol. 2: **Cleaner Technologies and Cleaner Products for Sustainable Development.** Edited by H. M. Freeman, Z. Puskas, and R. Olbina. 1995

Vol. 3: **Remediation and Management of Degraded River Basins.** Edited by V. Novotny and L. Somlyódy. 1995

Vol. 4: **Earthquakes Induced by Underground Nuclear Explosions.** Edited by R. Console and A. Nikolaev. 1995

Vol. 5: **Transportation Infrastructure.** Edited by R. M. Gutkowski and J. Kmita. 1996

Vol. 6: **Sustainable Development of the Lake Baikal Region.** Edited by V. A. Koptyug and M. Uppenbrink. 1996

Vol. 7: **Transboundary Water Resources Management.** Edited by J. Ganoulis, L. Duckstein, P. Literathy, and I. Bogardi. 1996

Vol. 8: **Urban Air Pollution.** Edited by I. Allegrini and F. De Santis. 1996

Vol. 9: **Advances in Groundwater Pollution Control and Remediation.** Edited by M. M. Aral (Kluwer Academic Publishers). 1996

Vol. 10: **Assessing the Risks of Nuclear and Chemical Contamination in the Former Soviet Union.** Edited by E. S. Kirk (Kluwer Academic Publishers). 1996

Vol. 11: **Ventilation and Indoor Air Quality in Hospitals.** Edited by M. Maroni (Kluwer Academic Publishers). 1996

Vol. 12: **The Aral Sea Basin.** Edited by P. P. Micklin and W. D. Williams. 1996

Vol. 13: **Radioecology and the Restoration of Radioactive-Contaminated Sites.** Edited by F. F. Luykx and M. J. Frissel (Kluwer Academic Publishers). 1996

Vol. 14: **Economics of Atmospheric Pollution.** Edited by E. C. van Ierland and K. Górka. 1996

Vol. 15: **Water Supply Systems.** Edited by C. Maksimović, F. Calomino, and J. Snoxell. 1996

Vol. 16: **Bioindicator Systems for Soil Pollution.** Edited by N. M. van Straalen and D. A. Krivolutsky (Kluwer Academic Publishers). 1996

Vol. 17: **Remediation of Soil and Groundwater – Opportunities in Eastern Europe.** Edited by E. A. McBean, J. Balek, and B. Clegg (Kluwer Academic Publishers). 1996

Vol. 18: **Environmental Engineering and Pollution Prevention – European Network on Excellence and Partnership.** Edited by J. Wotte, W. A. Halang, and B. J. Kraemer (Kluwer Academic Publishers). 1996

Vol. 19: **East-West Life Expectancy Gap in Europe – Environmental and Non-Environmental Determinants.** Edited by C. Hertzman, S. Kelly, and M. Bobak (Kluwer Academic Publishers). 1996

Vol. 20: **Regional Approaches to Water Pollution in the Environment.** Edited by P. E. Rijtema and V. Eliáš (Kluwer Academic Publishers). 1996

Vol. 21: **Global and Regional Mercury Cycles: Sources, Fluxes and Mass Balances.** Edited by W. Baeyens, R. Ebinghaus, and O. Vasiliev. (Kluwer Academic Publishers). 1996

Vol. 22: **The Inter-Relationship Between Irrigation, Drainage and the Environment in the Aral Sea Basin.** Edited by M. G. Bos (Kluwer Academic Publishers). 1996

Vol. 23: **Biodegradability Prediction.** Edited by W. J. G. M. Peijnenburg (Kluwer Academic Publishers). 1997

Vol. 24: **Protecting Danube River Basin Resources: Ensuring Access to Water Quality Data and Information.** Edited by I. L. Murphy (Kluwer Academic Publishers). 1997

Vol. 25: **Disturbance and Recovery in Arctic Lands: An Ecological Perspective.** Edited by R. M. M. Crawford (Kluwer Academic Publishers). 1997

Vol. 26: **Cytotoxic, Mutagenic and Carcinogenic Potential of Heavy Metals Related to Human Environment.** Edited by N. D. Hadjiliadis (Kluwer Academic Publishers). 1997

Vol. 27: **Sensitivity to Change: Black Sea, Baltic Sea and North Sea.** Edited by E. Özsoy and A. Mikaelyan (Kluwer Academic Publishers). 1997

Vol. 28: **Historical and Prehistorical Earthquakes in the Caucasus.** Edited by D. Giardini and S. Balassanian (Kluwer Academic Publishers). 1997

Vol. 29: **Scientific, Environmental, and Political Issues in the Circum-Caspian Region.** Edited by M. H. Glantz and I. S. Zonn (Kluwer Academic Publishers). 1997

Vol. 30: **Restoration and Forests. Environmental Challenges in Central and Eastern Europe.** Edited by R. M. Gutkowski and T. Winnicki (Kluwer Academic Publishers). 1997

Vol. 31: **Integrated Approach to Environmental Data Management Systems.** Edited by N. B. Harmanioglu, M. N. Alpasian, S. D. Ozkul, and V. P. Singh (Kluwer Academic Publishers). 1997

Vol. 32 · **Decommissioned Submarines in the Russian Northwest Assessing and Eliminating Risks.** Edited by E. J. Kirk (Kluwer Academic Publishers). 1997

Vol. 33: **Conflict and the Environment.** Edited by N. P. Gleditsch (Kluwer Academic Publishers). 1997

Vol. 34: **Biotechnology for Waste Management and Site Restoration. Technological, Educational, Business, Political Aspects.** Edited by C. Ronneau and O. Bitchaeva (Kluwer Academic Publishers). 1997

Vol. 35: **Atmospheric Nuclear Tests. Environmental and Human Consequences.** Edited by C. S. Shapiro. 1997

Vol. 36: **Nuclear Tests. Long-Term Consequences in the Semipalatinsk/Altai Region.** Edited by C. S. Shapiro, V. I. Kiselev and E. V. Zaitsev. 1997

Vol. 37: **Environmental Infrastructure Management.** Edited by J. J. Boland, M. E. Bell, and Eugene Z. Stakhiv (Kluwer Academic Publishers). 1997

Vol. 38: **Biosensors for Direct Monitoring of Environmental Pollutants in the Field.** Edited by D. P. Nikolelis, U. J. Krull, J. Wang, and M. Mascini (Kluwer Academic Publishers). 1997

Vol. 39: **Restoration of Degraded Rivers: Challenges, Issues and Experiences.** Edited by D. P. Loucks (Kluwer Academic Publishers). 1998

Vol. 40: **Air Pollution in the Ural Mountains. Environmental, Health and Policy Aspects.** Edited by I. Linkov and R. Wilson (Kluwer Academic Publishers). 1998

Vol. 41: **Actinides and the Environment.** Edited by P. A. Sterne, A. Gonis, and A. A. Borovoi (Kluwer Academic Publishers). 1998

Vol. 42: **Management of Lakes and Reservoirs during Global Climate Change.** Edited by D. G. George, J. G. Jones, P. Puncochar, C. S. Reynolds, and D. W. Sutcliffe (Kluwer Academic Publishers). 1998

Vol. 43: **Mineral Processing and the Environment.** Edited by G. P. Gallios and K. A. Matis (Kluwer Academic Publishers). 1998

Vol. 44: **Hydroinformatics Tools for Planning, Design, Operation and Rehabilitation of Serwer Systems.** Edited by J. Marsalek, C. Macsimovic, E. Zeman, and R. Proce. (Kluwer Academic Publishers). 1998

Vol. 45: **Environmentally Devastated Areas in River Basins in Eastern Europe.** Edited by A. G. Buekens and V. V. Dragalov. 1998

MIX
Papier aus verantwortungsvollen Quellen
Paper from responsible sources
FSC® C105338

If you have any concerns about our products,
you can contact us on
ProductSafety@springernature.com

In case Publisher is established outside the EU,
the EU authorized representative is:
Springer Nature Customer Service Center GmbH
Europaplatz 3, 69115 Heidelberg, Germany

Printed by Libri Plureos GmbH
in Hamburg, Germany